Electronics

Project Management and Design

Second Edition

D. Joseph Stadtmiller
Mohawk Valley Community College

PEARSON
Prentice
Hall

Upper Saddle River, New Jersey
Columbus, Ohio

Library of Congress Cataloging in Publication Data

Stadtmiller, D. Joseph.
    Electronics : project management and design / D. Joseph Stadtmiller.– 2nd ed.
      p. cm.
    Includes bibliographical references and index.
    ISBN 0-13-111136-1
    1. Electronic apparatus and appliances--Design and construction. 2. Project
management. I. Title.

TK7836.S73 2004
621.3815'4--dc21

2003065820

Editor in Chief: Stephen Helba
Editor: Dennis Williams
Development Editor: Kate Linsner
Production Editor: Rex Davidson
Design Coordinator: Diane Ernsberger
Cover Designer: Kristina D. Holmes
Cover Image: Corbis
Production Manager: Pat Tonneman
Marketing Manager: Ben Leonard

This book was set in Century Book by PublishWare. It was printed and bound by
R. R. Donnelley & Sons Company. The cover was printed by Phoenix Color Corp.

Pearson Education Ltd.
Pearson Education Singapore Pte. Ltd.
Pearson Education Canada, Ltd.
Pearson Education–Japan

Pearson Education Australia Pty. Limited
Pearson Education North Asia Ltd.
Pearson Educación de Mexico, S.A. de C.V.
Pearson Education Malaysia Pte. Ltd.

10 9 8 7 6 5 4 3 2 1

ISBN 0-13-111136-1

To the memory of my father and mother,
Dale and Frances Stadtmiller

Preface

Since the release of the first edition of *Electronics: Project Management and Design*, much has changed. The economic boom of the 1990s has turned to the bust of the new millennium, and the view of the world situation and globalization have been altered drastically. The telecommunications revolution has sputtered, at least temporarily, but technological advances continue. In this second edition, the concepts that promoted the original book have become even more critically important: to better prepare electronics professionals to begin their careers by giving them the preliminary tools needed to be productive in their first week on the job.

The foundation strategies used to accomplish this remain as follows:

1. Review the operation of a company.

2. Discuss teamwork and the role of the electronics professional.

3. Present methods of project management.

4. Define an engineering problem-solving process.

5. Discuss the practical aspects of an electronic project.

This second edition release has been reviewed completely, and general improvements, additions, and deletions have been made to all chapters. The most significant additions are:

Chapter 2: The definition of Total Quality Management (TQM) has been expanded and improved and a more complete discussion has been added.

Chapter 3: References to the new National Electrical Code 2002 are included, as well as an expanded discussion of ISO 9000 certification and its newly released requirements. The section on UL approval reviews various UL approval categories.

Chapter 8: An updated and expanded discussion of circuit simulation software elaborates on many of the data analysis capabilities of the latest versions of these software packages.

Chapter 9: A new section on digital displays has been added that discusses all types of LED, LCD, and VFD display devices. The basic theory of operation of each is included along with key design considerations.

Chapter 10: A discussion on the limitations of various breadboard techniques has been added.

The first edition included a six-month evaluation copy of *Microsoft Project 2000* software. Because evaluation software is available from many suppliers of project management software, a sample software CD has not been included with this edition.

The topics are presented in the simplest possible terms as this book is intended for use at many levels, but a basic understanding of electronics is assumed. The broad subject area addressed limits the depth to which any concept can be explored. Each chapter starts with an introduction that highlights the topics to be covered. In each section, examples are provided, wherever practical, to enhance topic discussions. Each chapter concludes with a summary and exercises that will vary depending on the chapter topic. This information is discussed as an actual design project is completed in an ongoing example throughout the book. The project example is included as a separate section at the end of each chapter, starting with Chapter 4.

The book is targeted as a textbook for electronic project management, senior project, or applied electronics courses. It can also serve as a resource and reference that can be used as an accessory text throughout a complete curriculum. It is most applicable to four-year programs in electrical engineering and electrical engineering technology but is also well suited for use in the second year of two-year programs in these areas of study. The first three chapters are a preliminary to electronic project discussions. The Six Steps, a process for solving engineering problems, are presented in Chapter 4. The Six Steps are then applied to electronic projects, one step at a time, in the ensuing Chapters 5 through 13.

Three appendixes are supplied as a reference to the project information covered in the main body of the book. Appendix A includes component reference information such as color codes and packaging information, Appendix B is a general reference for test equipment, and Appendix C includes information for professional organizations, periodicals, and approval agencies.

An instructor's resource manual is available with answers/solutions for the end-of-chapter exercises.

As we pass through the first decade of the new millennium, there will be increased pressure for adequately trained electronics professionals and high requirements for completing successful projects on time. There are many important problems that our society must resolve, and electronics technology will continue to be part of the solution. Raising the bar even further are issues such as diminishing natural resources, increasing world population and pollutants, a widening gap between the haves and the have-nots, the ongoing threat of terrorism, and diminished trust in big business. As always, the continuing challenge is to take a small step in the right direction. To accomplish this, all we need is a defined problem, a plan, project leadership, a team of accomplished players, and the fortitude to get it done.

Acknowledgments

I would like to thank all those who sent in comments and ideas about what they liked in the first edition of *Electronics: Project Management and Design* and ways to improve it. Many thanks to Acquisitions Editor Dennis Williams, Production Coordinator Tim Flem, and Production Editor Rex Davidson.

Contents

7   *Step Three: Develop a Solution Plan*
     (Project Scheduling)   *117*

8   *Step Four: Execution (The Preliminary Design)*   *143*

9 *Step Four: Execution (Component Selection)*   *183*

10 *Step Four: Execution (The Design Breadboard)*   *233*

11 *Step Four: Execution (Prototype Development)*   *269*

12 *Step Five: Verify the Solution (Design Verification)*   *315*

13 *Step Six: Conclusion (Design Improvements
and Project Performance Monitoring)*   *343*

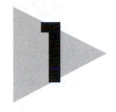

The Project Environment

▶ Introduction

In this book we will explore an electronic project from its inception to its completion. All projects begin with a desire to fulfill an identified need. They are usually worked on by a group of people while a project manager coordinates their activities. Before we begin a discussion of electronic project management, it is important to discuss the environment in which most projects are initiated and completed: a typical company. The company environment is where most project activities are performed and where the project results are realized. This chapter begins with an overview of a typical company and how it is managed and directed. This is followed by a discussion of the roles and concerns of the chief executive officer and each operating department. Then the role of the electronics professional is covered as it relates to development projects and the general operation of the company.

The concept of *teamwork* is as important to any organization as it is to a college basketball team. Realizing its importance, many corporations focus on improving teamwork within their organizations. The promotion of teamwork is discussed in general but is applied specifically to the electronics professional. Next, the inception and selection of projects are discussed as a prelude to the discussion of project design and management. The specific topics covered in this chapter are as follows:

- ▶ Company structure and overview
- ▶ The chief executive officer
- ▶ Marketing and sales
- ▶ Engineering
- ▶ Manufacturing
- ▶ Quality assurance

- ▶ Finance
- ▶ Human relations
- ▶ The role of the electronics professional

1–1 ▶ The Company Structure and Overview

This discussion covers the operation of a typical company, but the key points apply to any organization that might undertake a project. The typical company is a business that performs various combinations of operations, including the development, manufacture, and sale of goods or services. While some companies may perform all of these functions, others choose to perform only a few. At the very least, a company must sell a product or service.

Most companies perform these activities in the pursuit of making a *reasonable* profit. While the definition of reasonable profit varies, the owners of the company are taking a higher risk with their investment in the company. They are normally looking for a better return on investment that could be realized by simply putting the money in the bank. Of course, there are organizations deemed "not for profit" that exist to provide a product or service at the break-even level of operation.

A company can be owned by an individual, a small group of stockholders, or publicly owned by many stockholders. In publicly owned companies, stockholders who possess 51 percent of the stock have controlling interest of the company. Those in possession of controlling interest can determine the company's direction and, therefore, control its operation. This is how one publicly held company acquires another, a trend that increased in the 1980s and continues today. Stockholders with controlling interest direct the operation of the company by selecting a Board of Directors that operates the company through its appointed *chief executive officer*.

For income tax purposes the federal government classifies companies as a sole proprietorship, a C corporation, or a subchapter S corporation. In a sole proprietorship, there is only one owner, and any income is treated as the owner's income. The owner is legally liable for any results of the sole proprietorship. This means that the owner's personal assets are at risk if the business fails or is sued. With a corporation the company becomes an entity separate from the owners and they are protected from legal action taken against the company. In a C corporation the company's profits are taxed as well as the dividends paid to stockholders. In effect, the profits of the C corporation are taxed twice: first as profits to the corporation and second as dividends to the stockholder. The subchapter S corporation is a special type of corporation in which there are a small number of owners and the profits are only taxed once, as income to the shareholders.

Companies in general sell products or services for the purpose of receiving a reasonable profit. This is true for sole proprietorships as well as corporations. The president of a sole proprietorship will be in business as long as it is profitable. In a corporation the CEO and the Board of Directors will remain in power as long as the stockholder's interest in a reasonable return on investment is being met. If the primary goal of any company is to make a reasonable profit, then what is the secondary goal of a company? Usually it is to continue to make a reasonable profit and to improve profits year after year. The owners of the company desire a profit this year and will continue to desire a reasonable profit in the future. To accomplish this, the company must develop intricate plans for the future assuring its viable operation and profitability. This is called *long-term planning*. Strategic planning is the most popular method of long-term planning being used today. The strategic planning process creates a plan for the future based upon specific company goals and strategies.

Strategic Planning

Strategic planning is a method for companies or organizations to outline their strategic goals and objectives for the long and short term. Strategic planning begins with the development of a mission statement for the company or organization. The mission statement should be short, simple, and define the specific purpose for the company. Next comes the *situation analysis*, which includes a self-evaluation of the company's strengths and weaknesses as well as a list of the current trends likely to affect the company in the future. The self-evaluation is an objective analysis of what the company does well and not so well. The result is a list of strengths and weaknesses shown in the order of their priority. The list of important trends includes technical, cultural, legal, and geographical issues that the company will have to deal with.

Top-level goals are then identified, listed, and prioritized for the entire company. If possible, these goals should be *specific* and *measurable*. "Specific" means that the goals are identified very accurately, while "measurable" means that there is an obvious way to tell if the goal is achieved. With the goals defined, strategic objectives are developed that support achieving these goals. These must be specific and measurable. Each company department will review the list of top-level goals and strategic objectives and determine the specific tasks to be completed by that department in support of those goals. Each task will be listed with the person responsible for completion of the task and the required date of completion. The result of the strategic plan will be a document that includes all of the components listed in Figure 1–1.

It is important to understand the perspective of each company department in order to determine the best way to interact with the people from those departments. Individuals from the sales department have a very different perspective from those in the quality or finance departments. In pursuit of the *strategic plan objectives*, it is common for each department to become an encampment, fortified and isolated from the others. This occurs when the priorities of the departments oppose one another. For example, if manufacturing tries to keep inventories too low, the impact might affect delivery lead times and negatively affect sales. Many of the individual performance measurements of an operating department affect those of other departments. Upper management sometimes reinforces areas of conflict with aggressive promotion of goals for department managers, in effect

1. Mission statement

2. Self-evaluation

3. Future trends

4. Top-level, long-term (five-year) goals

5. Top-level, short-term (annual) goals

6. Strategic objectives, short-term (annual)

▶ **FIGURE 1–1**
Strategic planning components

causing the departments to compete instead of work together. Also, the personalities that characterize many departments sometimes clash with those of the others. The following discussion is an attempt to develop an understanding of the perspective and priorities of the operating departments for a company structure as shown in Figure 1–2.

1–2 ▶ The Chief Executive Officer

The chief executive officer (CEO) or company president is responsible for the complete operation and financial performance of the company. The operating department managers usually report directly to the CEO, although there are as many different structures as there are companies. The ideal CEO has broad knowledge

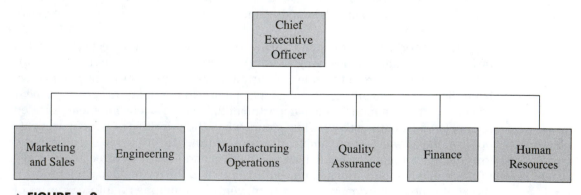

▲ **FIGURE 1–2**
Typical company structure

and strong leadership capabilities and is a good communicator and visionary. He or she must be a mentor to each of the department managers although he or she may not share their expertise. The performance of any company depends greatly on the people who make it up and their feelings about the company and their job function. The CEO can impact company morale more than any other individual. Typically, the CEO comes from a marketing and sales or finance background. However, a certain percentage come from manufacturing or engineering. The background of this kind of CEO will depend on the type of product that the company manufactures. If a company produces highly technical products, this increases the chances for an engineering-oriented individual to be its CEO.

The day-to-day activities for the CEO focus on operations and the ongoing development of the strategic plan for the future. The CEO periodically reviews key problems or projects and follows up and supports the department managers in reaching identified strategic goals. After developing the strategic plan, the CEO is responsible for its completion. The priorities of the CEO reflect the view from the top of the company and are shown in Figure 1–3.

1-3 ▶ Marketing and Sales

Imagine that you are vice president of marketing and sales for a small- to medium-sized electronics corporation. Try to think about what your top priorities would be and what a typical day would be like. To begin with, the marketing and sales department is likely to have a structure and responsibilities as shown in Figure 1–4. For this discussion, the department is broken into its two key components: marketing and sales.

▶ **FIGURE 1–3**
CEO priorities

1. Meeting strategic and profit plan goals

2. Continued development of strategic direction

3. Supporting each operating department in achieving identified strategic goals

4. Improving the image of the company

5. Improving company morale

6. Total quality management

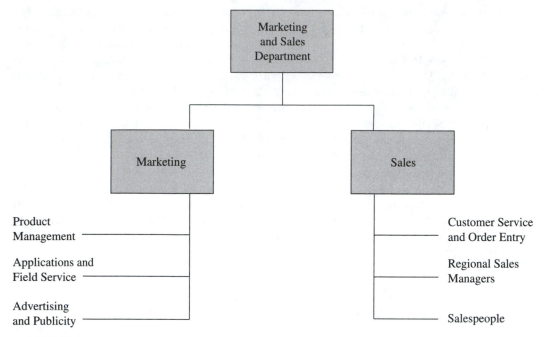

▲ FIGURE 1–4
Marketing and sales department structure

Marketing

Marketing leads the company strategically by determining the products, markets, and ways the products are sold. The definition for Marketing's function is as follows:

Product—Marketing determines the product or service to be sold by a company.

Price—Marketing will set a price for the product or service.

Promotion—Marketing determines the method and the amount of promotion provided for a product (i.e., magazine, radio, and television advertisements; trade shows; and so on).

Sales distribution—Marketing determines how the product will be sold (manufacturer's reps, distributors, direct sales, direct mail, and the Internet).

The primary concerns of Marketing's functions are:

1. Providing the basis for a strategic direction for the company.

This involves connecting the company's past with its future. What changes are looming in the marketplace and how will they affect the company's market? What new competitors will develop and what will their strategies be? Manufacturers of cassette tape decks had to reconsider their future after the development of compact discs (CDs), just the way the manufacturers of record turntables did after the introduction of cassette players. Marketing must be in tune with the fast-paced world in which we live to stay informed about market conditions and changes. Ultimately, Marketing will determine which new products to develop in order to pursue which markets.

2. The definition and development of new products.

    In pursuit of the company goals, specific products must be defined so that the engineering department can develop them. The marketing manager will be continually involved and interested in the status of new product development projects. This includes not only the definition of the product but issues such as internal and field training, promotion material and budgets, pricing strategies, sales and profitability forecasts, and, most importantly, the product's formal introduction.

3. The profit and sales performance of each current product line.

    The marketing group supports the sales group in the sale of existing products and also monitors their sales, profitability, and customer satisfaction. Sales and profitability levels of a product line are totaled and compared to previous month and year levels. Profitability is determined by the cost of manufacturing the product, sales, and overhead costs combined with the customer discounts offered. Customer satisfaction includes the product function, field failures, price competitiveness, ease of ordering, and the receipt and use of the product. Marketing will monitor and take action on all negative performance issues.

4. The definition of strategies for existing products.

    As market conditions change and new products are developed, the strategies regarding existing products must be reviewed. These strategies include pricing, promotion, expenditures, investment in product improvements, cost reduction, and sales emphasis. If an existing product is scheduled for replacement by a new product, then investments for improvements in the existing product will only be made if reliability is in question. At the same time, a price increase in the old product will make the new product even more attractive.

In order to address the marketing concerns just discussed, Marketing will have product managers for each major product line, an advertising and promotion manager and an applications/field service manager. Of course, the size of these departments will depend greatly on the size and diversity of the particular company. The product managers will be involved in monitoring and supporting existing products and the development of new products. The advertising manager will supply the required promotion materials and promotion as required for all products. The applications and field service manager is responsible for all field service and application problems.

Sales

The requirements for sales professionals have changed greatly over the years. Today's salesperson, selling technical products or otherwise, must be professional and capable in many areas in addition to possessing solid interpersonal skills. The sales manager's primary concern is current and future sales levels. Sales levels of the past are history, only significant as a reference for measuring the trend of current sales. There is always pressure on the sales manager to increase sales levels.

Sales usually includes a customer service group, regional sales managers, and, finally, the salespeople. The customer service group takes orders and communicates order and billing information, prices, discounts, and delivery dates. The regional sales managers are responsible for a large territory and a number of salespeople. The salespeople can be direct employees of the company, independent manufacturer's reps, or distributors. Direct salespeople usually receive a base salary and a bonus that depends on achieving sales goals. The regional and national sales managers also receive this type of compensation. Independent manufacturer's representatives receive a commission on all sales made into their territory. Distributors can buy product at a discount and sell it at whatever price the market will bear.

The primary concerns of the sales manager in order of priority are as follows:

1. The current sales performance level.

    Most important is whether sales are above or below the forecast for the year and the current month. The level of sales performance has a significant impact on the company. If the year-to-date sales are much lower than expected, there is much pressure on the sales manager to explain why and take corrective actions. The repercussions throughout the company can be the deferment of planned capital expenditures, budget cuts, and even layoffs. On the other hand, if sales are close to or exceed forecast levels, the sales manager can concentrate on the task of managing sales.

2. Improving sales performance.

    Whether sales are up or down, the sales manager must always think about improving sales. The ways to do this include better training, incentives, sales promotions, market coverage, and the replacement of ineffective salespeople. To improve sales, the sales manager must constantly monitor the performance of the department in these areas and take appropriate corrective actions. Each year the sales manager should identify a list of the highest priority programs to improve sales.

3. Communicating customer satisfaction/dissatisfaction.

    The customer service group, the salespeople, and the regional managers converse with customers on a daily basis. It is essential that customer comments and complaints be gathered and channeled to top management. The sales manager must have some method of gathering this information and communicating it to the company president and the other operating departments.

4. Communication of customer desires and market conditions and changes.

   The salespeople are also most able to sense changes in the market and communicate customer desires. This information is vital to Marketing and is why many marketing and sales functions are included in one department.

Marketing and Sales Summary

In addition to the specific marketing and sales concerns discussed previously, the marketing and sales manager must prepare budgets and sales forecasts annually for the upcoming year. Budget performance is monitored by the finance department and is usually discussed on a monthly basis. Sales forecasts must also be prepared for the upcoming year and sometimes three to five years for long-range strategic planning. Figure 1–5 shows a summary of the general priorities of the marketing and sales manager.

People often confuse the roles of the marketing and sales functions. Marketing determines what to sell, how to sell, the promotion, and the price at which to sell. Sales actually does the selling. Each member of Sales and Marketing is focused on serving the customer and promoting sales in one way or another. All other concerns and issues are secondary.

1–4 ▶ Engineering

The engineering department also covers a broad range of responsibilities. Today, the head of the engineering department is called the *engineering manager* or, sometimes, vice president of engineering. Thirty years ago the title most often used

▶ **FIGURE 1–5**
Priorities of the marketing and sales department

1. Providing strategic direction

2. Increasing sales levels

   • Defining new products

   • Strategies for old products

3. Customer satisfaction

4. Customer needs

5. Changing market conditions

was "chief engineer." The engineering manager is usually responsible for research and development, product support, and documentation development and control. In other words, the engineering manager manages all research and the design of new products, provides technical support for existing products, and develops and maintains all of the documents that define products. The structure for the engineering department is likely to be as shown in Figure 1–6. Many companies do not separate the development and support engineering functions. These companies are usually less active in the development of new products. Each of the three engineering functions will be discussed separately.

Research and Development

The research and development (R&D) group is the elite technical group in the company. In larger companies the research and development functions may be completely separate from the engineering department. It is important to understand the difference between research and development projects. A research activity is a pursuit or study for which the end result is not a particular system or product. A development project is the design of a specific device for use as a product or internally by the company. Research activities have defined goals. The difference is that the goals are not the development of a specific product. The R&D

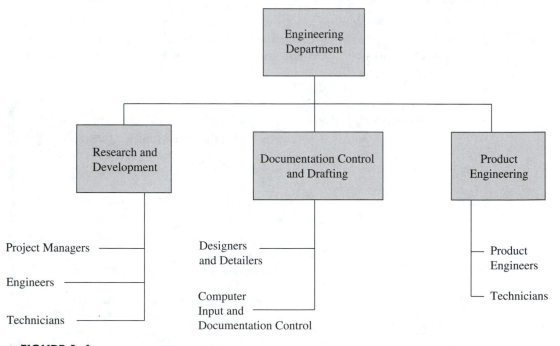

▲ **FIGURE 1–6**
Engineering department structure

group will have many projects occurring at the same time. Some will be the development of specific products while others will be research studies of some particular idea. A research project can include the development of an idea into a working prototype; then a specification is completed and the project can become a product development project. The R&D group consists of engineers, scientists, and technicians. The primary concerns of the R&D manager are listed in order of priority as follows:

1. The successful completion of projects on schedule and within budget.

   The R&D group has a very specific job to do. Its primary goals are to complete projects as scheduled and within the budgeted dollar amount. For a research project, success is achieved by completing the exploration of the defined topic. A research project may result in a new concept that is useable in the company's products. Success for a development project is the completion of a design that meets all aspects of the specifications.

2. Identification of areas for future research.

   The R&D group is constantly looking at competitors and keeping up with developments in all technological areas in order to make recommendations for future research.

More than any other department, the R&D group has a narrowly defined function. This is because R&D requires people with top-level technical and creative skills. The R&D group should be provided the time and equipment to bring the company's dreams into reality. In order to be successful, it will need entrepreneurial direction.

Product Support

The product support group provides technical support to products that have already been released. Product support people also get involved with new product development as those projects are developed and phased into manufacturing. (Other common names for the product support group are *product engineering* and *sustaining engineering*.) In many ways it appears that the product support group is assigned the more detailed and mundane of the engineering department's responsibilities. The R&D group is left to concentrate on the "fun" projects. However, working in this group is just as challenging and requires as much technical know-how and creativity as working on R&D projects. This group consists of manufacturing, mechanical and electronic engineers, and technicians. The concerns of the product support group listed in the order of priority are as follows:

1. Provide technical support for released products to the other company departments.

   The provision of technical support for existing products is a broad statement that includes the following:

a. *Providing training and answering technical questions about the product.* The product support group usually provides technical training for the company's operating departments. Whenever problems occur with a product that cannot be resolved, the product support group is the next level of support called on to find a solution. The product support group was first established as a buffer between the other operating departments and the R&D group. The daily occurrence of product problems in a manufacturing environment is usually a nuisance to the R&D group. Constant interruption precludes the timely completion of important R&D projects.

b. *The design, maintenance, and modification of test fixtures for the manufacturing, quality control, and marketing and sales departments.* The product support group designs custom test equipment for new products as they are developed. This function enhances the range of responsibility of the product support group with a creative design challenge that is often as imposing as the product design challenge.

c. *The development and maintenance of test procedures for all manufacturing, quality, and field test operations.* These procedures define the required tests, equipment, and the process for verification of products after their manufacture.

d. *Verifying that released products meet specifications.* This function includes completion of the product assurance testing described in Chapter 12. Product assurance testing verifies that the product meets the design specifications.

e. *Locating and approving the substitution of components.* The variation in lead time and the price of components and the problem of discontinued components generate many requests to find and approve acceptable substitutes. The product support group is responsible for locating and approving acceptable substitutes for all components.

2. Provide technical support for the phase-in of new products into production.

A member of the product support group is assigned to each new product development project to provide the support functions listed here: test fixture design, test procedure development, product assurance testing, and training.

3. Recommend and orchestrate changes to improve all aspects of existing products.

Continuous improvement is a way of life in the industrial world. The product support group is the focal point for requesting and implementing changes that improve efficiency, decrease costs, and maintain or improve quality levels. All of the operating departments can suggest and request changes that the product support group must explore, possibly approve, and implement.

Documentation Development and Control

The documentation development and control group—"documentation group" for short—of today's company only slightly resembles the drafting department of yesteryear. The drafting tables, with their precise drafting machines, have been replaced with personal computers, computer-assisted drawing (CAD) software, and plotters. CAD systems have increased the throughput and capabilities of the documentation group immensely. The greatest benefit of any CAD system is the ease of modification provided and the ability to take any part of a drawing and use it for some other purpose. The documentation group has the important and tedious responsibility of developing and maintaining all of the manufacturing documentation for the company. Manufacturing documentation includes all custom part drawings, standard part definitions, bills of material, assembly drawings, schematics, artworks, test fixture documentation, and test procedures. All of these documents will be described and discussed in a later chapter. The documentation group also has the responsibility for two important company-wide procedures: engineering changes, which are implemented with the *engineering change notice* (ECN) system, and the part number definition system which specifies the company part number system. The outputs of the documentation group are drawings and bills of material. The bills of material are loaded in the database of the central operations software that controls all manufacturing materials, purchases, and costs. The documentation group consists of mechanical and electrical designers as well as computer input personnel. The concerns of the documentation group are listed in the order of priority as follows:

1. The maintenance and distribution of accurate manufacturing documentation for all products.

   The documentation group must orchestrate the review and approval of all documents and respond to requests for engineering changes in a timely manner.

2. The development of new documentation.

   The mechanical and electrical designers within the documentation group work hand in hand with the R&D and product support groups to refine design details as they complete the actual design drawings. The electrical designers in the documentation group usually complete the design of the printed circuit board layout.

Engineering Department Summary

In addition to the responsibilities discussed, the engineering manager must prepare and monitor an annual budget and develop the capital expenditure estimates required to support planned product development activity. He also supports the development of strategic plans and strives for continuous improvement of the department's performance. Figure 1–7 shows the complete list of priorities for the engineering department.

1. Completing development and
   research projects

2. New technology developments

3. Providing technical support

4. Maintaining product documentation

5. Improving quality

6. Lowering costs

▶ **FIGURE 1–7**
Engineering department
priorities

1–5 ▶ Manufacturing and Operations

The manufacturing and operations department is the core of the company. The manufacturing and operations department deals more directly with the other company departments than any other. In addition to the normal range of management skills, the manufacturing and operations manager must be well organized and adaptable. This department is structured as shown in Figure 1–8 and includes material control, production control, manufacturing and plant management groups. There are many variations in manufacturing department structures as managers try to improve the efficiency of all operations. A separate discussion of each of the major functional groups within the manufacturing and operations department follows.

Material Control

The material control group is responsible for the procurement and storage of all material requirements for the company. This includes components and materials for manufacturing products as well as all other company operations. Within the material control group there are the following subgroups: purchasing and inventory control.

Purchasing is responsible for the purchase of all goods bought by the company. Most companies utilize a purchase requisition system to allow the other operating departments to request the purchase of goods. The purchase requisition is completed by the requestor and approved by the department manager. The dollar amount of the requisition will determine if any other approvals are required. The purchase of components and material to support the manufacturing of the company's products is usually driven by the manufacturing and operations software that generates purchase orders based upon the current inventory, product forecasts, and

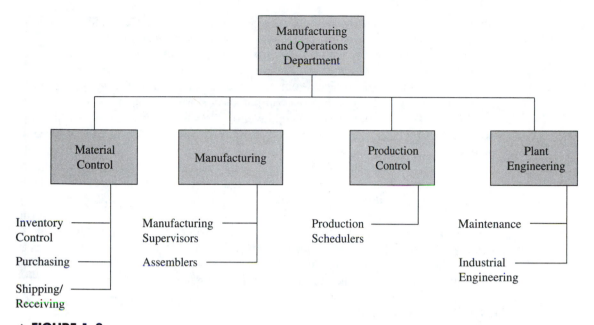

▲ FIGURE 1–8
Manufacturing and operations department structure

component lead times. These systems are called *material resource planning* (MRP) systems and are in use by most companies today. The buyers, who are usually organized into groups that specialize in types of commodities (electronic components, mechanical parts, plastic parts, and so on), follow through with the purchase by contacting the supplier and submitting a purchase order. A purchase order is a legal commitment requesting the purchase of a quantity of a particular item and guaranteeing payment at the stated price.

One other type of purchase that requires special handling is a request to purchase capital equipment. *Capital equipment expenditures* are items defined in principle as having a high value and a long life. High value is defined as over $1000 for most companies, and five years of useful life or longer is considered a long life. Sizable capital expenditures have high visibility and must usually be approved by the CEO. If the company is part of a larger corporation, other approvals may be required.

The inventory control group is responsible for receiving all shipments to the plant, storing all purchased components and materials, and completing all shipments outgoing from the company. Inventory control includes receiving, stock room management, and shipping subgroups. As goods are received, the receiving group unpacks them, notifies Purchasing that they have been received, and distributes them as required. Goods used for the manufacture of products are placed in stock under the appropriate company part number. The components and materials will be drawn from stock as required for the manufacture of products. Goods that are for use by the operating departments are forwarded to their requestor.

The components in stock comprise the company's *inventory*, which is the current investment in parts and materials for the purpose of manufacturing products.

The stock room management subgroup is responsible for maintaining efficient and organized storage of the inventory and maintaining an accurate count of each item. This data is maintained on the central computer—but the data is based upon manual or scale counts. In order to maintain an accurate count of inventory, the stock room subgroup will perform planned "cycle counts" on various categories of components to verify the current inventory count. Most companies perform a complete inventory count annually. Most companies wish to keep inventory levels as low as possible while supporting the timely manufacture of products to meet customer needs. This is because inventory has a cash value that could be invested in some other way. To minimize inventory, many companies utilize *just-in-time* (JIT) processes, which means that suppliers provide weekly shipments of components in quantities required to meet one week's worth of production. Ideally, JIT means components are available just in time for their use in the product. JIT requires a good working relationship between the supplier and customer, high quality levels, accurate forecasts, smaller and frequent deliveries, and overall flexibility. "Inventory turns" is the measurement most often used to show how well a company minimizes—or turns—its inventory. The higher the inventory turns, the better the company is minimizing and using inventory efficiently. The total inventory purchases, the inventory total, and cumulative inventory turns are reviewed on a monthly basis.

After components are drawn from stock to manufacture a particular product, they become *work in process inventory* (WIP). When the manufacture of the product is complete, it becomes *finished goods inventory* (FG) until it is shipped. The shipping group packages and ships all items from the company. This includes packages and mail outgoing from the other departments as well as products being sent to customers. As a product is shipped, FG inventory is relieved of the item.

The key concerns of the material control group listed in the order of their priority are as follows:

1. The procurement of quality components and material to support the manufacture of product to meet customer orders.

2. The quality and timely shipment of goods to the customer.

3. Selection of quality suppliers with short delivery times.

4. Minimization of inventory levels and maximization of inventory turns.

5. Maintenance of an accurate inventory count.

6. Supporting the development of new products by participating in new product phase in process.

7. Fulfilling purchase requests for requisitions from other departments in a timely manner.

Production Control

The production control subgroup performs all the scheduling of production builds for the manufacturing and operations group. As customer orders are received, Production Control will review them. Based on the customer's requested ship date, production backlogs, and component availability, the production planner or scheduler determines when the product will be built and its target ship date. The customer is then notified, usually with a customer acknowledgment or form sent via fax, mail, or even e-mail, that the order has been received and that it will ship on the designated ship date. The production control subgroup consists of planners or schedulers that usually handle a similar group of products. The planner must be aware of the availability of parts and materials, production backlogs, sales forecasts, and the current performance to sales forecasts. After reviewing all of this data, the planner schedules the manufacture of the product.

The priority concerns of the production control subgroup are:

1. Scheduling customer orders.

2. Adjusting purchase quantities to meet current order levels while minimizing inventory.

3. Requesting updated sales forecasts and modifying production forecasts accordingly.

Manufacturing

The manufacturing subgroup is usually the largest group in the company. It includes all of the people who manufacture the company's products. There are two primary ways to organize the manufacture of products, by *product line* or by the type of *assembly level*. A product line organization divides manufacturing up into manufacturing cells that complete one line of products exclusively. Organization by assembly type separates manufacturing into groups by subassemblies (i.e., low- and middle-level assemblies), such as circuit board subassemblies, mechanical subassemblies, and final assemblies. Most companies today prefer the manufacturing cell approach but, in fact, use a combination of the two different concepts, as shown in Figure 1–9. Manufacturing cells are used wherever possible, but when expensive capital equipment is required for an assembly operation (i.e., circuit board component insertion equipment), then a separate group is usually created to perform this assembly operation for all the products that require it. Throughout the manufacturing subgroup's organization, there are supervisors, foremen, or group leaders who head up either the manufacturing cells or the different assembly groups.

There is one priority for the manufacturing subgroup:

Manufacture quality products to meet customer orders at the required ship dates at the lowest cost possible.

In order to meet this goal, the manufacturing subgroup is measured in the areas of *quality level*, *schedule performance*, and *cost effectiveness*. Quality measurements are in the form of *product failures* in production and product failures at customer sites.

Schedule performance is shown by a running total of the number of orders that are shipped later than promised and later than requested. The *promise date* is the date that the company indicated to the customer that the product would be

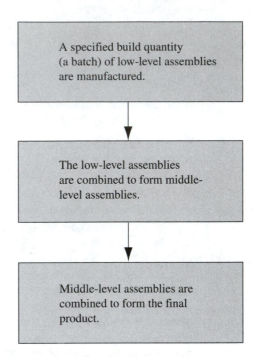

Cellurized Manufacturing—
One Piece Flow

Assembly Level Manufacturing—
Batch Processing

All final products are manufactured in one cell, one piece at a time, in a continuous one-piece flow operation.

A specified build quantity (a batch) of low-level assemblies are manufactured.

The low-level assemblies are combined to form middle-level assemblies.

Middle-level assemblies are combined to form the final product.

▲ **FIGURE 1–9**
Manufacturing organization cells or assembly type

shipped. The *request date* is the ship date requested by the customer. Orders shipped later than promised are called *overdue*. The number of times that the promise date given did not meet the customer's request date is called *later than requested*. Every effort is made to meet the customer's request date. The rate at which a company is unable to accomplish this is the most important measure of its schedule performance.

Cost effectiveness is shown by the current cost to manufacture the product. These numbers include all component and labor costs applied to the product and are developed on a monthly basis.

Plant Management

Plant Management is completely responsible for the grounds and facilities of the company. It oversees or performs all new construction and installs and maintains all plant equipment. The plant management group will include people with the following skills:

Electrical, electronic, and control technology

Mechanical and machine maintenance

Construction and carpentry technology

Industrial engineering technology

The primary concerns of the plant management group are listed in the order of priority:

1. To maintain all plant equipment and facilities in good working condition with a professional and neat appearance.

2. To plan and provide for expanded facilities and new equipment in support of the improvement of company efficiency and the introduction of new products.

Manufacturing and Operations Summary

The manufacturing and operations manager must also develop an annual budget for the department and monitor the status of expenditures. A capital equipment budget must also be developed. The priorities of the manufacturing and operations manager are shown in Figure 1–10.

1-6 ▶ Quality Assurance

The quality assurance (QA) department is the watchdog of company operations and the corporate police that blows the whistle on the "less than quality" actions of the other departments. It is natural for the manufacturing and operations manager and the quality assurance manager to have an adversarial relationship, which is why the two departments must be kept separate, each with its own manager who reports to the CEO. QA will have most of its dealings with the manufacturing

▶ **FIGURE 1–10**
Manufacturing and operations managers' key priorities

1. Manufacturing quality products
   to meet customer requested ship dates

2. Minimization of manufacturing costs

3. Minimization of inventory

4. Maintaining product documentation

5. Improving quality

6. Lowering costs

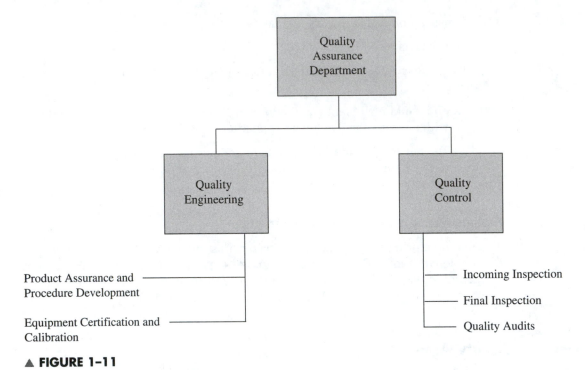

▲ **FIGURE 1–11**
QA department structure

and operations and engineering departments. The quality assurance department includes the quality engineering and quality control groups with a structure as shown
in Figure 1–11. Each of these groups will be discussed separately.

Quality Engineering

The quality engineering group develops the rules for acceptable quality. The quality engineers develop procedures and policies that should guarantee the achievement of the company's quality goals. This group usually has an overall group leader
and is separated into various product groups, consisting of engineers and technicians who focus on the quality level of their particular product line. The key concerns for the quality engineering group, listed in the order of priority, are as follows:

1. The development of policies and procedures that support the company's
   strategic quality goals.

   This concern deals with the process for verifying that a product is free of
   defects and ready for shipment to the customer. It includes the specific
   procedures to be performed on the product as it passes through the manufacturing process. The quality engineers may also review and recommend
   changes to key operating procedures for other departments that impact
   quality levels.

2. The development and implementation of periodic, quality-level reporting for each product line and the company overall.

   On a monthly basis, the quality engineering group will complete reports that indicate current quality levels for each product line. These include manufacturing failures, out-of-box failures, 30-day failures, warranty failures, and out-of-warranty failures for each major product line. These terms will all be discussed and defined in a later chapter. The purpose of these reports is to determine the current quality level, how it is changing (increasing or decreasing), and which products and areas need action to be taken. Other data that might be presented are customer complaints and the results of customer surveys. After a review and discussion of any quality problem areas, there are many actions that the quality assurance department and other departments can take to resolve them.

3. The completion of product assurance testing that assures that new products meet all published specifications.

4. The support of the ongoing improvement of quality levels.

The quality engineers continually address quality problems or the improvement of quality levels.

Quality Control

The quality control group performs the inspection of products and assemblies and thereby enforces the rules developed by the quality engineers. This group can be broken down into three different functions: incoming inspection, WIP (work-in-process) inspection, and final inspection. Inspection is viewed as a nonproductive (no value added) use of labor and equipment, so there is always pressure to minimize the amount of inspection. Without some level of final inspection, the customer will experience firsthand any quality problems. The *only* concern of this group is to perform inspection and review of the company's components, subassemblies, and procedures in support of the company's quality and efficiency goals.

Incoming Inspection

Incoming inspection is an attempt to verify the quality level of suppliers and weed out problems before they ever get into manufacturing. Companies today take a dim view of performing incoming inspection to verify the quality level of a supplier. The lowest-priced supplier can lose a customer due to poor quality levels. Also, incoming inspection does not necessarily improve the quality level of the product shipped to the customer. It can, however, improve the efficiency of manufacturing products with poor quality components. Ultimately, incoming inspection is best used to identify suppliers of poor quality components. Once identified, corrective actions can be taken.

One popular way to implement incoming inspection is to randomly audit shipments of categories of components and materials. If excessive levels of nonconforming goods are detected, then larger samples are inspected on future shipments and the issue can be addressed with the supplier. Another method is to perform incoming inspection after problems are detected in WIP or final inspection. Many types of components can be inspected on an incoming basis. Standard components such as mechanical hardware (screws, nuts, bolts, and the like) or electronic components (resistors, capacitors, ICs, and so on) rarely require incoming inspection. The more critical items are those custom made for the company, plastic molded parts, castings, stampings, and labels. Custom parts are treated differently than other components because they are not readily obtained from another supplier. Also, any rejected custom components become scrap material to the supplier because they are unable to sell them to other customers. Incoming inspection of custom components is more common and critical because they are naturally more susceptible to variations in tolerances.

One special type of incoming inspection is called a "first article inspection." This type of inspection is usually performed on the first part received after new tooling has been completed for a custom part. Tooling is defined as a mold, fixture, or other special device that allows the manufacturer to produce the custom component per the drawing and specifications supplied by the customer. A first article inspection includes the measurement and recording of all key dimensions shown on the part drawing for the component inspected. The results are used to determine if the tooling has successfully reproduced the component as specified. Any inconsistencies result in either a change to the drawing or the tooling. Quality control people must be knowledgeable and adept at making a variety of mechanical measurements.

WIP Inspection

WIP inspection includes the inspection of subassemblies that will eventually become part of a final product. WIP inspection is performed when there is a significant chance for an error to occur and the result of not inspecting or testing results in higher costs and less efficiency at the final assembly and test level. The best example is circuit board assemblies. Circuit boards, manually assembled with many components, have a higher probability of being assembled incorrectly. It probably makes sense to test this type of board before assembly into the final product. The decision to test or inspect a circuit board will also depend on how easy it is to isolate a failure at the final product level and how easy it is to remove and repair it. The decision whether to test any WIP subassembly is based upon its complexity and the impact of a failure at higher levels of assembly.

Quality Assurance Summary

The quality assurance manager is focused on measuring, maintaining, and improving the quality levels of the company. This includes every aspect of the company's products. The QA manager also must submit and maintain an operating budget and is concerned with improving the department efficiency, the cost of quality procedures, and inspections. The QA manager and the department are typically people who "go by the book." That is the nature of their job. The QA department's main purpose is to set up rules that promote quality in the company's products and to enforce those rules. QA people have a difficult time with "gray areas" because their rules must judge the result as black or white, pass or fail. The summarized priorities of the QA manager are shown in Figure 1–12.

1. Measuring and reporting quality levels

2. Maintaining and improving quality levels

3. Minimization of the cost of quality

▶ **FIGURE 1-12**
QA manager's key priorities

1-7 ▶ Finance

The finance department is another company watchdog and its special concern is the use of company money and capital. Its primary purpose is to control, disperse, and report on the company's funds and its financial performance. The finance department is usually structured as shown in Figure 1–13, with groups for accounts receivable, accounts payable, and cost accounting, which all report to the controller. The controller and computer information systems group reports to the vice president of finance.

Accounts Receivable

The accounts receivable group receives all incoming payments to the company. The payments received are usually for goods and services provided by the company. As payments are made, the customer account is updated and the invoice is shown as fulfilled. The real work for this group involves those invoices to customers where payments have not been made per the terms of the sale. The terms of sale are always stated on the invoice and are usually "net 30 days." This means that payment of the net amount shown on the invoice must be received within 30 days. When the

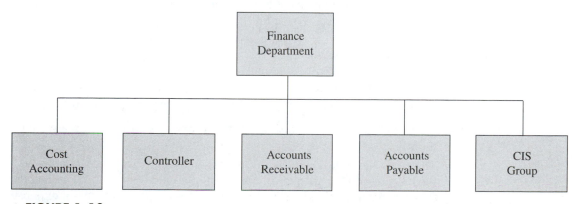

▲ **FIGURE 1-13**
Finance department structure

terms of sale have not been met, the customer's account is placed on overdue status and special measures are taken to procure payment. The company may place all other orders for the delinquent customer on hold, revoke the customer's open account status (*open account* means to receive the company's standard terms of sale), or forward the account to a collection agency. Loss of open account status would require the customer to pay on a *cash on delivery* (COD) basis. The key concerns of this group are as follows:

1. The receipt and recording of all payments made to the company.

2. The procurement of payments from customers with overdue accounts in the minimum time period.

3. The calculation, reporting, and minimization of overdue accounts receivable and receivable days.

   Receivable days are a measurement of how well the company is performing at receiving payments within the company's terms of sale. Receivable days are the average number of days within which customer payment is received. If a company has net 30-day terms of sale, and receivable days for a one-month period are 45 days, the company's terms are being exceeded on average by 15 days. This can represent a significant amount of money and can have a serious impact on the company's cash flow. The finance department will publish the total accounts receivable and the receivable days on a monthly basis.

Accounts Payable

The accounts payable department is responsible for payment of all of the company's liabilities. This includes the payroll, invoices, and other monthly expenses. The finance department will plan the payment of all liabilities on a priority basis. Before payment can be made, there must be some indication to the purchasing department that all of the goods or services were received in an acceptable manner. If so, the purchasing department will submit the invoice for payment. The key concerns of the accounts payable group are as follows:

1. To make payment for all of the company's liabilities in as long a time frame and with the least expense possible.

2. To calculate, report, and maximize accounts payable days.

   Accounts payable days are similar to the accounts receivable days calculation and are the average number of days within which an invoice is paid.

3. To adjust payments according to the company's cash flow in any given month.

   The sales, liabilities, and cash flow of the company is forecast and scheduled to provide for adequate cash for all planned operations. The finance department will sometimes accrue money for an upcoming major expense in a month where sales exceed forecast levels.

Cost Accounting

As its name implies, the cost accounting group of the finance department determines the cost associated with products and other company expenditures for the purpose of calculating company profits. Most of the cost accounting group's time is spent itemizing the current product costs. Each month this group will total material, labor, and overhead costs against the net sales for a particular product group and calculate the various types of profit by product line: operating profit, net profit, and earnings before income taxes. This information is critical for the company as it identifies the progress made toward meeting profit goals. These numbers also identify the most profitable products and the largest opportunities for improvement.

To ensure that the profit calculations are correct, it is important that the bills of material for the products are correct. Also, time studies that determine the allocation of labor to the product cost must be accurate. The cost accounting group will take steps to monitor and verify this data. The key concerns of the cost accounting group are as follows:

1. Determine accurate material, labor, and overhead cost data for all company product categories.

   This entails making sure that the bills of material are accurate and that the current cost data for a component is accurate. Time estimates of labor consumption for various product lines are studied and overhead allocations are verified.

2. Calculate and report operating profit, net profit, and earnings before income taxes.

   This is totaled for the company and broken down into the major product categories.

3. Review and study the costs of other company operations on request.

   These are special situations that involve significant expenditures to the company.

The Controller

The controller may be just one individual in a small company—or a management-level employee with a staff of accountants reporting to him or her in larger companies. In any case, the controller usually oversees the cost accounting, accounts receivable, and accounts payable groups and primarily manages the cash flow of the company.

Well before the company begins its fiscal year, the CEO and department managers are busy preparing operating budgets and sales forecast numbers that will determine the company profit plan. The profit plan starts as a forecast of the company's expected profit, but it becomes one goal of the company and thereby a measure of its performance. In order to accomplish this, the controller determines a working budget forecast for the company based upon the approved budgets for all of the operating departments. The initial budgets are plugged into the profit plan and combined with the sales forecast and expected manufacturing costs to calculate profits. If the profit levels are agreeable to upper management, the departmental budgets are accepted as is. Otherwise budget reductions will ensue until acceptable profit plan numbers are achieved. Every month each department manager receives a report of his or her department's monthly and year-to-date expenditures compared to the forecast.

The controller will develop a complete balance sheet, income statement, and report on the status of department budgets and capital expenditures. Literally every category of income and expenses will be compared to the profit plan numbers for the current month and the year to date. Once a month the company reviews this data at a special staff meeting. At this meeting the reports generated by the controller—or the finance manager—are first on the agenda. The primary concerns of the controller are as follows:

1. To achieve the current profit plan forecast.

   In order to achieve the profit plan, the sales level must be equal to or above the forecast number. If not, the controller must take action to reduce company expenditures so the profit plan can be maintained. Sometimes budget cuts are not practical or in the best interest of the company long term. Upper management may accept operation at the lower profit level in this instance.

2. To orchestrate sound financial short- and long-term planning for the company.

   This involves leading the development of annual budgets, forecasts, and the profit plan, as well as long-range strategic planning.

Computer Information Systems

The typical computer information systems (CIS) group has grown consistently over the past 40 years. This group is responsible for the company's central computer system and its hardware, software, and operating procedures. The use of mainframe computers, distributed processing, and now *local area networks* (LANs) have changed the skills and knowledge required by this group as much and as quickly as the computers themselves. In response to this, many companies have created a separate telecommunications department that manages all telephone and computer communications as well as the primary computer system. The telecommunications manager usually reports directly to the CEO, increasing the visibility and status of this important group. We will discuss the function of the more predominant CIS group.

There is usually one software package central to all company operations that is used by each department. This software will use many databases for the processing of customer orders. Some of the operations and the databases, starting with the receipt of a customer order, are described as follows:

1. *Order entry function.* Orders for products are entered utilizing customer and final product databases. Orders will be entered only if acceptable terms

of payment are established between the buyer and seller (i.e., net 30 days, major credit card, and so on.)

2. *Manufacturing shop orders*. These schedule the manufacture of the product by scheduling the manpower and allotting the parts as called for on the bills of material (the product database) for the product.

3. *Purchase orders*. These are generated to replace components consumed in manufacturing and provide for meeting future orders. These orders are generated from the forecasts for products and the bills of material for those products.

4. *Receipt of goods*. As goods are received, there are two types that are tracked by the computer system: returned products and items that have been ordered. Returned products are returned for repair, replacement, credit, or scrap, and there is usually a separate system that will handle these situations. Items that have been ordered to manufacture product are checked into inventory.

5. *Inventory totals*. As components and materials are checked in and out of stock, the computer database is updated.

6. *Shipping orders*. When products are complete and ready for shipment, all of the shipping documents (packing list, invoice, and labels) are printed out by the computer.

7. *Payments for products*. These are processed through accounts receivable.

These are just some of the central operations usually implemented on the central operating system. The computer information systems group will often utilize the same central software package to generate reports of manufacturing costs, profits, department budgets, product failures, and much more.

The prioritized concerns of the computer information systems group are as follows:

1. To provide an accurate database and an efficient computer system to handle day-to-day company operations.

2. To provide efficient and appropriate security for the company's operating systems and databases.

3. To provide for future expansion in efficiency and capabilities in order to meet future needs of the company.

Finance Department Summary

The finance department manager, who usually carries a title of vice president, has an operating budget to forecast and control—and also monitors the budgets of all the operating departments. He or she must coordinate the efforts of the controller's group and the CIS group. Like other managers, the finance manager continually looks for ways to improve the operation of the finance department. He or she also looks

1. Maintaining secure financial records

2. Meeting profit plan objectives

3. Identifying and reducing key company costs

4. Provide adequate cash flow for company operations

5. Accurate reporting of the company's financial transactions

▶ **FIGURE 1–14**
Finance department manager's priorities

for significant cost improvement opportunities throughout the company. The finance department will champion cost reductions because of their unique perspective and ability to see the relative costs of the company. The finance manager will spend much time completing monthly reports for upper management and the CEO while making sure that the company meets all current tax laws and directives. The summarized priorities of the finance department manager are shown in Figure 1–14.

1–8 ▶ Human Resources

The human resources department is saddled with some of the more difficult company responsibilities: those that involve people. The human resources manager (also called the personnel manager) is responsible for finding qualified people, determining fair and competitive compensation, improving personnel performance, developing and maintaining company policies, and promoting the company's morale. If you will be graduating soon and will be searching for employment, the HR manager is whom you should be contacting to first inquire about employment opportunities.

To help find qualified people, the HR manager will use employment agencies, online recruiting, or a general advertisement. The HR manager must become knowledgeable of the various positions throughout the company and develop a feeling about the qualities of the right person for a given job. In support of this and the determination of fair compensation, complete job descriptions are developed that categorize job functions by job title. The HR department is responsible for generating these and keeping them up to date. The HR manager will take the lead in bringing in candidates for an interview and usually interviews the candidate first—and then last. He or she is the primary point of contact for the job candidate until reporting to work. The HR manager is the person who will greet you on your first day of work and will meet with you on your final day, typically in an *exit interview*.

For each job description, the HR department performs surveys on the salary levels being paid within the geographic area of the company to determine the range of salary being paid by other companies for that particular function. These become the salary ranges the company is willing to pay for specific job descriptions. As the cost of living rises, these ranges are usually shifted upward accordingly.

Once joining a company, an employee's pay increases should be determined by performance and be the result of performance reviews. These should be done at least annually and could be as frequent as every six months. The HR manager will orchestrate the completion of the performance review, which is actually completed by the department manager or direct supervisor. The HR manager and possibly other management will discuss the review with the department manager before it is approved and discussed with the employee. The performance review should always be honest and direct. The emphasis should be on constructive criticism, the positive reinforcement of solid performance, and the improvement of weak areas. The ways to improve performance start with the recognition of weak areas. Through work experience, self-improvement books, special training, and academic courses, weak areas can be improved.

The other compensation offered by most companies typically includes health insurance, life insurance, and retirement plans. The HR manager must budget and coordinate these programs. These are considered to be budget costs allotted to the HR department, and there is always pressure to reduce the cost of these programs or, at least, minimize their increase in cost.

Every company must have an overall set of rules, and the HR department must generate and maintain these in the form of a company policy manual. It includes the company's position on issues such as vacation time, sick and bereavement time, leave of absence, job posting, and many more. The most difficult and unpopular functions of the HR manager are the enforcement of these policies or the termination or layoff of an employee. The priorities of the HR manager are shown in Figure 1–15.

1–9 ▶ The Role of the Electronics Professional

The primary role of any technical professional is to provide sound technical solutions to technical problems. If something does not work, the electronics professional will be asked to make it work. In a design project, a design engineer creates a technical solution to solve the design problem. A design modification is suggested by a technician to remedy an ongoing field failure problem. In all cases, technical solutions should meet the overall objective of the design and promote general efficiency (time and money), quality, and reliability. These are all prime examples of the results most expected from the electronics professional.

The secondary role of most technical people is to provide a vision of future technological trends. Electronics professionals are exposed to these trends more than any other group within the company, and, as such, provide a better vision of coming technologies. This requires that all technical professionals keep current with new technologies as they become available.

1. Procuring qualified personnel

2. Improving personnel performance

3. Developing, maintaining, and enforcing
   company policies

4. Evaluate and recommend compensation levels

5. Manage and minimize cost of employee benefits

6. Maintaining and improving company morale

▶ **FIGURE 1–15**
HR manager's priorities

The final role of any technical group is to provide creative thinking to general problem solving. Because technical professionals practice creative thinking more often than most other groups, they usually have better-developed creative thinking skills. They are a source of creative thinking and are often sought after for suggestions in solving nontechnical problems. The roles of the electronics professional are summarized in Figure 1–16.

▶ Summary

In Japanese companies most new employees rotate through job assignments in each of the company's departments solely for the purpose of gaining an appreciation of each department's perspective. This is just one of many different practices that illustrate the difference between U.S. and Japanese companies—but it

▶ **FIGURE 1–16**
The role of the electronics
professional

1. To provide sound technical solutions to
   technical problems

2. To provide a vision of future technological trends

3. To provide creative problem solving for general
   problems

exemplifies the importance the Japanese firms place on understanding the roles of each company department.

Both U.S. and Japanese companies are owned by stockholders who desire a return on investment. If a company is profitable today, then the owners and managers will desire to increase profits at a rate greater than the rate of inflation. In order to achieve these goals, sales levels must increase and relative costs must decrease. These will usually be the top two goals of the corporate strategic plan: to increase both sales and profits to particular levels. The strategic plan will identify many other objectives designed to support the sales and profitability goals. Many of these strategic objectives become projects that the company will undertake and fall into one or more of the following categories:

1. *New products and services.* The development or acquisition of new products or services for sale.

2. *Research.* The research of new technology or other areas new to the company.

3. *Improving Efficiency.* The development of systems, procedures, or equipment that improve the efficiency of some company operation.

4. *Improvement of customer satisfaction.* The development of systems, procedures, or equipment that improve the quality, reliability, ease of use, and delivery of the company's products or services.

In this book we will discuss the completion of electronic-oriented projects that will determine whether significant strategic objectives of the company are met. The success of a project will depend on its viability and how well it is defined, planned, and implemented. In the next chapter, the method of selecting projects is discussed along with general project management practices.

▶ Exercises

1–1 List the benefits of becoming a corporation as compared to a sole proprietor.

1–2 Explain the difference between an *S Corporation* and a *C corporation*.

1–3 Define the term *strategic planning* and list the steps required to complete one.

1–4 What is the primary difference between the priorities of the marketing and sales functions?

1–5 What are the four primary functions that define marketing?

1–6 Which department is most concerned with the long-term strategy and sales level of a particular product line? Why?

1–7 Which department is most concerned with the sales level of the company overall? Why?

1–8 What is the primary reason for separating the product and support engineering and development engineering groups?

1–9   What is the primary difference between a research project and a product development project?

1–10   Which department is usually responsible for approving the substitution of components used in the company's products?

1–11   Define what is meant by an ECN. What is the purpose of an ECN, and which department is responsible for issuing and controlling an ECN?

1–12   Describe what is meant by an MRP system. What are its primary functions?

1–13   List the differences between what are called *capital expenditures* and *normal component expenditures*.

1–14   Describe the concept of *just-in-time* (JIT). What is the primary goal of a JIT system? What is necessary to make JIT work well?

1–15   What does the term *inventory turns* mean and of what is it a measure?

1–16   What department is responsible for scheduling the assembly of products within manufacturing?

1–17   What are the two basic ways of organizing a manufacturing operation that were discussed in this chapter? List an advantage and disadvantage of each.

1–18   As discussed in this chapter, which department is responsible for maintaining the company's facilities?

1–19   Which department is responsible for reviewing company procedures that affect the company's overall product performance?

1–20   Describe what is meant by the term *first article inspection*.

1–21   Which specific group within the company is responsible for overseeing invoices where payment is overdue?

1–22   Which specific group within the company calculates and reviews the allocation of labor to various product lines?

1–23   What is meant by a profit plan? Which part of the finance department develops the profit plan?

1–24   Which department is responsible for budgeting and coordinating health plan benefits for the company's employees?

1–25   Apply the process of developing a strategic plan to your personal goals by developing a *personal strategic plan*.

2 ▶ Managing Electronic Development Projects

▶ Introduction

In Chapter 1 the basic purpose and operation of a typical company were discussed. The process of strategic planning was introduced as a method for determining long- and short-range strategies and objectives to provide future growth and profitability for the company. A team from top management reviews and reissues the strategic plan on an annual basis. It usually includes a short-range plan (one year) and a long-range plan (three to five years). The key part of the strategic plan is the listing of *goals* and *objectives* for the company for each year of the plan. The strategic plan gives the entire company a vision of the company's game plan for the next few years.

The annual profit plan discussed in the last chapter is based on the strategic plan. Each year company management determines specifically how the company will achieve the objectives listed in the strategic plan. The result is the identification of many different types of projects. Developing new products, new systems for use internally, or computer system upgrades are a few examples. For the remainder of this book, the discussion will center on completing the electronic projects that result as objectives to support the strategic plan.

A key part of managing electronic development projects is the concept of *concurrent engineering*. In this chapter concurrent engineering is explained as a complementary process with strategic planning—and a concept called *total quality management* (TQM). The implementation of concurrent engineering is explored as well as the expected results. The specific topics for this chapter include the following:

- ▶ A history of engineering project management
- ▶ What is concurrent engineering?
- ▶ The results of concurrent engineering
- ▶ Methods of project management

2–1 ▶ A History of Engineering Project Management

The operation and methods used to operate companies has changed significantly over the course of the twentieth century. Advances in technology and population growth have driven most of these changes. All operational areas have been affected. In manufacturing the first major change came with Henry Ford's implementation of the assembly line in the early 1900s. The creation of the assembly line was a key innovation that changed manufacturing processes used by companies around the world. The following is a brief history of the trends that have changed the company environment and have ultimately affected the way engineering projects are managed.

The Early 1900s

Before the assembly line, products were manufactured manually by one or more skilled craftsman, usually one item at a time. Raw materials and components were purchased as required to meet customer orders. The product was manufactured per customer order and was often customized to meet customer specifications. Each craftsman paid attention to the quality of the product. The product's designers usually worked their way into their positions through years of experience on the production floor, so they possessed a solid understanding of the manufacturing process. Consequently, the products were designed with the knowledge of how they would be manufactured. Product documentation was minimal, as the craftsman would habitually perform the same operations daily. The volume was low, product lead times were relatively long, the quality level was very good, and the customer could buy custom product variations. There was minimal, localized competition between companies. Product life cycles—the length of time a design is produced before being replaced with a new design—were very long. Projects to develop products or systems were fairly simple, and the designers had a thorough understanding of how the product would be manufactured and used.

Pre-World War II Era

The implementation of the assembly line changed all this as the product moved through the plant on a main assembly line through the various assembly stations in sequence. At each assembly station, a component, subassembly, or some operation was performed on the product until it was completed. Many subassemblies were preassembled in batches, ready to be applied to the final product as it moved down the assembly line. The assembly line promoted the manufacture of large volumes of the same product. Product variations were difficult to implement and became unattractive to most companies. The low costs and high efficiency of the assembly line fueled mass marketing and the consumer society that still exist today.

Many companies built product to forecast instead of responding to customer orders. The product designers lost touch with the now complicated and spread out assembly process—and they usually had no experience on the assembly line. Improved product documentation was required to define the assemblies and to make sure that everyone was using the proper revision levels of components and procedures. Communication of design changes became critical. *Work-in-Process* (WIP) inventory was needed to support the assembly line concept for subassemblies that were preassembled and other unfinished assemblies. Also, finished products not required to meet customer orders were placed in *finished goods* (FG) inventory until their shipment to customers.

The quality of the end product was the result of many operations performed by different individuals located throughout the plant. It became necessary to inspect subassemblies and the final product to verify the quality level.

Competition was minimal and less localized. Product life cycles declined slightly during this era. Engineering projects to develop projects and systems were usually conducted informally, without a detailed understanding of the requirements for the product or system or how it would be manufactured. Engineering would complete development projects internally and just pass them on to manufacturing when complete. After a while, manufacturing would determine how to assemble the product efficiently and marketing would be able to sell the product at a profit.

Post-World War II Era

Automation 1945–1975

During this era many changes took place in the business environment as companies grew and expanded their operations. Increasing population and the application of automation fueled this expansion. Automation was used on the main assembly line or to complete subassemblies to feed the main assembly line. Automation was applied to a whole range of applications as companies increased production volumes and batch sizes and lowered costs to meet increasing competition from U.S. companies. The type of automation applied was more mechanized than computerized at this stage. More and more companies *built to forecast*, which reduced lead times to customers at the expense of growing inventories.

Product designers were increasingly distanced from the assembly process as new manufacturing plants were built, often in separate locations from the engineering department location. Documentation requirements were expanded as required to attain uniform quality levels of final and subassemblies possibly assembled at different locations. Product life cycles continued to decline and companies became increasingly departmentalized. The increased use of automation continues today, driven more by computers and telecommunications.

The methods of managing engineering projects began to change over this period. Many companies realized the importance of formally documenting project specifications before implementation. Projects were completed on a more formal basis and under more scrutiny from management. There were efforts to involve manufacturing earlier in the project to promote a smoother phase in to manufacturing of the new product.

The Computer Revolution 1975–1995

During this era, the computer was increasingly applied to solve company problems, improve efficiency, and reduce costs. Mainframe computers were first applied that performed mostly accounting operations such as payroll, accounts receivable, and accounts payable. Next the personal computer (PCs) literally invaded every department as PCs monitored sales data, created product literature, helped design the products, and created higher levels of automation for many assembly and test operations. Eventually the PC could interface with the mainframe, and local area networks (LANs) and wide area networks (WANs) became commonplace.

Companies continued to grow and spread out as computers and telecommunications made communication inexpensive and easy. Designers were increasingly distanced from the manufacturing floor that now utilized more sophisticated machines and computers than ever before. Company acquisitions increased as businesses chose to increase sales by acquiring new products and markets instead of developing them.

Integrated circuit technology, which had fueled the development of the computer industry, also picked up speed as new technology and components were applied to new products. Product life cycles began to decrease significantly. Increased emphasis was placed on short-term profit as stockholder interests became paramount to employee considerations. Corporate downsizing and efficiency improvements to increase profits and be competitive were the overwhelming strategies in place in the 1980s. Literally, all the trends initiated with the introduction of the assembly line continued on through most of this era. Volumes and batch sizes increased as customization of the product to meet specific customer needs decreased. Documentation increased as all bill of material information was maintained in the central computer. See Figure 2–1 for an outline of the key historical changes in the ways companies operate.

At the end of this period (1985–1995), a significant trend developed as the *World Trade Organization* (WTO) and other trade alliances were successful in eliminating tariffs and promoting global competition between businesses. Global competition combined with high levels of automation, increased technology, large departmentalized companies, high volumes, and short product life cycles have all served to change the way projects are managed and the way companies operate. The level of competition was greater than ever before, and most companies began to think on a global basis. This created more pressure to improve efficiencies and performance even further with quality improvements, a reduction of product lead times, reduced inventory levels, shorter development project times, and an improvement in their results. Literally every aspect of company operations was scrutinized for improvements in overall performance, quality, efficiency, and the reduction of costs.

At many companies this resulted in fewer managers, leaner departments, and a strategy of revamping all company processes. The development project became a key target for improvement because of its importance in meeting the company's strategic objectives. The following is a list of the key objectives for improving the development process—and results from the problems experienced—of most engineering projects during the computer revolution era:

1. Reduce project completion time

2. Improve success rate of sales and profits and meeting customer needs

3. Minimize changes required after the project release

Craftsperson manufactured	Assembly line manufactured	Assembly line manufactured with automation	Batch processed with automation and computerization
Built to customer order	Built to forecast	Built to forecast	Built to forecast
Customized per order	Decreasing customization	Decreasing customization	Little customization
Minimal documentation	Better documentation	More documentation	Computerized material lists required
Low volume	Higher volumes	Higher volumes	Very high volumes
Little competition	Increasing competition	Acquisitions and increasing competition	Acquisitions and global competition
Good quality	Average quality	Average quality	Average quality
Long product life cycles	Declining product life cycles	Further declines in product life cycles	Short product life cycles
Companies had a local perspective	Increasing national perspective	Predominant national perspective	Companies have a global perspective
Early 1900s 1900–1930	Pre-World War II 1930–1945	Post-World War II 1945–1995	Computer Revolution 1975–1995

▲ **FIGURE 2–1**
Outline of historical company trends

4. Improve the rate at which projects are completed on schedule.

5. Reduce the *ramp-up* time of manufacturing

6. Improve the reliability and quality of the released project

7. Reduce the cost of development

2–2 ▶ What Is Concurrent Engineering?

Concurrent engineering (CE) is the logical result of the company environment that promoted improving the engineering project process. It is a direct response to the objectives listed in the previous section for improving the performance of company projects. It can be defined this way:

> Concurrent engineering is an engineering project management concept that promotes the consideration of all project requirements from the beginning of the project to the end. These requirements include performance, reliability, quality, customer use, marketing and sales, manufacturing, and financial issues.

1. Develop complete specifications

2. Consider manufacturing, customer use, and quality up front

3. Utilize a multifunctional team

4. Develop detailed schedules

5. Include a project verification stage in the project schedule

6. Involve key suppliers

7. Envision the goal

8. Promote continuous improvement

▲ **FIGURE 2–2**

Summary of strategies for improving engineering projects

Concurrent engineering is not a complete project management system but a concept for improving traditional methods. The strategies for improving the performance of engineering projects were developed to achieve the improvement goals outlined previously (and summarized in Figure 2–2) as follows:

1. Develop complete specifications defining the project and minimize changes to the specifications after the project is started.

2. Consider manufacturing, quality, customer use, field service, and disposal issues at the very beginning of the project.

3. Utilize a project team that has representatives from all affected departments. Promote innovation in the team. Empower them and give them incentives to promote their success.

4. Develop detailed schedules with distinct project phases. Deliverables should be defined for the end of each phase that must be complete before moving on to the next phase.

5. Include a *project verification stage* in which the quality, performance, and reliability of the project is verified independently and the project is utilized in the intended environment.

6. Involve key suppliers very early in the project.

7. Maintain a perspective of the overall project goal and all of the different design issues on a concurrent (simultaneous) basis.

8. Provide for continuous improvement by evaluating the performance of the project when complete and reflecting on what went well and the areas that need improvement.

The concept of concurrent engineering complements the strategic planning process. Strategic planning is a long-range planning method accomplished by the development of key company objectives and strategies to meet those objectives. Concurrent engineering provides an improved method for completing the engineering projects that result from the strategic plan.

Total Quality Management (TQM)

TQM is another concept that evolved about the same time as concurrent engineering and for the same reasons: global competition, demand for higher quality, profitability, and improved efficiency. The attempts during the post-World War II period and computer revolution eras to improve quality focused solely on product quality while other aspects of the company were not addressed. Improvements in product quality were attempted by incoming inspection, the inspection of subassemblies, and the final inspection of products before shipment to the customer. These methods succeeded in weeding out inferior products before reaching the customer but did little to resolve the underlying reason for most quality defects. The quality and manufacturing departments became armed camps, where there was little teamwork applied to resolving quality and manufacturing problems. The responsibility for improving quality was given to the quality department and its focus was on product quality.

TQM proposes a different approach. TQM places the responsibility for quality on all company departments and provides for the continuous improvement of quality in all areas of the company. This includes the following areas:

1. Customer operator manuals

2. Advertisement and sales promotion material

3. Company facilities

4. Packaging materials

5. Processes, policies, and procedures

6. Company documentation

The key aspects of TQM include the following: customer focus, continuous improvement, and total company involvement. Focusing on the customer includes both internal and external customers of the company. Internal customers are employees who are recipients of services of the various company departments (for example, employee benefits administered by the human resources department). External customers are the actual purchasers of the company's end product.

Continuous improvement promotes the realization that we can never improve enough; we can always get better. In TQM, certain output parameters are targeted for improvement. Methods are developed to measure the values of these parameters. The parameter is measured and reported monthly as strategies are continually applied to improve its performance.

Total company involvement simply means that all areas of the company—employees, managers, supervisors, suppliers, and customers—are involved in the quality improvement process.

The quality assurance department promotes TQM by developing methods of monitoring the quality level in all of the areas listed and working with all of the company departments to resolve quality problems.

2–3 ▶ The Results of Concurrent Engineering

When the principles and strategies of concurrent engineering are applied to the management of engineering projects, good things happen. The most important result is the *synergy* gained from the team approach to project management. Concurrent engineering is the application of good common sense to engineering projects to rectify the situations that evolved as companies grew in size and become departmentalized. The specific results of concurrent engineering are as follows:

1. The multifunctional team ensures that all project issues are addressed up front. Customer, manufacturing, quality, financial, and field service issues are considered equally with other design issues.

2. Project activities are scheduled in an orderly way, taking advantage of parallel paths wherever possible and resulting in minimal project linear time. The project team believes the schedule can be achieved.

3. When involved early in the process, key suppliers can supply new and different perspectives that can improve the project results.

4. Project delays are identified and addressed as they are realized.

5. Delays resulting from incomplete or changing specifications are eliminated.

6. The project visibility is very high throughout the company.

7. The ramp-up time for starting manufacturing is planned and minimized.

8. Quality is measured independently.

9. Creativity is maximized.

10. The team feels a joint ownership in the project (i.e., they become stakeholders) and its results.

The results of concurrent engineering, when properly applied, address all the areas of improvement previously noted. They are summarized in Figure 2–3. However, implementing concurrent engineering in many company environments is difficult as existing departmental structures try to maintain their power base and naturally oppose the application of these concepts. The implementation of concurrent engineering is discussed in the next section.

2–4 ▶ Methods of Project Management

The implementation of concurrent engineering is usually where a company falls short in meeting its goals. The natural structure and political environment of many companies is sometimes so strong that concurrent engineering concepts can be applied with disappointing results. It is important to completely apply CE principles and let the team do the rest. Project failures are most often caused by defi-

1. Manufacturing, customer use, and quality are considered up front.

2. Project schedules are minimized.

3. Key suppliers provide new ideas.

4. Project delays are identified and addressed quickly.

5. Project delays from specification changes are minimized.

6. Project visibility is high.

7. Short ramp-up time to manufacture.

8. Quality is improved.

9. Creativity increases.

10. Team members feel ownership.

▲ **FIGURE 2–3**
Summary of the results of concurrent engineering

ciencies in the areas of project definition, leadership, time, or money. Shortcomings in these areas bring about negative results with or without the use of CE concepts. Some specific reasons for project failures follow:

1. The project is not properly reviewed and defined initially.

2. Ineffective project management.

3. A lack of accountability from the team members to the project manager—when team members are not accountable to the project manager in any way.

4. Lack of focus by team members—when team members are assigned to the team in addition to all of their normal responsibilities. Team members are expected to be productive and active team members while also performing their full-time responsibilities.

5. Political maneuvering—constant positioning and meddling by top management to force project decisions in favor of their departments. Certain managers consistently intervene to make project decisions follow their department's agenda.

6. Unreasonable project schedules—project schedules dictated by top management that are not practically achievable with the allotted budget. In this

case the project team seldom believes that the project can be accomplished in the time frame—and it seldom is.

7. Inadequate investment in the project—a project budget dictated by top management that is unreasonable.

Figure 2–4 shows a summary of the reasons for project failures.

The Project Manager

For every major project someone must be responsible for its completion. The project manager provides direction and pulls together all aspects of the project, the technical disciplines, logistics, financial issues, and most important, the people. The project manager develops the project schedule in conjunction with the project team. The project manager has a significant impact on the project and ultimately determines the degree to which it is completed on time and meets the original goals. Consequently, he or she should be on board, leading the project from step one: gathering information. Ideally, the project manager should have experience in project management and be someone that is close to the *center* of the project. The center is the company department where most of the project activity takes place. For a product development project the project manager usually comes from engineering or marketing. Someone from Manufacturing or Engineering usually heads up a project to develop a new manufacturing system.

Good project managers are hard to find and develop. There are few ways to really learn the requirements for good project management other than going through the experience. The best way is to learn by experience as part of a well-managed project team. There are certain qualities that foster the management of

▶ **FIGURE 2–4**
Summary of reasons for project failures

1. Projects that are not properly defined

2. Ineffective project management

3. Lack of team member accountability

4. Lack of team member focus

5. Political maneuvering

6. Unreasonable project schedules

7. Inadequate project investment

1. Possesses good project vision

2. Technically skilled

3. Broad knowledge type

4. Proactive

5. Lead blocker

6. A complete communicator

7. A task master

8. A team player

9. A mentor

▶ **FIGURE 2–5**
Qualities of the project manager

a project that resemble the qualities of a good athletic coach. These qualities are stated and discussed as follows and are summarized in Figure 2–5:

1. *Good project vision.* The project manager must have the ability to step back and see the big picture for the entire project and at the same time be able to focus meticulously on every project detail.

2. *Technical skills.* The project manager should be proficient technically.

3. *Broad knowledge type.* Preferably, the project manager should be the type that knows a little about a lot of different areas (broad knowledge) versus one that knows a lot about a little (a narrowly focused specialist).

4. *Proactive (linebacker) mentality.* As problems develop, the project manager should be the type that operates proactively and reacts to problems as they develop. Project managers should not be afraid to get involved with project activities and fill in with project tasks as appropriate and possible.

5. *Lead blocker.* The project manager should look for and remove bureaucratic obstacles for the design team. This promotes a "get it done" attitude among the other team members.

6. *Complete communicator.* The project manager must possess articulate written and verbal communication skills.

7. *Taskmaster.* The project manager must be demanding of the team members and others within the company. At the same time the project requirements must be achievable and reasonable.

8. *Overall team player.* The project manager must be an example of a good team player by never losing sight of the ultimate goal. The vision of the ultimate goal must be constantly made visible to the team.

9. *Mentor.* Overall, the project manager must be a mentor for the team and a meticulous planner. He must also be a sound thinker and decision maker.

If you see yourself as having a good mix of these skills, give some consideration to project management as your career develops. Good project managers are worth a premium on the job market.

Twelve Methods for Managing Projects

Project management is an extremely difficult job. Good project managers are rare because they must have such balance in their personalities, leadership, organizational, creative, technical, and communication skills. There are distinct methods that successful project managers utilize. Twelve distinct methods are identified here that serve to combine good project management techniques with concurrent engineering concepts. (These are summarized in Figure 2–6.)

1. Select an overall project group that includes at least one person from each involved department.

   This typically would include team members from marketing, engineering, manufacturing, quality assurance, and finance. In some cases it may be necessary to have multiple team members from a particular area. This usually occurs with the operations department when both the manufacturing and material control groups need representation on the team. It is important to keep the team down to a reasonable size. The project team should represent a focal point for the directing, monitoring, and tracking of the progress of the project. It should not be where all of the work is completed. Strive for *quality* and not *quantity* in the project team. Team members are expected to represent the department and communicate project activity and concerns to and from their department.

2. Conduct regular and efficient project meetings.

   Determine the frequency of the meetings by the phase and nature of the project. During the design phase, for example, certain members of the design team may get together often, but the overall project group will not need that type of frequency at that stage of the project. If meetings are too frequent, everyone's time is wasted. If the meetings are not frequent enough, issues will not be addressed in as timely manner as they should. When conducting project meetings, realize that the purpose of the project meeting is

1. Select a multifunctional project team

2. Conduct regular and efficient project meetings

3. Develop timely and accurate minutes

4. Manage by walking around

5. Involve the project team in all major decisions

6. Address project slippage immediately

7. Focus on the goal

8. Promote interdepartmental communication

9. Value the time of each team member

10. Be obsessive about quality

11. Lead by example

12. Enjoy the process

▶ **FIGURE 2–6**
Summary of project management methods

to direct, follow up, and track project performance and issues. When complex issues come up that involve only a few team members, assign someone the task to set up a separate meeting to discuss and resolve the issue, and then follow up at the next project meeting. Every effort should be made to make meetings as efficient as possible.

3. Develop timely, consistent, and accurate minutes to all project meetings.

The minutes should be brief and to the point, summarizing discussions and always concluding with some action to address key points. When problems are discussed at the project meetings, summarize the discussion and determine a specific action item to address the issue, noting who is responsible for completing the action and the due date. (See Figure 2–7, a sample meeting report.) The minutes should be distributed within one day of the project meeting. While everyone will take their own notes, the meeting minutes will be the one document that everyone will see and respond to. The minutes should be distributed to all top management affected by the project.

Date: 1/1/00
Subject: Meeting Report—Digital Thermometer Development Project
Attendees: *Engineering—Jeff Way*
               Finance—Gary Swift
               Manufacturing—Gene Former
               Quality Assurance—Clarence Wood
               Marketing/Sales—John Espisito
               Field Service—Jim Field

Completed By: John Espisito

The subject meeting began with the introduction and review of the Digital Thermometer Project business proposal. The proposal was given to all attendees and is available to any interested parties that may not have received a copy.

John Espisito is the Project Manager for this project and the attendees are the primary project team.

There was a long discussion about many aspects of the project, but there was general agreement that the project appears to be reasonable and within each department's capabilities. The biggest concern was caused by the cost goals combined with the need for SMT technology to meet the physical size criteria. In order to address these concerns, the following action items were discussed and agreed upon:

*   **Action Item:** Develop a complete engineering specification for the Digital Thermometer based upon the marketing specification and business proposal for the project.*
*   **Person Responsible:** Jeff Way*
*   **Required Completion:** Next meeting*

*   **Action Item:** Determine a list of five possible SMT board suppliers for the project.*
*   **Person Responsible:** Gene Former*
*   **Required Completion:** Next meeting*

*   **Action Item:** Procure literature available on SMT quality issues and report key areas that should be emphasized.*
*   **Person Responsible:** Clarence Wood*
*   **Required Completion:** Next meeting*

*   **Action Item:** Review market forecast with sales force to verify that the forecast numbers are achievable.*
*   **Person Responsible:** John Espisito*
*   **Required Completion:** Next meeting*

The next meeting will be held in two weeks at the same time and place. Please review this report and let me know of any discrepancies.

▲ **FIGURE 2–7**
Sample meeting report

4. Manage by walking around.

Get out of the office and see firsthand what is going on in different project areas. One can determine what is really going on in the project by periodic and impromptu visits to the various people involved in the project. This shows a sincere interest in project activity and provides an opportunity for discussion about the project on a more informal basis.

5. Involve the entire project team in all major project decisions.

Nothing can destroy the team makeup more than removing team members from the decision-making loop. Ultimately a decision will have to be made that not all team members will favor, but it is important for all to have their say about major project issues.

6. Address project slippage immediately.

Whatever the cause of the slippage, some action must be taken to make up for the slippage immediately, or the project end date will have to change.

7. Keep the team focused on the ultimate goal of the project.

As the project progresses, it is always important to keep sight of the original goal. With the incredible detail that comprises most projects, it is very easy to lose sight of the overall goal of the project.

8. Promote communication between the departments and all members of the project team.

Promote the sharing of project problems, activities, ideas, and solutions between the various departments and other project teams. Try to leverage the creative power available in the company. *Good communication* is critical between all team members.

9. Place a *high value* on the time of each member of the project team.

Never waste time by having meetings that are too long or too frequent or having team members perform activities that are wasteful. Make-work projects that attempt to make someone look good are just another source of declining team morale. When time is wasted, it gives the team members the idea that their time has little value and they become less concerned about wasting it on their own.

10. Be obsessive about the quality level of every aspect of the project.

This means to apply the concepts of total quality management to the entire project. This sets an example that encourages each team member to do the same. The result will be an emphasis on quality through all aspects of the project.

11. Lead by example and respect the perspective of each team member.

In order for the team to function well, each member of the team must respect the perspective of each other team member. Review the priorities of each department discussed in Chapter 1 to get a fresh perspective for each department. The project manager must exemplify this concept and expand on it. The project manager should understand the perspectives of the other departments more than anyone else on the project team.

12. Enjoy the process—feel the excitement.

The project team is giving birth to something wonderful. It is very hard work but very rewarding at the same time. Take many deep breaths to counteract the natural tendency to "wind up" too tightly. The best work will flow effortlessly. At times in the project, have lunch brought in for the team or have them go out to dinner together. The best projects are the ones that are the most fun.

The Team Player

The most important aspect of being a good team player is understanding the position you are playing. If you are an offensive tackle on a football team, you will not be running with the football. This section discusses the general role of the electronics professional in order to develop a better team attitude.

Since our review of the differing departmental perspectives in Chapter 1, you should have a better understanding of the interests of each department manager and the issues most important to the operating departments. It is important to develop an understanding about the needs of other departments while placing the goals of the company first and foremost. This is why it is so important for a company to prioritize its goals and to make them practical. The results of being unrealistic occur when a company states competing goals, such as a significant increase in market share and profits in the same year. In competitive markets price concessions may be required to successfully increase market share, making it difficult to increase profits at the same time. The company, as well as any project team, must clearly identify the goals before starting a new fiscal year.

Have you ever thought back and wondered why the concept of teamwork was always promoted by youth organizations while you were growing up? Little league baseball, scouting, music, and dance programs all promote teamwork. While we all know that teamwork is important, do we really understand the depth of its importance? During youth, our performance levels are monitored for our pursuits individually and as a group. Many times it is the individual measurements that become the most recognized and praised. In a team environment, there is always a conflict present within us, working for ourselves or to achieve the goals of the team, whether to sacrifice bunt or try for the game-winning hit. When teamwork is promoted, the words "put the team first, let's win the game" are often used. On the other hand, human nature wants us to succeed individually more than as a team. It is part of our survival instinct. We naturally put number one, ourselves, first, and we should not feel guilty about it. It is part of our makeup. The most important thing we need to realize when we are part of a team is that:

All of the team members' individual goals will be better served by a team that works together to achieve the team's goals.

We need to make the connection between the team and individual goals, while realizing that our primary instinctual goals for ourselves are best served if the team is successful.

Electronics professionals should realize that they are typically not experts in marketing, manufacturing, quality assurance, and finance. They should realize that other departments do have functions and needs that are not clear or easily understood by technical people, much the same as other departments have difficulty understanding the needs of technical people. This is why it is important for each to understand and respect the perspective of the other departmental areas. Many times in a design project, Marketing or Manufacturing will request that the design possess some feature or function that goes against the design philosophy of the engineering group. It is important to have an open mind in these situations and consider the legitimacy of the request. Sometimes design engineers will indicate that a design feature is not technically feasible, just because they do not agree with the request. This is not the attitude of a team player.

Good team players are not usually the standouts on athletic teams—or project teams. We all recognize, however, that without them there would be no successful teams. How many times have there been teams that were not as exceptionally skilled as other teams but were successful because of how they worked together? In the workplace and on project teams, the project leader is an important ingredient in determining how well a team works together, but each team member must take some responsibility for fostering a positive team attitude. Good team players will most often exhibit the following traits:

1. *Focuses on the goal.* Understands the goal of the project and strives to attain it.

2. *Looks for the best way to contribute.* Finds the unique talents that he or she can best contribute to achieving the project goals.

3. *Keeps an open mind.* Is open to all ideas and suggestions. Hears people out and does not try to second-guess them.

4. *Possesses good listening skills.* Listens and concentrates on ideas and concepts. Consequently develops a strong understanding of most concepts.

5. *Communicates needs well in advance.* Coordinates and plans needs that come from other departments and co-workers. Accordingly, these needs are usually completed on time and with little disruption to other company operations.

6. *Quietly gets the job done.* With little fanfare completes required project activities on time and at a high quality level.

7. *Is considerate of co-workers.* Is responsive to the needs of co-workers.

It is not a coincidence that people that are good team players can also develop into good project managers. Many of the characteristics of the good team player foster good project management.

Some working relationships in a company are more important than others. The most critical and problematic is the relationship between marketing and sales and engineering. The level of cooperation between these two areas can mean success or failure to most companies. When working together, these two groups can develop very creative functional products that meet customer needs. When they operate separately, the product is developed in a void and seldom meets its strategic objectives.

The reason this relationship is so important is that the marketing and sales person is the representative of the customer on the development team. The engineer is the person that has the technical know-how to resolve customer issues and desires in the design of the product. These two groups must work closely and use creativity to resolve design issues. The engineer must try hard to understand the perspective of the marketing and sales person and the customers he or she represents. The marketing and sales person must realize that the engineer should understand everything possible about the customer situation in order to provide good technical solutions. All too often, marketing and sales people will demand certain performance improvements in new products simply to make the product appear better than a competitive product, without understanding the actual benefit to the customer. The engineer, on the other hand, finds the increase in performance difficult to achieve and unjustified as a benefit to the customer. There are many reasons and situations that make it difficult for these two groups to work together, but in the long run it is critical that they do. The best way for this to happen is for the engineer to travel and visit customer sites with the marketing and sales person. There is no better way to build the relationship of these two areas. The engineer develops a better understanding of the customer situation as well as the world in which the marketing and sales person must operate.

The engineering and manufacturing relationship is also critical to the company's success. The level to which these two groups work together determines the overall efficiency of the manufacture of the product. The relationship between these two groups is often strained by the immediate needs of the manufacturing department to resolve problems on the manufacturing floor. Here too it is important for the engineer to understand the situation that Manufacturing must deal with. The best way is to experience the situation firsthand. The manufacturing department must thoroughly communicate its needs and ideas to the engineering department. When these two departments work well together, the manufacture of the product can be refined down to a science.

Finally, the relationship between Marketing and Sales and Manufacturing is also crucial, as the members of the manufacturing department must know well in advance the needs of the customer in the form of forecasts and other information. This relationship does not involve much creativity, but it is based more on consideration and good communication of customer needs in order to utilize the capabilities of the manufacturing department.

▶ Summary

Much of what we have discussed in this chapter has involved communication. Even in this information society, with elaborate and sophisticated ways of communicating globally, we are finding it increasingly difficult to communicate accurately and work together. Good project management and concurrent engineering are all about good communication.

When strategic planning is combined with concurrent engineering and total quality management, a company is able to focus on the most important issues that will support future growth and profitability:

1. *Strategic Planning.* Provides a planning road map for the future.

2. *Concurrent Engineering.* Provides a method for the completion of projects.

3. *Total Quality Management.* Provides a focus on continuous improvement of quality levels of all aspects of the company.

In the beginning of this chapter, we discussed the manufacturing methods of the early twentieth century. This era was characterized by craftspersons who manufactured quality products one unit at a time. These craftspersons focused on the quality and efficiency of everything they did and continually improved their process. The combination of strategic planning, concurrent engineering, and total quality management, working together in today's high-technology and segmented business environment, is an attempt to emulate the results of the craftspersons of the early 1900s.

Individuals who start a company on their own and attempt to develop, manufacture, and sell custom products often perform the functions of marketing, sales, manufacturing, quality control, engineering, and finance—all at the same time. Such people develop a new appreciation for the concepts of concurrent engineering because they experience them almost in perfect form. The ultimate goal of concurrent engineering is to consider these issues completely.

As the world and the business environment continues to change, so too will the manufacturing and engineering project management processes of the future. Technology, telecommunications, global competition, population, and natural resources will continue to lead the way. The concept of continuous improvement will direct the nature of these changes as the world continues to respond to change and continues to improve our way of life.

▶ References

Kolarik, W. J. 1995. *Creating Quality.* New York: McGraw-Hill.

Prasad, B. 1996. *Concurrent Engineering Fundamentals.* Upper Saddle River, NJ: Prentice Hall.

Turtle, Q. C. 1994. *Implementing Concurrent Project Management.* Upper Saddle River, NJ: Prentice Hall.

▶ Exercises

2–1 Identify the key innovation that changed the way products were manufactured between the early 1900s and the pre-World War II years. Also, list all of the results of applying this innovation.

2–2 List the key changes that were made to manufacturing processes from the pre-World War II to the post-World War II eras.

2–3 List the key changes that were made to the manufacturing processes from the post-World War II years to the Computer Revolution era.

2–4 Define in your own words what constitutes the successful completion of an engineering project to develop a new product.

2–5 Define the terms *WIP* and *FG* inventory. Which type of manufacturing process results in significant levels of WIP inventory?

2–6 What is meant by the term *project linear time*?

2–7 What are the undesirable results of design changes after a project has been released to production?

2–8 Define the concept of *concurrent engineering*.

2–9 In today's business environment, what are the key reasons for the application of concurrent engineering principles?

2–10 In your own words, define the term *total quality management*.

2–11 List all the reasons why a project team should include a representative from each major department.

2–12 Why is it important to distribute meeting minutes soon after the meeting has occurred?

2–13 This chapter discussed the typical reasons for project failure and included "lack of accountability" as one of the reasons. What do you think this means? How are the team members accountable to the project manager for project activity when they do not typically report to the project manager?

2–14 Which three concepts discussed in this chapter work together to promote the future growth and profitability of a business?

2–15 Between which two company departments is an attitude of working together most critical to the success of the project and the company?

2–16 Define the concept of *continuous improvement* and how is it implemented in most companies.

3 ▶ Approval Agencies

▶ Introduction

A major consideration of any new product design is the requirement for approval by an independent testing laboratory. The requirement for agency approvals should be stated in the specifications before the project is started. The marketing department, after analyzing the market requirements for agency approvals, determines if any agency approvals are required. The design team must obtain and review all the appropriate requirements of the approval agencies and factor them into the specifications and the project schedule. Some companies allow products to be sold without formal approval on an "approval pending" basis, while others require formal approval before product sales can commence. It may take a long time to obtain approval, so it is important to submit applications and pay any up-front costs as soon as possible in the project.

This chapter discusses the leading approval agencies and the process for getting approval. The specific topics of this chapter include:

- ▶ An overview of approval agencies
- ▶ National Electrical Code
- ▶ Underwriters Laboratories, Inc.
- ▶ CSA International (formerly the Canadian Standards Association)
- ▶ VDE and TÜV
- ▶ The European Committee (CE)
- ▶ International Standards Organization ISO 9000 and ISO 14000

3–1 ▶ An Overview of Approval Agencies

The foundation for most approval agencies has come from the insurance industry, which has supported the development of institutions to develop standards in product safety and performance. The approval agencies that are discussed in this chapter are the primary ones recognized around the world. These agencies function in a geographical area where they have gained acceptance by the government or a group of customers (considered a market segment) that use particular products. Sometimes an approval agency covers a very broad area of product types. Underwriters Laboratories, Inc., is a good example of an approval agency with broad coverage. Other agencies focus on narrow product segments, such as Factory Mutual, which deals primarily in industrial safety devices. Once an approval agency is accepted in a geographical area or a product category, it is important for the manufacturers of these products to be approved by that agency. In some cases approval is mandated. Such is the case in Europe, where acquiring CE (European Committee) certification is mandated as a legal requirement before products can be imported into any European Union country. In the United States, Underwriters Laboratories (UL) is an accepted authority for many product categories, but federal law does not require a manufacturer's product to be UL approved. In this case, the overwhelming majority of customers for these products in the United States will only buy products that have been UL approved. In the United States, the market for the product dictates the requirement for agency approval, not the government. On the other hand, there are certain categories of equipment, such as medical products sold in the United States, that must be approved before they can be sold. A government agency, the Food and Drug Administration (FDA), is the approval agency for all medical products. The FDA is also involved with the approval of equipment used to manufacture food and drugs as well as the approval of the food and drug products themselves.

The most well-known standard in the electrical industry is the National Electrical Code (NEC). The NEC is not a product standard but a standard on how electrical work should be performed. The NEC also specifies that certain types of equipment be used in certain situations. For example, the NEC may require the use of a UL-approved connector in certain situations. The NEC effectively becomes an approval agency, as it requires the certification by approval agencies of certain types of electrical components. Consequently, for many product categories it is important to be knowledgeable about current NEC requirements.

In the United States, approval agencies are recognized by a federal agency, the Occupational Safety and Health Association (OSHA), as a Nationally Recognized Testing Laboratory (NRTL) to certify products for the U.S. market. The address, phone number, and Web address for the major approval agencies are included in Appendix C.

It is important to understand the perspective of the approval agencies. These agencies are always concerned first and foremost with product safety. They are adamant about knowing what the product will do if it catches fire or if the mechanical or electrical ratings are exceeded. They are equally concerned with the personal safety of individuals using the product. They do not necessarily concern themselves with how well the product works. If the specifications indicate that the product is to be 1 percent accurate, they will probably not test the product to determine the accuracy or be concerned with that aspect of the product. Approval agencies are concerned with what happens when a technician inserts a screwdriver into a slot on the side of the product and the potential for electrical shock.

They are also concerned with what happens when the input power source exceeds the operating power rating by 100 percent: does the product fail and catch fire or simply fail with a blown fuse? The point is that obtaining an agency approval does not necessarily mean that the product is a good product, it means that these agencies have found the product to be a safe product.

3–2 ▶ The National Electrical Code

The National Electrical Code (NEC) is one of the most recognized electrical standards in the world. It was developed initially in 1897 by the National Fire Protection Association (NFPA) and is revised every three years. The current edition is the 2002 edition. The NEC defines the methods and the equipment that must be used for electrical work. The NEC is put into effect when some governmental agency, be it federal, state, city, county, and the like, requires electrical work performed within its domain to meet the NEC standards. When this is done, a code enforcer is retained by the governmental agency to inspect any work that is completed. The current NEC is an extensive document that includes more than 700 pages in its new 8.5″-×-11″-format. It is divided into nine chapters as follows:

Chapter One—General Information

Chapter Two—Wiring and Protection

Chapter Three—Wiring Methods and Materials

Chapter Four—Equipment for General Use

Chapter Five—Special Occupancies (i.e., hazardous locations)

Chapter Six—Special Equipment (i.e., industrial equipment, pools, and the like)

Chapter Seven—Special Conditions (i.e., emergency systems)

Chapter Eight—Communication Systems

Chapter Nine—Tables and Example Calculations

The NEC refers to many other approval agencies, as it requires certain approvals for products that can be used in electrical installations. If you are involved with the design and manufacture of products that are utilized in residential or industrial electrical systems in the United States, a good working knowledge of the NEC is vital.

3–3 ▶ Underwriters Laboratories, Inc.

Underwriters Laboratories, Inc. (UL), was established in 1894 for the purpose of certifying the safety of industrial, commercial, and consumer products. UL is an independent, not-for-profit organization. As one of the first organizations established for this purpose, UL has become a leading product safety certification organization in the United States and a common household name. Because of the large U.S. market for products, there is also much interest by foreign companies in UL standards. Consequently, there is a significant ongoing effort at Underwriters Laboratories to harmonize their standards with other similar standards around the world. There are five Underwriters Laboratories testing laboratories located strategically around the United States:

Northbrook, Illinois—Main Office

Research Triangle Park, North Carolina

Melville, New York

Santa Clara, California

Camas, Washington

Underwriters Laboratories functions in two different ways, the certification of a company (some process or capability of the company) or the certification of the company's product. Formerly, product certification was the only function of Underwriters Laboratories and is still the largest part of its operation. However, in the last 15 years a significant trend toward the certification of company operations and procedures (ISO 9000 certification and so on) has led to a significant growth of this type of activity at Underwriters Laboratories and other approval agencies.

UL Product Certifications

When working on design projects that require Underwriters Laboratories' approval, first determine the applicable UL specification that covers the product category being developed. UL has published more than 800 different safety specifications. Just a few examples follow: electrical equipment for laboratory use, electrical controls for households, electric oil heaters, electric ranges, and sensing controls. Many times the UL specification that applies is already known from past experience and is spelled out in the project specifications. Other times statements are made, such as "The product must meet applicable UL specifications." In this case a search of the possible applicable UL specifications is made to determine the appropriate category and specification number. Another way to determine this is to check on the UL listing of competitive products. If there is any question about the appropriate product category, consult with Underwriters Laboratories technical support staff.

Product certifications can be one of two types carrying either a "Listed" or "Recognized" mark. "Listed" products are approved as ready to use, complete products. "Recognized" products are certified as components that can be used to make other end-use products. Take the example of a digital panel meter (a simple digital indicator). The digital panel meter could be listed for use as a complete product or it could be recognized as a component for use in a UL-listed product such as an FM radio receiver. Using UL-recognized components in an end-use product greatly simplifies the task of receiving its listing approval. The specifications that a product must meet to receive recognized or listed status vary. In any case, a copy of the appropriate UL specifications for the product must be procured and available to the design team at the very beginning of the design process. The design team should review the specifications and highlight all areas that apply specifically to the product under development. Questions about the specifications, their meaning, and how to apply them must be addressed up front. Completing this step is critical and can save an entire project from failure or costly design changes later.

As the design proceeds, the schedule should be monitored for the planned date of application to Underwriters Laboratories. At this point Underwriters Laboratories should be contacted to request a quotation for the approval of the design project. The quotation will estimate the entire project cost, the required payment with application, the expected time to completion, and when testing can begin. Today the UL application can be made directly through the Underwriters Laboratories Web site (www.ul.com). The first time a company seeks certification of a product category, it is wise to have someone from the design team visit the Underwriters Laboratories testing facility. This is to determine how Underwriters Laboratories will perform the tests and what is required for documentation and samples. The submitting company should make every effort to complete similar tests before submitting the product to Underwriters Laboratories for testing.

When submitting the product to Underwriters Laboratories, all manufacturing documents that define the product should be included with any documents that are supplied to the customer at shipment, such as part number designations, specifications, operator manuals, and service manuals. Underwriters Laboratories will require at least one working sample of the product. If the product enclosures or other mechanical parts utilize materials that are not UL approved, a number of samples of these parts will be needed for flammability testing.

When Underwriters Laboratories completes its tests, it will submit a report indicating the results of product conformance testing. In the case of nonconformance, the deficiencies will be listed along with recommended corrective actions. The design team must review the UL report and implement corrective changes to the product or its documentation and then resubmit for testing. Once the product has been certified by Underwriters Laboratories, it can carry a mark that indicates UL approval. The product is also listed as an approved device in the appropriate UL listing catalogs published annually by Underwriters Laboratories. The most useful of the UL catalogs are published consistently with distinctive cover colors with the following titles:

Electrical Construction Equipment Directory ("The Green Book")

Electrical Appliance and Utilization Equipment Directory ("The Orange Book")

General Information of Electrical Equipment Directory ("The White Book")

These catalogs can all be purchased from Underwriters Laboratories. The size, shape, and content of the UL marking is designated by the UL specifications and is shown in Figure 3–1. In some cases Underwriters Laboratories will require a site visit to the facility to certify the manufacturing process of the product. In these cases Underwriters Laboratories may revisit the site periodically to verify certification. It is also important to realize that when design changes are made to

▶ **FIGURE 3–1**
The registered UL mark of
Underwriters Laboratories

a product that has UL approval, it may be necessary to have Underwriters Laboratories review and possibly retest the product with the changes incorporated. Failing to do so can result in the UL approval being rescinded by Underwriters Laboratories.

Underwriters Laboratories and CSA International now have an agreement that allows each to test and certify products and components for the other. When UL performs testing for CSA listing and the UL and CSA standards differ, Underwriters Laboratories will certify marking the product with a "C" to the left and outside the circle as shown in Figure 3–2. This indicates that the product has been tested by Underwriters Laboratories to the CSA standard only.

In North America, UL and CSA codes and standards are becoming uniform as part of the North American Free Trade Agreement (NAFTA). When complete, these codes should be common for the United States, Canada, and Mexico.

UL Company Certifications

Company certification from Underwriters Laboratories involves the certification of a service performed by a company. Alarm systems are a good example of this type of certification. Underwriters Laboratories offers an alarm system certificate to qualified companies. To be certified, the company applies and requests the appropriate approval certificate. Underwriters Laboratories will qualify the performance of the company by inspecting alarm systems installed by them using UL specifications that cover the application. If acceptable, Underwriters Laboratories certifies the company's work and allows them to issue UL certificates for any system that it installs that is covered in the UL specification. The UL-approved company is subject to periodic inspections of its operation, practices, and current installations.

▶ **FIGURE 3–2**
CUL listing

3–4 ▶ CSA International

CSA International is the new name for what had previously been called the Canadian Standards Association or simply CSA. CSA International was established in 1919 and is the equivalent of Underwriters Laboratories in Canada, with a growing international presence, as their new name implies. CSA International performs identical functions as Underwriters Laboratories and, in fact, can approve products for the U.S. market as well. This is due to the agreement between Underwriters Laboratories and CSA International described in the preceding section. One primary difference between CSA International and Underwriters Laboratories is that CSA International does not differentiate between a component (recognized mark) or end use of equipment (listed mark) certifications. There is a growing trend for all of the national approval agencies to work toward common specifications and testing requirements. The head office for CSA International is located in Toronto, with other offices in Canada, the United States, and Asia.

3–5 ▶ VDE and TÜV

VDE, the German abbreviation for the Association for Electrical, Electronic & Information Technologies, develops standards and provides certification testing in Germany. VDE can be considered the equivalent of Underwriters Laboratories or CSA International in Germany and is one of many agencies that are authorized by the German government to issue the German "GS" mark, which signifies compliance with their safety standards. Prior to the development of the CE (European Committee) approval mark, VDE was a recognized certification by many other European countries and was viewed as critical for accessing the European market overall.

TÜV, the German abbreviation for Technische Überwachungs Verein Rheinland, is composed of 14 independent testing laboratories in Germany. Unlike VDE, TÜV does not develop standards, but it is authorized to issue the GS marking. There are several TÜV offices in the United States to help expedite the approval of European approvals for U.S. manufacturers.

3–6 ▶ CE Approval

The European Union (EU) consists of 15 countries that have joined together to create one European market and eliminate trade barriers between themselves. Part of this process has been the development of uniform product standards. European Committee approval requirements are the result of many years of discussion and represent one set of product standards for all products imported into any EU country. The European Union has required that *common European standards* must be implemented at the national level of each EU member. Any national conflicting standards must be withdrawn. The development of the European standards was completed by the following three agencies:

European Committee for Standardization (CEN)

European Committee for Electro-technical Standardization (CENELEC)

European Telecommunications Standards Institute (ETSI)

The most important of the European standards are known by the prefix "EN," which are published by the *Official Journal of the European Communities*. The development of these standards can be tracked in a publication called the *European Standards Bulletin*. Some of the more important standards follow:

EN 60601-1 Medical Electrical Equipment, Part 1: General Requirements for Safety

EN 60950 Safety of Information Technology Equipment, Including Electrical Business Equipment

EN 61010-1 Safety Requirements for Electrical Equipment for Measurement, Control and Laboratory Use, Part 1: General Requirements

EN 50081-1 Emission Requirements for Light Industrial and Household Equipment

EN 50081-2 Emission Requirements for Heavy Industrial Equipment

EN 50082-1 Immunity Requirements for Light Industrial and Household Equipment

EN 50082-2 Immunity Requirements for Heavy Industrial Equipment

EN 61000-4-2 Electro-Static Discharge up to 30 kV

EN 61000-4-3 HF Immunity up to 30 kV

EN 61000-4-4 Surge

EN 61000-4-5 EFT Burst, up to 4.4 kV

EN 61000-4-6 Current Injection up to 50 V RMS or 150 ma

A number of directives have been developed for various types of products that refer to the standard specifications just listed. A listing of the directives is shown in Figure 3–3. In order to determine the requirements of carrying the European Committee's CE (Conformité Européene) mark, one must determine the directive that applies to the product in question and review the directive for the standard specifications that must be met.

As of January 1, 1997, any products covered by the standard product directives must carry the CE mark to be imported into any European Union country. This was a significant event for any manufacturer who had been selling products into Europe. There are two primary requirements for CE approval: meeting the safety requirements and the required noise immunity and radiation testing. Meeting the CE safety requirements is typically not difficult for most manufacturers because these are similar to other international standards. However, the noise immunity

1. Safety of toys — 88/106/EEC

2. Construction products — 89/106/EEC

3. Electromagnetic compatibility — 89/336EEC, 92/31/EC

4. Machinery — 89/392/EEC, 91/368/EC, 3/44/EEC

5. Personal protective equipment — 89/686/EEC, 93/95/EEC

6. Telecommunications terminal equipment — 91/263/EEC,
   93/97/EEC, 98/13/EC

7. Medical devices — 93/42/EEC

8. Low voltage — 73/23/EEC

9. Automotive — 95/54/EEC

10. Energy efficiency for household refrigerators
    and freezers — 96/57/EC

▲ **FIGURE 3–3**
European Committee CE directives listing

and radiation test specifications pose a much more difficult problem. (Both noise immunity and radiation testing are described in Chapter 12.) Prior to the CE requirement, there were no legal or formal requirements to perform noise immunity testing. In fact, neither noise immunity nor radiated noise testing is required for UL or CSA certification. Radiated noise testing has been required for some time in the United States by the FCC on specific types of products, but no agency had previously required passing a noise immunity specification. Consequently, the requirement for CE approval forced many manufacturers to complete noise immunity tests that they had never performed before. Accordingly, significant interest developed regarding the procedures and equipment needed to complete this type of testing. The result has been a significant market for agencies that can perform testing per CE standards and many companies now offer these services.

The application of the CE mark involves the completion of a document called a "Declaration of Conformity." An authorized person from the manufacturer and the importer must sign the declaration of conformity. This document contains a unique identifier of the manufacturer and the importer, a list of the directives to which

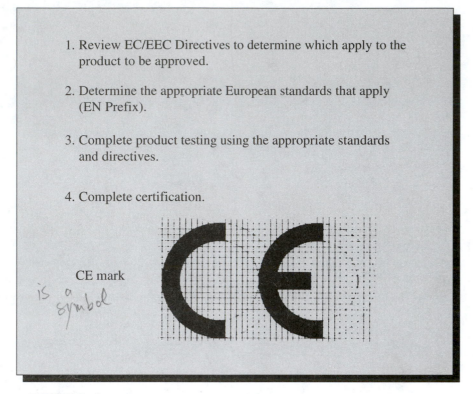

1. Review EC/EEC Directives to determine which apply to the product to be approved.

2. Determine the appropriate European standards that apply (EN Prefix).

3. Complete product testing using the appropriate standards and directives.

4. Complete certification.

CE mark

is a symbol

▲ **FIGURE 3–4**
The European Committee CE mark and process

the product complies, and a dated signature. As such, the CE marking process is a self-certifying procedure, and the manufacturer or importer is liable for the product in any case.

If you take note of the approval markings on various items and equipment you purchase, you will see the CE mark on many of them. It is a sign of the growing trend toward globalization. The formal CE mark is shown in Figure 3–4 along with the process for declaring conformity.

3–7 ▶ International Standards Organization— ISO 9000/14000

The International Standards Organization (ISO) is an international group that focuses on quality and environmental safety. The ISO group has been active in developing standards to promote and improve the quality of goods and services provided by companies, as well as their environmental performance. The ISO has published standards on each: ISO 9000 and ISO 14000. ISO 9000 is a standard on good company operating practices, while ISO 14000 deals with the environment and certifies that a company is a good environmental steward, complying with all applicable environmental laws and regulations.

ISO 9000

ISO 9000 is an approval standard that has experienced tremendous acceptance and growth since its initial publication in 1987. ISO 9000 is an approval standard for a company's operational procedures. There are many examples of companies that proudly state their achievement of ISO 9000 approval in advertisements and on letterheads. The general acceptance of the ISO 9000 concept by industrial customers has lead to near universal requirement for ISO 9000 by companies selling into industrial markets.

Within each company, there are operational procedures that define the proper way to perform critical activities within the departments. Many times these activities involve interaction with other departments. Accordingly, all other departments must be made aware of these procedures. Employee turnover and training also requires that operational procedures be documented. For an engineering department, the following are examples of typical department procedures:

Engineering change notices

Part number designations

Drawing system designations

Engineering test procedures

Engineering project numbers

Engineering test fixtures

Engineering field alerts

Field failure monitoring

Departmental operational procedures are usually developed and modified over time as improvements and changes are made to the company's operational systems. It is simply good practice to document, maintain, and distribute these procedures as they are developed and modified. It is these procedures, their development, maintenance, and implementation with which ISO 9000 is most concerned.

When a company decides to obtain ISO 9000 approval, it must make application to an approved ISO 9000 auditing organization. There are many groups that are authorized to audit for ISO 9000 approval. Underwriters Laboratories is a primary ISO 9000 auditor in the United States. The selected auditor will conduct a plant visit to determine the status of the company's operational procedures. After a review of a list of deficiencies, the company puts together a project team to bring the company into compliance. The auditor will then perform a formal inspection with a final report that the company must act on within a specified time period. It may take several inspections before a company receives approval after its preliminary review. Once approval has been achieved, the auditing agency will perform annual inspections of the company thereafter.

There are two levels of the ISO 9000 approval standard. ISO 9001 is the highest level and includes both the manufacture and sale of standard products as well as the custom design of products and systems. ISO 9002 is a little less rigorous because it covers only the manufacture and sale of standard products. The primary difference between the two is the emphasis placed on the development of design documentation as it applies to custom-designed products and systems included in ISO 9001.

The general acceptance of ISO 9000 is a positive statement for quality. There are many companies that now have documented processes where previously they had none. The emphasis of the ISO 9000 auditing agencies has been to "Document your operational procedures and do what you document."

The latest update to the ISO 9000 specification was completed in the year 2000. While significant changes have been made to the standard, its goal remains essentially the same. The previous version of the ISO 9000 standard included more than 20 separate defining documents. The new standard streamlines these down into the following four primary standards:

ISO 9000: Quality management systems—Fundamentals and vocabulary

ISO 9001: Quality management systems—Requirements

ISO 9004: Quality management systems—Guide for performance improvement

ISO 19011: Guidelines on quality and/or environmental management systems auditing

The primary goals of the new standards are to relate current quality management thinking to the actual processes that occur in the organization, with a focus on customer satisfaction and continuous improvement. The ISO 9000 standards are intended to apply globally.

The main changes within the body of the standard have to deal with increased attention to top management, its commitment to the standard, and the use of process improvement to achieve quality objectives. The standard is centered around the following eight quality management principles:

1. Customer focus

2. Leadership

3. Involvement of people

4. Process approach

5. System approach to management

6. Continuous improvement

7. Factual approach to decision making

8. Mutually beneficial supplier relationships

The resulting standard is segmented into the following five topical chapters:

1. Quality management system

2. Management responsibility

3. Resource management

4. Product realization

5. Measurement, analysis and improvement

These new standards are a significant improvement over the previous ones, because they focus more on the process rather than on simply having a process and documenting it.

ISO 14000

The ISO 14000 standard establishes the international requirements for being classified a good corporate environmental citizen. It is important for every company to consider this standard, but it carries more weight with companies that have visible environmental concerns, such as oil companies. ISO 14000 applies to the company's products as well as the company's operations. The packaging a product uses is one example that applies to most products. ISO 14000 calls for all packaging to be recyclable. Again, there is no current legal authority that requires a company to meet ISO 14000. The customers (the market segment) for the company's products will determine the viability of this standard by the level to which they demand ISO 14000 from their suppliers. ISO 14000 certification is obtained in a similar manner as ISO 9000. An application must be made to an organization certified to audit for compliance with ISO 14000. The certifying organization performs a preliminary inspection to identify the effort required to comply with the standard. A timetable is established that culminates with a final inspection and, hopefully, certification. Follow-up inspections will occur annually thereafter.

▶ Summary

In this chapter we have discussed many of the approval organizations used to verify the safety and quality of products and services. These organizations reflect the ongoing trend toward globalization of all goods and services. This process is amplified by the revolution in the telecommunications industry. A company that develops products today can meet most global standards by designing their products to meet the appropriate specifications published by Underwriters Laboratories, CSA International, and the European Committee. In the future, it is likely that global standards will be developed for product categories—and that many independent agencies will be authorized to certify them. Development of the ISO 9000 and ISO 14000 continues to promote good business practices that are not mandated by law but by the consumer.

▶ **Exercises**

3–1 Define what is meant by the initials *NRTL* and who approves them in the United States?

3–2 What is the primary concern of most approval agencies?

3–3 How do you determine which approvals are required for a particular product?

3–4 How can you determine the requirements for approval by an agency for a particular product category?

3–5 How does a manufacturer indicate their certification once a product has been tested and accepted by an approval agency?

3–6 What is the difference between the *listed* and *recognized* status for products that have been approved by Underwriters Laboratories?

3–7 What mandates the requirement for Underwriters Laboratories approval in the United States?

3–8 What is the most significant difference between Underwriters Laboratories and CSA International discussed in this chapter?

3–9 Which marking will Underwriters Laboratories specify for use after completing certification testing on a product using CSA International specifications only?

3–10 Which approval organization develops specifications for the German market?

3–11 Describe the concept of *self-certification*. Which of the specifications described in this chapter uses self-certification?

3–12 Which type of certification legally requires noise immunity testing?

3–13 In the United States, which organization requires radiated electrical noise testing on some product categories?

3–14 What is the purpose behind the ISO 9000 standard? Who developed it and how is it attained?

3–15 ISO 9000 has two levels of requirements: ISO 9001 and ISO 9002. In general, what is the difference between the requirements for ISO 9001 and ISO 9002?

3–16 What is the purpose behind the ISO 14000 approval?

3–17 What mandates the requirement for ISO 9000 or ISO 14000 approval in the United States?

4 ▶ The Six Steps

▶ Introduction

So far we have discussed the typical structure and operation of a company. We have reviewed the role of the electronics professional, the importance of teamwork, and how to manage projects. In the last chapter we reviewed why agency approvals are important and how to obtain them. Before beginning the discussion of specific project activities, it is important to develop a logical process to solve the design problem.

At this point in your academic career, you have probably solved many problems. Many of them were homework problems done quickly after skimming the problem write-up and frantically paging through the book to find an appropriate formula. After plugging in the numbers and performing the calculations, the answer is compared with those in the back of the book. If the answer checked out, then on to the next problem. If not, the process was repeated again, with perhaps a different formula. Only after numerous failures to achieve the book's answer will the average student reread the problem to make sure of the facts given and verify what is required in the form of an answer. Very often at this juncture comes the realization that the problem write-up is asking for something completely different from what you initially thought. A simple change in direction will often lead to the successful solution. And so it is with the educational process and life—as the pace of our lives continues to go faster and faster.

The primary role of engineers, technologists, and technicians is to supply technical solutions to problems, design problems, operational problems, and failure problems. As you enter a career in the electronics industry, the pace will continue to be rapid. Today's industry requires fast and correct answers to a complex set of problems. In this chapter we will discuss a simple process of solving problems that works.

This chapter includes only one section other than the introduction and summary. Within that section each of the following Six Steps to solving problems are discussed:

67

- ▶ Research, gather information
- ▶ Define the problem
- ▶ Developing a solution plan
- ▶ Execute the plan
- ▶ Verify the solution
- ▶ Develop a conclusion

4–1 ▶ The Six Steps of Problem Solving

The Six-Step process is not new. It is a basic engineering problem-solving process that has essentially been around for a long time. Many similar processes have been presented in the past, one of which was called the *professional method*. The professional method included the latter five steps of the Six-Step process presented here. In today's high-tech society, we sometimes look on older methods as being outdated and not applicable today. We also have a tendency to overcomplicate things. The Six Steps that follow are simple, easy to use, and will solve many types of problems:

Step One—*Research* the problem by gathering information.

Step Two—Completely *Define* the problem. Study the problem and list as many facts as possible to fully define the problem.

Step Three—*Plan* the solution. Develop a plan, a list of steps that are intended to solve the problem.

Step Four—*Execute* the plan. Simply perform the plan outlined in Step Three.

Step Five—*Verify* the results. Check the results achieved in Step Four, making sure that they do solve the original problem.

Step Six—*Conclude*. Develop a set of conclusions. Take note of what is learned in the process.

In our educational systems the focus is on Step Four, the execution stage of problem solving. This chapter will emphasize each of the other steps. Figure 4–1 shows a summary of the entire process. Let's look at some examples.

Example 4–1

Let's start with a very simple problem that does not involve electronics: the need to define a file system—hard copy and computer disc—for laboratory experiments that would be used each semester as part of some technical courses. Using the Six Steps, one can perform each step as follows:

Step One—Research the problem: Gather information about the problem.

This is done by discussing the problem with fellow teachers to see how they solved the problem.

Step One: Research and gather information

Step Two: Define the problem

Step Three: Plan the solution

Step Four: Execute the plan

Step Five: Verify the solution

Step Six: Develop a conclusion

▶ **FIGURE 4–1**
The Six Steps

Step Two—Problem definition: Develop a list of facts about the problem.

After a cursory review of the problem, the following three facts should be obvious:

1. The laboratory experiments will be developed and maintained on a word processor, and a computer disc file is needed.
2. A paper file is needed.
3. The experiments will be used semester after semester.

Looking deeper, these additional requirements can be seen as well:

4. Current copies of the experiments to be utilized will need to be kept in a file for use during the semester.
5. The experiments will be improved and modified on a regular basis.
6. There will be a need to note the revision level on the experiment.
7. Master copies of the current revision will need to be maintained.
8. Revisions to the current master for the following semester will need to be maintained.
9. A simple file-naming scheme will need to be developed for the disc files.

Step Three—Plan the Solution: Develop a plan, a list of steps intended to solve the problem.

The solution plan should be simple. As the experiments are developed for a course, some mechanism is included to address each aspect of the problem highlighted in the problem definition.

Step Four—Execute the plan: Simply perform the plan outline in Step Two.

The process is performed as outlined in the solution plan.

Step Five—Verify the results: Check the results achieved, making sure that they do solve the original problem.

At the end of each semester, both the experiments are reviewed and the file system, noting any problem areas and ways that the system could be improved.

Step Six—Develop a set of conclusions: Take note of what is learned in the process.

After completing verification, conclusions about the process are made while taking into account the problem areas noted, and then modifications to the system are developed.

When faced with this problem in reality, you would not sit down and write out all the steps exactly as shown here. This is a simple problem; and after many years of applying this process, it is natural for one to think and operate this way. But it is important to write out all the aspects of the problem definition.

As a result of the first cycle of this process, the following system was developed:

Each experiment was keyed into a word processor, noting the revision date on the title page. Copies were made of the master and kept in one file for use during the semester. The master was kept in a separate file, along with a marked-up copy of the master that included any changes to be implemented the next time that the course was taught. The file-naming system for the discs had to be short, because the operating system was Windows 3.1. This necessitated using the five-character course name followed by LB# and the lab number (e.g., ET265LB4).

After the first semester, the system worked well. But one modification was needed: simplify the paper filing system. There was no need to keep two separate files, one for the master and the other for lab copies. Now only one file needed to be kept. The first sheet in the file is the current master copy. The second is the current markup of the master followed by any copies for use during the semester.

This simple problem was resolved with a simple and timely solution. Yet if the effort had not been made to completely define the problem, it would have taken much longer to reach a successful solution. If the problem had been defined as simply the need to develop experiments on a word processor and keep a paper file of them, one would have realized later the need to note revision levels on the documents. This realization would have occurred when editing the files along with an appreciation for a file-naming system.

Defining the problem is everything. Understanding the depths of the actual situation to be dealt with has everything to do with the quality and timeliness of the solution.

Example 4–2

This problem was posed to a product engineer whose job is to support the phase-in process for new products as they are turned over to manufacturing. It involves a project to promote the manufacturability of a new capacitance fuel gauge system designed for small aircraft:

> The capacitance fuel gauge system consisted of a circuit board, a capacitance sensor, and interconnecting wires all mounted to a stainless steel plate that mounted the unit into the gas tank of small aircraft. The circuit board was totally potted (encapsulated, molded into an epoxy potting compound) to form the shape of a small cube. Where the interconnecting wires exited this cube, there was a need to protect the wires from bending or, in engineering terms, provide strain relief for the wires. On the prototypes rubber boots had been glued on that were shaped like a small cylinder to perform the strain relief. A better way must be found, one that was more effective as a strain relief and more manufacturable.

Step One—Research the problem: Gather information about the problem.

The assembly process is reviewed and the installation of the fuel gauge system in an aircraft is observed.

Step Two—Problem definition: To provide a functional and easily manufactured strain relief for the capacitance fuel gauge system.

Develop a list of facts about the problem.

1. The interconnecting wires needed mechanical strain relief.
2. The term "easily manufactured" meant a simple, fixturable process that did not add significant assembly time to the product.
3. The current method was messy, time consuming, and not reliable.

Step Three—Plan the solution: Develop a plan, a list of steps intended to solve the problem.

The following steps for a solution were developed:

1. Develop two alternative solutions that address all the issues brought up in the problem definition.
2. Prototype and test each of the solutions.

Step Four—Execute the plan: Simply perform the solution plan.

The process outlined in the product engineer's solution plan is performed.

Step Five—Verify the results: Check the results achieved, making sure that they do solve the original problem.

Review each solution with engineering and manufacturing and choose the preferred solution.

After analyzing the problem, a couple of solutions were devised. One alternative involved changing the tooling (the mold for applying the potting compound) for the potted assembly to include the strain relief; and the other utilized an "off-the-shelf" strain relief. The tooling change was the most attractive solution, because it required no additional parts or assembly. The strain relief became part of the main assembly and was formed at the same time as part of the potting process. Both solutions were built up and tested and the tooling change was the selected alternative. The benefits of this solution were:

1. No additional parts were needed.

2. No additional labor time was needed.

3. The strain relief was very secure and functional.

Step Six—Develop a set of conclusions: Take note of what is learned in the process.

What the product engineer learned from this problem was, if possible, to design all requirements of the project into any tooled parts.

Example 4–3

In this example, a project manager for a team of engineers has attempted a very difficult product design. The product is a 32-channel strip chart recorder that prints out in six colors on 10-inch-wide paper. In the later stages of the project, the team tested eight prototype units at locations in their plant. During their tests, most of the prototypes at one time or another would simply stop running and lock up, not responding to any key depressions except for a manual reset of the entire system. The product design utilized three microprocessors. The software was very complicated. The electrical and mechanical designs were intricate and new to the team. The project was well over budget and behind schedule. This is how the team approached this problem:

Step One—Research the problem: Gather information about the problem.

Additional information on noise immunity and the software operating system was obtained.

Step Two—Problem definition: Develop a list of facts about the problem.

1. The problem was intermittent. The team did not know which conditions instigated the failure. They had no way to make the product fail so that they could analyze it before, during, and after.

2. After the failure, the main processor would be locked up, out of its normal program, and would respond only to a hard reset.

3. The team deduced that the problem could be the result of external electrical noise, internal electrical noise, software problems, or electronic or mechanical hardware problems.

Step Three—Plan the solution: Develop a plan, a list of steps intended to solve the problem.

The most difficult problems to resolve are those that are not readily induced or simulated. In this case the problem occurred apparently at random. After reviewing the list of facts about the problem, the project manager and the engineers realized that they did not know enough about it to solve it. So, their first step toward solving it would be finding out more about the nature of the problem. The team set up a plan to do this by attempting to qualify the source of the problem. They needed to know the basic nature (i.e., electrical noise, software, and so on) of the source of the failure. They hoped to accomplish this by setting up environments for testing their prototypes where only one of the potential failure sources existed. Using this method, the team hoped to isolate the nature of the failure mechanism.

1. To verify the software, the team utilized an earlier version that had been heavily tested and debugged. They ran this software on emulation computers for 48 hours to verify that the software would at least run under emulation for that long, and then loaded this same software revision into each of the seven prototypes.

2. To eliminate external noise sources, the team set up one prototype to operate in a noise-free environment without any external signal connections.

3. The team set up two prototypes operating in an environment with a normal level of external electrical noise. One had input signals connected to transducers located outside the prototype, while the other had no external signal connections.

4. A fourth prototype was set up in the team's electrical noise-testing lab, where an excessive level of external noise was induced.

5. The remaining three prototypes were set up in the team's main laboratory, all under the same conditions with no signal inputs connected.

6. The prototypes were checked every hour during normal working hours, which were much extended considering the urgency of the situation. A checklist of tests to be performed and information to be recorded in the event of a failure was developed.

7. After one week of testing, the team regrouped and analyzed all the data relating to any failures observed. This was done in order to decide whether there was enough information to pursue a solution or to discuss how to improve the test process.

Step Four—Execute the plan: Simply perform the plan outline in Step Two.

The test plan was implemented for one week. This time period was chosen because it often took up to two days before a failure would occur.

Step Five—Verify the results: Check the results achieved, making sure that they do solve the original problem.

The results of the first week indicated that the software seemed to be all right. The program running under emulation did not fail during the entire week. However, all of the other prototypes failed at random and without any other common factor except that they all were printing when the failures occurred.

Step Six—Develop a set of conclusions: Take note of what is learned in the process.

The team decided to modify their test program and have half the prototypes set up to operate normally, but with the print mechanisms disconnected so that printing did not actually occur.

After the second week of testing, the data showed that the prototypes that were not printing never locked up. The team now had enough information about the nature of the problem to solve it. They added the following fourth item to their previously developed facts about the problem:

4. The lockup of the recorder microprocessor is related to the actual printing process.

The team's implementation plan was modified as follows:

1. Determine the relationship between the printing process and the ultimate failure.

2. Eliminate the root source of the disruption.

After one more week of testing and attempted solutions, the problem was resolved. As it turned out, the print mechanism, as it traveled across the paper, would build up a charge of static electricity on the metal part of the print head. Under some environmental conditions the charge would be sufficient enough to arc over to the recorder's metal chassis, creating a large electrical noise spike that sent the main microprocessor well out of its normal program loop looking for its next program instruction. A simple ground wire was added to the print mechanism to prevent it from building up a charge. That simple change solved this problem permanently.

We have reviewed three examples of problems that were resolved using the Six Steps: a simple filing system problem, a packaging problem, and an electrical noise-induced failure problem. Each had its own unique set of facts and methods of execution. In the last example, initially, there were not enough facts known about the nature of the problem in order to solve it. The first order of business was to set up a plan to get more facts about the problem. The complexity of the examples we have looked at varied significantly, but each would be considered a short-term or small problem when compared to developing a new product or a large project. Let us review one more example that involves the development of a product discussed in outline form.

Example 4–4

A company in the music electronics business desires to develop and market a new innovative electronic guitar tuner. A project team has been put together to study the project and, if feasible, complete it. A project manager is selected and a multifunctional project team is put together. Here is how the project would be completed using the Six Steps discussed in outline form.

Step One

The project team will gather all the information needed to consider this project. The exact information that will be needed is discussed in detail in Chapter 5 but generally it will include technical, market, and financial information that will be required to determine if the project is feasible and can proceed onto Step Two. The completion of Step One is realized with a project proposal that will include all of the technical information gathered on guitar tuners, information on the market for these products, and financial data that will estimate the cost and profitability of the project if it were completed.

Step Two

With management approval of the Step One proposal, the project team proceeds to Step Two, where the design specifications are completed. Completing design specifications is discussed in Chapter 6. Design specifications define the problem for a product development project. The result of Step Two is a set of specifications for the guitar tuner that will allow the project team to determine a solution plan and to develop the product using concurrent engineering concepts.

Step Three

With the design specifications complete, the project team develops a solution plan that is better known as a *project schedule* for product development projects. Developing project schedules is discussed in Chapter 7.

Step Four

Now the project team can begin implementing the development of the project. The project will begin with the preliminary design stage, where initial design ideas will be generated, explored, and simulated. Next components will be selected, procured, and breadboarded to test the design concepts. After testing and modifying the breadboarded circuits, prototype circuit boards and a prototype guitar tuner are assembled and tested. Step Four includes the preliminary design, component selection, breadboarding, and prototype development, which all involve the implementation of the design.

Step Five

The Project Team will now verify the design to make sure that it meets the original specifications. Product assurance tests are performed on the guitar tuner as well as field tests and a financial analysis. If approved at the completion of Step Five, the guitar tuner will be ready to be released for sale to customers.

Step Six

This step will be completed some time after the project has been released as the performance of the project from sales, profitability, quality, and customer use issues are reviewed on a monthly basis. The results of the project are summarized and reviewed to determine what can be done better on the next project.

▶ Summary

At the end of this chapter and every chapter for the remainder of the book, a running example project is discussed relative to the particular topic covered in the chapter. Please refer to this ongoing example for additional insight as to how the chapter topics are addressed in an actual project. The example project is the development of a digital thermometer. This project was chosen because of its relative simplicity, the fact that it utilizes a variety of different technologies, and its ultimate function is easy to understand.

We have completed our general discussion of the Six Steps. Consider the importance of each, but it is most important to understand the significance of Step Two, "Defining the problem." Poorly defined problems are the major reason for poor or incomplete problem solutions. It is also important to emphasize Step Six, "Develop a conclusion." All too often our desire to complete a project prevents us from taking the time to review and reflect on what was learned from it. Developing a conclusion promotes the concept of continuous improvement: the ongoing learning that should be part of our everyday life.

All that is necessary now is to learn by doing. There will be more examples as we proceed through our project example and the additional problems at the end of this chapter. However, the best way to develop the use of the Six Steps is to begin using them for all the types of problems on an everyday basis. In the beginning, it is better to be overly aggressive and really pick apart the problem definition. Take your time before starting to solve a problem. When you fail, go back and see why you failed. Most often it will be because the problem was not defined correctly up front or not enough information was gathered. After a while, you will begin to believe in the power of the *problem definition*. Consider this to be a tested theory of engineering problem solving:

> The quality of a problem solution is most dependent upon and directly proportional to the quality and the depth of the problem definition.

Digital Thermometer Example Project

In this chapter an example project, the design of a digital thermometer, will be initiated, and it will be continued at the end of each of the following chapters. In this way each of the chapter topics can be experienced in an actual project example.

Background Information: The digital thermometer project is being initiated by a company called Incontrol, Inc. Incontrol is a startup company that desires to be a manufacturer of custom electronic temperature indicators and controls. The founders of the company possess years of experience in this area and see an opportunity to develop custom temperature indicators and controls for sale to original equipment manufacturers (OEMs) that manufacture industrial ovens. The primary *strategic objective* of this company is to develop a standard temperature indicator and control product line that they can advertise, sell, and provide a basis for customization. This product line will be a vehicle for Incontrol's entry into the custom, OEM industrial, oven-control market. The founders possess backgrounds in engineering, manufacturing, and marketing, and they will be the primary project team working on this project.

▶ References

Ver Planck, D. W., and Teare, B. R. 1954. *Engineering Analysis*. New York: Wiley.

▶ Exercises

4–1   List all of the benefits that result from a complete definition of a problem as described in Step Two of the Six-Step process.

4–2   List all of the benefits that result from the conclusion stage of problem solving as described in Step Six of the Six-Step process.

4–3   You are given a full-wave bridge rectifier that has four connections, all of which are unmarked. Define the problem and develop a solution plan for determining the two AC and the DC plus/minus connections.

4–4   You wish to determine the key parameters and connections for a transformer that includes six unmarked connections. Define the problem and develop a solution plan.

4–5   You are given a seven-segment LED display that includes a decimal point. The device is completely unmarked. Define the problem and develop a solution plan to determine all of its ten connections. There are five dual in-Line connections on the top and bottom of the LED display. The display's individual LEDs are available wired in either a common anode or common cathode configuration.

4–6   A logic circuit has four switches—A, B, C, and D—as inputs. The output of the circuit is supposed to be high if A and B or C and D are depressed. One of the switches, A, B, C, or D, is not functioning properly. The truth table shown below lists the actual output of the circuit. Define the problem and develop a solution plan to determine which of the switches is not functioning. Execute the solution plan to determine which switch has failed. Has it failed open or closed?

Switch Truth Table   *0 = Off (open)*   *1 = On (closed)*

Switch →	A	B	C	D	Light
	0	0	0	0	0
	0	0	0	1	1
	0	0	1	0	0
	0	0	1	1	1
	0	1	0	0	0
	0	1	0	1	1
	0	1	1	0	0
	0	1	1	1	1
	1	0	0	0	0
	1	0	0	1	1
	1	0	1	0	0
	1	0	1	1	1
	1	1	0	0	1
	1	1	0	1	1
	1	1	1	0	1
	1	1	1	1	1

4–7   You are asked to determine the equivalent electrical circuit for a potted electronic assembly. Previous testing has been performed using the circuit shown in Figure 4–2. When Switch S1 is in the ON position, the 90 V DC supply is connected to the unknown circuit. When Switch S1 is initially closed, the current goes immediately to 30 ma and then exponentially declines to 10 ma with a time constant of 20 ms. At time T1, the switch position is changed to the OFF position, and the circuit is shorted out with the DC supply disconnected. At this time the current changes instantaneously to –20 ma and decays to 0 with the same time constant. The reading of the ammeter is as shown on the graph of current versus time in Figure 4–2. Define the problem, develop a solution plan, and execute the plan to determine the equivalent circuit for the unknown potted assembly. Is your circuit the only one possible?

4–8   You are asked to evaluate the purchase of two electric motors: Motor A and Motor B. Both motors are the same size and require 50 hp. The application requires the continuous operation of the motor, 24 hours a day and seven days a week. The two motors being compared are the same in every regard except for efficiency. Motor A is 80% efficient and Motor B is 88% efficient. Motor A has an installed cost of $800 and Motor B can be purchased installed for $700. Define the problem, develop a solution plan, and execute the plan. Estimate any other parameters required to complete the analysis.

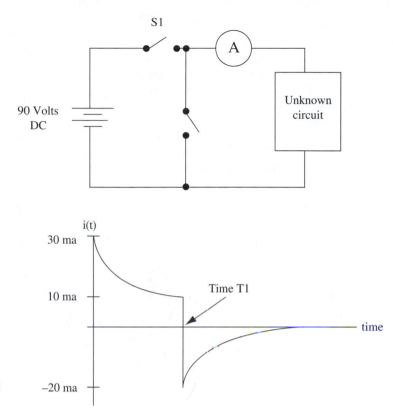

▶ **FIGURE 4–2**

Potted assembly test circuit and graph

4–9 Your company is considering investing in surface-mount technology for the production of printed circuit board assemblies. Consider this investment by completely defining the problem that the study should consider. In other words, what are the general topics that the study should address?

4–10 As a product engineer for a line of electronic patient monitoring devices, you are asked to find an acceptable substitute for a special connector that connects an electronic data recorder to monitor patient ECG data. Either a substitute must be found, or a lifetime purchase of the current connector must be made, because the current connector is being discontinued. Completely define the problem that must be resolved.

 Step One:
Research and Gather Information

▶ Introduction

The efficient collection, use, storage, and distribution of information are increasingly characteristics exhibited by successful people and companies. In the Information Age, there is more information available, and it is more accessible than ever before. The Internet is the key to accessing much of this data so computer and Internet skills must be well developed, especially if the Internet is your only source. For example, many companies have discontinued printing product catalogs and other related materials altogether and only make them available as downloads from the company Web site. The Internet, however, is not the only source of information and should not be viewed as such. There is still plenty of information that is only available firsthand by talking to competitors, customers, and co-workers. Many marketing studies are published periodically and are available for a fee. A lot of marketing information is available on the Internet if you are willing to pay for it as well. Indeed, there are many types of information that must be gathered and many sources for it. This chapter discusses the types of information needed to begin development projects in general, where to look for it, and how to use it. The specific topics of this chapter are as follows:

- ▶ Researching general failure problems

- ▶ Researching design problems—What information is needed?

- ▶ Researching design problems—Where to gather the data?

- ▶ The project proposal—Deliverables from Step One

- ▶ The project approval process

5–1 ▶ Researching General Failure Problems

This section covers researching general failure problems. These problems involve some operation, system, or device that is in place and is not functioning as desired. In this case the electronics professional is asked to provide a technical solution for a general failure problem. The approach used here is somewhat different from researching design problems, which are covered next. The first step in solving a general failure problem is to determine exactly what the device is supposed to do and how it functions. This information can be found in operator manuals, service manuals, block diagrams, schematic and assembly drawings, and parts lists. Usually all of this information is not available, but it is imperative to have at least a schematic or block diagram. Study the information available and develop a working knowledge of the device to be analyzed.

Next, the actual failure must be qualified. Application and service engineers too often simply hear "the product doesn't work" when people report product failures. In response, the first question posed by the service engineer is to explain exactly what is not working. Many times the product was never intended to perform the operations being attempted. When a true failure exists, the specific details must be determined. This is achievable with a working knowledge of the system.

When the nature of the failure has been determined, there is one last bit of information that should be collected before moving on to Step Two (Defining the problem) of the Six Steps. Gather information about the history of the system by posing the following questions:

How long has the system been in operation?

Is this a new system startup? Has the system ever functioned properly?

How often has the system failed in the past and what was the nature of those failures?

What recent changes have been made that involve the system? Be sure to note details that may seem trivial, such as personnel, system modifications, and changes in the environment.

With this information in hand, the general failure problem can be accurately defined. (See Figure 5–1 for a summary of these steps.)

5–2 ▶ Researching Design Problems— What Information Is Needed?

Design problems require that significantly more information be gathered in order to provide a good definition of the problem. In Step Two of the Six Steps, this information will be used to develop the product specification. But before Step Two can begin, management will want to review the viability of the project. The information gathered at this point will also be used to evaluate the feasibility (and therefore the risk) and financial impact of the project. A formal project proposal will be put together at the completion of Step One.

1. Determine normal device function

2. Fault identification

3. Gather background information

  • Is the installation new?

  • How long has the system been in operation?

  • What is the system's failure history?

  • What recent changes have been made to the system?

▲ **FIGURE 5–1**
Summary of gathering information for general failure problems

Before we get too much further into detail, it is important to stand back and take note of the real reason the project is being considered. This should be made very clear to everyone involved in the project. One suggestion is the development of what could be called the *strategic objective* of the project. The strategic objective is the overriding reason for implementing the project or, in other words, the ultimate corporate problem that is to be solved. It should be highlighted on all project documentation and communication. It should be the banner that gives focus and drive to the project.

There are usually many reasons to complete a project. The strategic objective is the number one reason for completing it and is the strategy behind it. For a new product project, the company wants to increase sales and profits—which is what is meant by "the ultimate corporate problem." Indeed, in the long term, all new product development is to accomplish this goal. But why does a company choose to develop a particular product instead of an alternative one? Sometimes a company wishes to develop a position in a new market in which it can continue to grow in the future. The primary reason for developing the product here is to gain market share, not to maximize profits (at least in the short term). Surely, the company will want the product to be profitable, but it will likely have to accept lower profits to develop market share in a new market. Later on, when a strong market position is established, prices can be increased to maximize profits. In this case the strategic objective is to develop a market share in the new market.

In another situation, a company may have a large market share in a particular market segment. Company management desires to increase overall profits for the company and has chosen this market segment as a good opportunity to do so. Management's strategy is to develop a high-end product with which significant profit levels can be achieved. In this case the strategic objective is to maximize profits for the company.

It is also important to identify the strategic objective for projects to develop systems for use internally by the company. The overall goal of this type of project can be to reduce costs, improve quality, or reduce manufacturing area. It is important to identify which is the most important and why. The strategic objective should be in the forefront of the minds of all company managers and the project team.

Example 5–1, Compressor Tester System

At an air conditioning company, a research engineer developed specialized automated test equipment for the company's own internal use. One strategic objective of the company is to reduce the number of incorrect warranty returns of compressors. These objectives evolved as the company realized that the number of compressors returned under warranty was growing every year. Furthermore, when the compressors were cut apart for inspection, many were found to be in good working order. Whatever fault had caused the air conditioner to require service had been incorrectly diagnosed as the compressor. A good compressor was replaced and returned for warranty credit while the fault that initiated the problem remained unresolved. The company had no easy way to check out the compressor upon its receipt from the field and routinely cut apart each returned compressor for failure analysis. After the compressor was cut apart, it became apparent after inspection that it was in good working order.

The company developed a strategy in their annual strategic plan to reduce warranty credits provided incorrectly for compressors returned in working order. To address this strategic objective, a project was initiated to develop a computerized compressor tester system. The system specifications called for the completion of tests on returned compressors in a period of five minutes. With this tester in place, returned compressors could be verified quickly and warranty credit could be denied without having to cut the compressor apart. An aggressive $200,000 project was defined to develop and manufacture four of these testers, locating one at each of the company's independently owned and operated remanufacturing centers. The project was a long and difficult one but technically a success. A compressor tester system was developed that would completely evaluate the compressors in under five minutes, and four of them were assembled, delivered, and installed at the remanufacturer locations.

Unfortunately, the remanufacturers had not been included in the project discussions and were not happy about their installation and use. They felt that the level of warranty returns was not their problem, and they were unwilling to invest their time, effort, and space in resolving the issue. Each remanufacturer demanded additional money for the storage, use, and maintenance of the compressor tester system. The amount of money requested by the remanufacturers was excessive and was greater than the amount of loss the company felt it was incurring by the invalid warranty returns. Unfortunately, the compressor tester systems were never used by the remanufacturers and were eventually shipped back to the main plant and placed in a warehouse. In this example, the strategic objective was not attained because it was not extended completely to resolve and include a key part (the remanufacturers) of the original problem. The remanufacturers should have been part of the project team. If the remanufacturers had been included in the project at the beginning, then either the project would not have been undertaken, or it would have evolved differently: the costs of using the tester would have been planned and more readily accepted by the remanufacturers.

There are three general types of information that must be gathered before starting to define the problem: technical, market and application, and financial data. First we will discuss the types of information that are needed—and then how it can be collected.

Technical Issues

On a new design, the innovation level required to solve the design problem must first be determined. To resolve this issue, develop a list of the unique aspects of the design problem by posing the following questions:

Has a similar design been completed before?

> If so, then all information available about the design should be gathered: the design documentation, evaluation of system performance, and suggestions for design improvements. If the design was completed within the same company, then this information should be available. If this information is not on paper, then obtain it from talking to the designers themselves.

What is the closest, similar design completed previously?

> If the design problem is unique, then the most similar design should be identified and, if available, its design information gathered.

What areas of the design problem does the design team have the least experience?

> The answer to this question will identify the key weak areas of the design team relative to this particular design project. Be sure to address all aspects of the design, including any processes involved in its manufacture. All the available information about each of these areas should be collected.

Is the feasibility of the design based upon the availability of some new type of component?

> If so, then all the information about the component or components needs to be gathered.

Are there any extraordinary manufacturing or quality issues expected as part of the project?

> Any issues that would require significant changes in the operation of these departments should be researched and included. Let us say a new

plant is required to manufacture the new product. This becomes a major part of the Step One project proposal.

Market and Application Issues

Market and application issues address what will be done with the project after its completion. For a new product project, these are product and market issues. For a project to develop a system for use by the company, these are application and use issues. First let's discuss the product and market issues for a new product development project.

Market and Product Information

The market information is developed and gathered by the marketing department, usually under the leadership of the product manager for the product line in question. It is important for the electronic professional to be aware of the market information and the issues that it affects. If an opportunity develops to visit customers or get direct field experience in other ways, it is wise to take the opportunity. The experience and knowledge that is developed in these situations is priceless. All too often a design engineer is satisfied and relegated to staying in the main office. The creative solution of the customer's problem is much better served if the designer has a first-hand view of how the product is applied. These are market and product issues for a product that should be well understood before Step One can be completed:

Market segment and size

Classification by market segments is a way of categorizing customers into a common group with similar needs. Most companies categorize and report their sales activity by market segments. Which market segments (application areas) will require the product and which one is the focus for this particular design? Also, what is the total size of the market segments and what market share is possessed by key competitors?

Competition

Who are the competitors for the new product and which specific products will be competitive? What features, benefits, and weaknesses do they have? What are their market positions?

The best way to compare the market position of a number of competitors is to use a *feature vs. price rating chart*. On this type of chart, price is plotted on the horizontal axis. A relative feature rating is determined for each competitive product and plotted on the vertical axis. A circle is drawn around the coordinate for the competitive market position. The size of the circle represents the relative market share of the competitive product. An example of a feature vs. price rating chart is shown in Figure 5–2.

Similar products

Who are the companies that offer similar but not directly competitive products? What are the features, benefits, and weaknesses of these products?

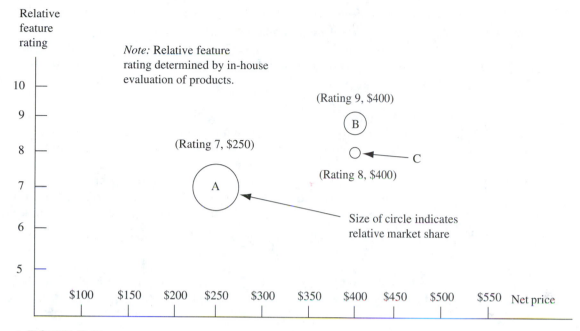

Relative feature rating

Note: Relative feature rating determined by in-house evaluation of products.

(Rating 9, $400)

(Rating 7, $250)

(Rating 8, $400)

Size of circle indicates relative market share

Net price

▲ **FIGURE 5–2**
Market position graph

Sales process

How are the products sold? Through a catalog, by sales reps, distributors, telemarketing, or the Internet?

Product support

If the customer requires service of any kind, how is that product support provided and what is its quality?

Customer profile

Develop a profile for the customer that represents the average customer for the product in each of the focus market segments to be addressed.

Customer use

How will the customer use the product and what are the important aspects of its installation, use, and service?

Physical environment

What is the range in environment that the product will see (i.e., ambient temperature, shock, vibration, moisture, electromagnetic interference, and the like)?

Market environment

Regulations, reliability, quality, pricing, and discounts—a profile for these issues should be put together for each market segment and competitor.

Developing priorities

> Which of the following issues mentioned are most important: customer use, method of sale, reliability, quality, price, and discount? Each market segment and potential customer may have a different mix of priorities.

Application and Use Information

This information is for a development project to produce some device or system for use internally by the company. It is equivalent to the market and product information discussed previously for a new product project. This information should be put together by the department that will use and operate the system. In this case they are the customer for this project. These are the application and use issues for a project that should be well understood before Step One can be completed:

Similar designs

> Research and gather data about any equivalent or similar systems that have been developed outside or by the company.

Competing alternatives

> What are the competitive alternatives to the design project? Repairing and continuing to use an existing system or, perhaps, purchasing a system? Obtain information about these alternatives.

Installation

> Where and how will the system be installed? What installation issues are important? Will the system have to be transported?

Application and use

> How will the system be applied and who will use it? Develop a profile of the application and end user.

Support

> How will the system be supported after installation? Will training need to be supplied? What will be the training and support issues?

Environment

> What is the physical environment where the system will be installed?

Regulatory issues

> Are there any government regulations that the new system will have to meet?

Priorities

> List the issues discussed previously in order of priority.

Financial Issues

Before any project or new product can be taken to the specification stage, preliminary financial information must be put together. In order for management to pass judgement on the project, some financial information is necessary to determine the project's fiscal viability. The project manager will be involved with the collection and dissemination of all of this information. It is most important for the electronics professional to understand the relationship between meeting the product development and manufacturing cost goals and the ultimate success of the project. The following financial information should be gathered to enable estimates of the project's costs and benefits to be determined:

Project and product development total costs

> This is a complete estimate of all expected costs needed to complete the project.

Net selling price

> What is the anticipated *average net selling price* for the product in order to be competitive and meet the sales forecast? What are the equivalent prices of all competitive products? Net selling price is the list price minus any discounts given for certain types of customers. The average net selling price is an average of all the selling prices and is closest in value to the discount price given to the largest number of sales.

Sales forecast

> A forecast in units and sales dollars should be forecast for the product over its expected life.

Capital expenditures

> What long-term capital expenditures must be made to support the product and what is the expected life of the product? These costs will be depreciated over the life of the product.

Manufacturing costs

> What estimates are projected for the manufacturing costs of the product?

Cost of sales

> Sales commissions and other direct sales costs combined with the cost of sales promotion (advertising, catalogs and the like) should be estimated and projected over the life of the product.

Administrative overhead

> Administrative overhead costs should be identified.

Profitability projections

> A profit estimate can be developed for the life of the product based upon all of the data discussed previously.

5–3 ▶ Researching Design Problems— Where to Gather the Data?

There are usually many sources, inside and outside the company, for the types of information described in Section 5–2 of this chapter. The sources for each type of information are discussed separately. The primary purpose for acquiring this information is to learn from what others have already done and to provide a basis to further evaluate and define the design problem.

Sources for Technical Information

Books, periodicals, and published papers

These are the most common sources for research, technical, and product information. It is important to receive all of the pertinent periodicals published in the general area of electronics but also those published in the market area of interest to the company. Keep a file on various technical subjects that are close to the technical area in which you are working. It is important to develop a good filing system for this information so that it is accessible when needed.

Products: Competitive and noncompetitive

Products of all kinds are an important source for technical information. Information should be acquired and filed for all of the company's competitive products. Competitive products should be purchased and tested using consistent procedures so that they can be evaluated on an equal basis. It is equally important to look at noncompetitive products, which may have some aspect or novel idea that might be adaptable to the company's products. Develop a habit of looking for creativity everywhere.

Colleges and universities

These are best used as a source for research into various technical areas, where a particular university has focused on a certain research topic.

Trade shows

General electronics trade shows are the best way to determine what is new in the industry. Most manufacturers save their major new product releases for the big shows. While there is a lot of hype, there is also a lot of good information. These shows also provide a sense of the direction that the industry is taking overall, and knowing this direction usually gets the creative thought process flowing.

Professional technical associations

These serve as ongoing sources for information. Their meetings and conventions can provide an opportunity to meet and discuss issues with other technical professionals.

Military and government agencies

> The military and other government agencies are a warehouse of information. Military specifications exist on many topics and are the de facto standard in many areas. Agencies such as the Food and Drug Administration (FDA) and the Federal Communications Commission (FCC) are important sources for specifications required for various products. The FDA issues many directives for both medical products and products used to process foods and drugs. The FCC issues the requirements for all types of telecommunications products.

Manufacturer's associations

> There are many industry groups where manufacturers get together and agree on common specifications to provide a common basis for products and systems to be developed. The Institute for Interconnection and Packaging Electronic Circuits (IPC) is a good example of this type of organization. This group has developed much information on important electronic packaging standards.

Consultants

> There are many consultants that specialize in a particular type of research or a specific design area. They can be of great help when addressing a difficult new area, where the company has no expertise or background. When dealing with consultants, it is important to define up front exactly what is expected, as well as their rate of compensation.

The Internet

> The Internet is the tool that will help you find all of the sources listed previously. Not only can you find, download, and print much of this information, you can also purchase the information as required in the form of books, specifications, and market studies.

Sources for Market Information

The sources for marketing information are similar in type to those listed for technical information. There may be market information available internally from previous projects, but make sure it is still valid.

Books, periodicals, and published papers

> Articles in periodicals are the best source of market information that can be used as project information. Often articles are written about types of products as well as their current trends, options, and features.

Products: Competitive and noncompetitive

> The marketing issues for a new product project are usually addressed by both competitive and some noncompetitive products alike. As competitive products are purchased for evaluation, all aspects of the purchase

should be reviewed and noted. This will serve to record the customer service performance of the competing company. When purchasing products directly from well-recognized competitors, it is advisable to have a friendly, low-profile intermediate company make the purchase for you.

Distributors and sales representatives

These professionals are both strong sources for market information in general. These individuals are visiting customers on a daily basis and are well aware of what is occurring in the field for a particular type of product. Because they often sell competitive products, distributors may be a good source of information and the products made by other manufacturers. This will depend on the type of relationship that exists between the distributor and the competitive company, as well as your own company's relationship with them.

Marketing studies

There are many organizations that develop and sell general market studies on particular market segments. They are usually published annually or every few years depending on their size and scope. The type of information included in these studies is data that is usually hard to find. Typically these studies define a market segment and list its total dollar value and the market share of all competitors. The current trends, technology, and customer application issues will likely be discussed as well.

Marketing consultants

If the market segment of interest is very specialized, market studies may not be available. In this case, market consultants are available that will research and develop specific market data for the company.

Market organizations

These organizations are specific to a market segment. When a company joins these organizations, it agrees to share key sales information with the other members. The market organization then develops a report with very accurate information about the particular market size and the relative market share of all competitive members.

Sources for Financial Information

Most of the financial information to be developed for the project is available internally from the company or can be developed by using the gathered market information. The only outside data required would be quotations for the projected capital equipment needed to complete the project. Cost accounting will be heavily involved in developing the project's financial data. They will use data from other equivalent product lines or will develop new estimates. See Figure 5–3 for a summary of the information sources described in this section.

Sources for Technical Information
    Books, periodicals, and published papers
    Other products
    Universities
    Trade shows
    Professional associations
    Military and government associations
    Manufacturer's associations
    Consultants
    The Internet

Sources for Marketing Information
    Distributors and sales representatives
    Published market studies
    Marketing consultants
    Trade organizations

Sources for Financial Information
    Potential suppliers
    Internal databases

▶ **FIGURE 5–3**
Summary of information
sources

Example 5–2

A company that manufactures temperature indicators has completed their strategic planning process and has identified a strategy to increase sales into new markets adjacent to the markets currently being served. One of their strategic objectives is to develop and market noncontact infrared (IR) temperature indicators. The following are the steps that the company might use to gather information for this project:

1. The current manufacturers of IR temperature indicators are identified by reviewing advertisements in periodicals for industrial instrumentation. Catalog information is requested and a number of the most viable products are purchased, tested, and reviewed.

2. Research studies on IR temperature measurement are obtained using the Internet.

3. A search for manufacturers of IR detectors and lenses is made using the Internet.

4. The marketing group determines the key market segments for IR temperature indicators and their size. They also identify the key competitors

currently selling into these market segments and estimate their relative market shares.

5. The marketing group further researchs each competitor and product and completes a weakness and strength evaluation and a market position chart for each.

6. Key marketing and engineering people get together to roughly define a product strategy and its intended market position on the market position chart. This idea is expanded to include an outline of the design specifications.

7. Marketing does field research to create a profile of the customer and to determine the typical product application environment. The product idea is field tested to determine a rough estimate of its sales potential. This is accomplished by talking to potential customers and salespeople.

8. A group, that will likely become the project team, will decipher the data and develop a project proposal using the gathered information combined with historical financial data, and create new estimates where no historical data is available.

5–4 ▶ The Project Proposal—Deliverables from Step One

The requirements for the completion of Step One will vary from project to project. Each company has its own system of requirements for authorizing a project. Usually the company will require a *project proposal* that is similar to the deliverables shown for Step One that is discussed in this section. Above all, the project proposal should estimate how well the project will achieve the desired strategic objective. If the strategic objective is to gain market share in a new market, then how much market share is expected? Figure 5–4 shows the components for a project proposal as a list of the deliverables from Step One for a new product project. Figure 5–5 shows an equivalent list for the development of a project for internal use by the company.

5–5 ▶ The Project Approval Process

Once the project proposal is complete, it is reviewed by upper management to determine whether to continue on to Step Two. There are usually many projects that have been outlined as potential projects by the company. As information is gathered for these projects and project proposals are completed, they are reviewed and compared. The areas of the project proposal that are reviewed are listed and discussed as follows:

1. *Strategic objective priority*—What is the relative priority of the strategic objective that this project will address?

> *Step One Deliverables*
>
> A Step One project proposal for a product development project should include the following deliverables:
>
> 1. The strategic objective
> 2. Technical information in similar designs or research
> 3. Data on key components
> 4. Technical data on competing products
> 5. Market segment and size
> 6. Competitive market data: price, discount, sales process, weaknesses, and strengths
> 7. Customer profile
> 8. Product environment
> 9. Regulatory environment
> 10. List of priorities
> 11. Projected development costs
> 12. Projected average net selling price
> 13. Sales forecast
> 14. Capital forecast
> 15. Manufacturing cost forecast
> 16. Profits and return on investment
> 17. Identify risks

▶ **FIGURE 5–4**
Deliverables for a product project

The company will have many strategic objectives. A project that addresses a higher-priority strategic objective will be favored over a project that addresses a lesser one.

2. *Strategic objective resolution*—How well will the project resolve the strategic objective?

The relative impact of the project as it resolves a particular strategic objective will be compared against other potential projects.

3. *Investment*—The total amount of investment required.

The total cost of the project, including material cost, capital equipment costs, and man-hours will be a key determination for approval of the project.

Step One Deliverables

A Step One project proposal for an in-house development project should include the following deliverables:

1. The strategic objective
2. Technical information in similar designs or research
3. Data on key components
4. Technical data on competing alternatives
5. End-user profile
6. End-use environment
7. Regulatory environment
8. List of priorities
9. Projected development costs
10. Capital investment forecast
11. Net savings and return on investment
12. Identify risks

▶ **FIGURE 5–5**
Deliverables for a project for use by the company

Because the company has a limit of these resources, the mix of projects that provide the best rate of return on investment will be the ones selected.

4. *Risk*—The amount of risk entailed in the project.

   There are many types of risks associated with a project. There is always the risk that the project will not meet the basic operational requirements. Other risks include exceeding the investment projected, profit levels that are much less than expected, or competitive actions that will render the project ineffective. The amount of risk will be a key determinant in its approval.

5. *Return on investment*—The rate of return on investment is the amount of return expected on an annualized basis over the life of the project.

   The calculation of this number is made by taking the investment, annual costs, and revenues and then spreading them out over the project's life. The result is an equivalent cash flow for the project that the company will receive each year of the project's life. The amount of that cash flow compared to the total investment is the rate of return on investment. Many companies have a *minimum acceptable rate of return* (MARR), which is required of all potential projects in order to receive management approval.

Figure 5–6 shows a summary of the project proposal review issues.

1. The priority of the project's strategic objective

2. The expected impact of the project on the strategic objective

3. Investment

4. Risk

5. Return on investment

6. Project description

▲ **FIGURE 5–6**
Project proposal review issues

▶ **Summary**

With the completion of Step One, enough information has been put together to review the projected results of the project with management. Also, enough information has been gathered to complete the project specifications that will be performed in Step Two. The project proposal should indicate to management the expected results of the project. They should be able to envision the project as complete and that the projections they are reviewing are real. The entire project team should be involved in developing the proposal and should feel confident in the projections. Each department will have data that they will supply, and the project team should have a feeling of ownership toward the project. It is important that all projections be as realistic as possible, perhaps even slightly pessimistic, to temper the tendency to overestimate the possibilities.

With the completion of Step One, a general statement of the problem has been made in the form of the strategic objective. The project's results have been projected internally, externally, and financially as well as how they meet the project's strategic objective. With the approval of management, Step Two can commence.

Digital Thermometer Example Project

The project team is put together to complete Step One for the digital thermometer project idea discussed at the end of Chapter 4. The team utilizes the process described in this chapter and proceeds to develop the following Step One project proposal.

Digital Thermometer Project Proposal

Strategic objective: Develop a standard digital thermometer product that can be sold directly to industrial environmental test chamber OEMs in the United States and provide a basis for advertising, sales calls, and further customization.

General description: The digital thermometer should be packaged in one of the industry standard DIN sizes and should be capable of measuring and displaying temperatures up to ± 200°F with a standard RTD sensor connected.
   Note: DIN stands for Deutch Industriale Norm which is a German standard for mounting and and package dimensions and configurations. It applies to industrial instrumentation products.

Technical information: The following are the key technical issues and the results of the research completed thus far.

   Sensor—The RTD sensor should be the .00392Ω/Ω/°C coefficient that is a standard in this industry.

   Electronics—The electronics should be a routine design with which the design team has had much experience. The challenge will be to minimize the size of the product using SMT components.

   SMT manufacturing—This is a weak point in the design team as there is minimal experience with SMT. A number of specifications, periodicals, and books have been obtained and are being reviewed in the initial stages of the project.

   Enclosure and packaging—A copy of DIN specifications and the requirements for SMT land patterns (minimum spacing requirements) have been obtained.

Market information:

   Market segment and size—The U.S. market size for digital thermometers sold into the environmental test chamber market has been estimated internally at about 12,000 units per year. The average net selling price for each is further estimated at $125. This brings the total market to $1,470,000. The market is expected to grow at about 5 percent per year.

   Competing market data—There are currently four major companies active in this market that comprise 85 percent of the estimated market. A profile of each follows:

   Temp1—Sales 4200 units at $115 each for a total market size of $483,000/$1,250,000 = 37 percent. Product feature rating is = 8. Average discount is 25 percent. They sell through a strong group of manufacturers representatives and provide good customer service and support. They focus on this market. Their only weakness is that their product line is getting old.

Temp2—Sales 2100 units at $135 each for a total market size of $283,500/$1,250,000 = 23 percent. Product feature rating is = 7. Average discount is 25 percent. They sell through manufacturer's representatives and provide average customer service and support. Their product line is up to date but harder for the customer to modify. This market is more of a sideline to them.

Temp3—Sales 1200 units at $200 each for a total market size of $240,000/$1,250,000 = 19 percent. Product feature rating is = 9. Average discount is 25 percent. They sell through a few direct sales representatives and provide average customer service and support. Their product is the premium product on the market and is used on customer request.

Temp4—Sales 1200 units at $183 each for a total market size of $219,500/$1,250,000 = 18 percent. Product feature rating is = 7. Average discount is 25 percent. They sell through a combination of direct sales representatives and manufacturer's representatives and provide less than average customer service and support. Their product line is old and was the initial digital thermometer sold into this market.

Projected market position—The preliminary market position identified for this product is to have a feature rating of 8 and average net selling price of $160. As a new company, Incontrol will not be able to be that price-competitive. The strategy for the product is to provide a starting point for the development of custom temperature indicators and controls.

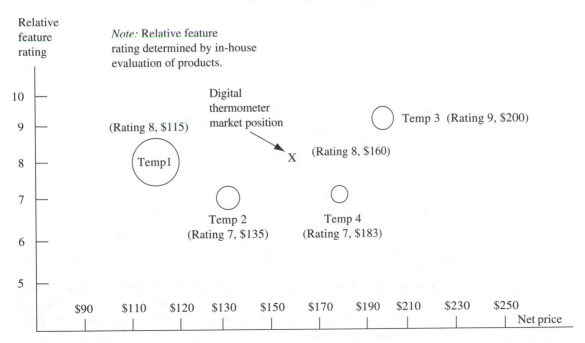

▲ **FIGURE 5–7**
Digital thermometer market position

Customer profile—The OEM customers are all sizeable modern companies that possess differing strategies and market segments. The consistent theme is reliability, price, ease of use, quality, and support, price being more important when trying to make the initial sale. Reliability and quality are more important for maintaining the business.

Product environment—In all cases the product is used in an average industrial environment.

Regulatory environment—All temperature indicators and controls sold into this market must be approved by UL specifications 1092 and equivalent CSA specifications. CE listing is also required as the OEM products are sold in Europe.

List of priorities:

1. Reliability

2. Price

3. Ease of use and modification

4. Quality

5. Customer service and support

Financial information:

Projected development costs—$50,000

Average net selling price—$160

Sales life—7 years

Sales forecast:

Year 1—500 units/$80,000

Year 2—1200 units/$192,000

Year 3—1500 units/$240,000

Year 4—1700 units/$272,000

Year 5—2000 units/$320,000

Year 6—2000 units/$320,000

Year 7—2000 units/$320,000

Capital equipment forecast—$10,000

Manufacturing cost forecast—$75

Profits and return on investment—Earnings before income taxes are projected at 20 percent of the average net selling price. At that rate, the total investment will be recouped after the sale of unit # 1875, which would be in the third year of sales.

Risk Identification—The risk for any startup company is extreme. The key risks relative to this project are:

1. Completing the project on schedule
2. Getting customers to give the new company a chance
3. Finding qualified sales representatives
4. Getting capital for large purchases

▶ Exercises

5–1 Define the term *market segment*.

5–2 What is the definition of the term *market share*?

5–3 What is a market position graph? List the information that is available from the review of a market position graph.

5–4 Define the term *strategic objective*.

5–5 List the information-gathering steps that should be performed when resolving a general failure problem.

5–6 Which are the Step One project proposal deliverables that would be developed mostly by the marketing department?

5–7 Which are the Step One project proposal deliverables that would mostly be developed by the engineering department?

5–8 List the key decision factors that management will review to determine whether or not to proceed with a project.

5–9 What are the types of risk that can occur on a new product development project?

6

Step Two: Define the Problem (Develop Design Specifications)

▶ Introduction

Having reviewed the Six Steps and Step One in Chapters 4 and 5, respectively, we are ready to look at Step Two, Define the Problem. For large development projects, defining the problem becomes the development of design specifications. Typically, development projects are complex, involving a number of people from different disciplines working toward the same goal. This chapter discusses the following topics that describe the process and the importance of developing design specifications:

- ▶ Specifications define the problem
- ▶ Specification format
- ▶ Specification development
- ▶ Software specifications

6-1 ▶ Specifications Define the Problem

What are specifications and why do we need them? Specifications are nothing more than a formal document listing the requirements of a project or product. They are a detailed list of the facts relating to the problem definition for a significant project. When you construct a house, there is a set of specifications that define details about the house that must be met upon its completion. The general contractor is responsible for satisfying these specifications, but all the subcontractors (i.e., carpenters, electricians, masons, and the like) must be aware of these requirements to ensure that all aspects of the specifications are fulfilled. Generally speaking, a specification is simply a nontrivial problem definition. We need specifications to define the problem for us in detail, just as we used the problem definition described in Chapter 4 on simpler problems.

The other benefit of using a formal specification is that a format can be developed that becomes a useful guide in completing specifications for future projects. This will help ensure that key aspects of a future project are reviewed before the design is started. When one small detail is not identified up front, it can result in the project being completely aborted. The better the specifications are developed, the better the design problem is defined, and the more likely the final design functions and meets the original need. At the end of this chapter, we will develop a specification for the example digital thermometer project to show the result of the specification development process.

Before we discuss the development of specifications further, let us again look at what precedes and drives their development. First, there is the need for whatever the design project represents. This results from a desire to meet the strategic objectives set by the company. During Step One, a significant amount of information is collected about a project and a proposal is put together. This information is developed to determine the viability of the project and provide for the development of the specifications. At this point the multifunctional team will get together and use the Step One proposal information to develop the specifications. This is a critical point for the application of concurrent engineering principles. The specifications must be developed jointly with all of the ultimate requirements of the project in mind.

The marketing representatives are the team members who are the experts on how the project is employed as well as the business aspects of its success. The marketing team members are usually not technical experts and are therefore not aware of the technical details, knowledge, and capabilities required to complete the project. Most people who have a house built for them are not experts on building houses, but they are most knowledgeable on what they want their house to look like and how it will be equipped. As they sit down with their contractor to discuss their requirements, they often find that aspects of the home they desire will present difficulties to the contractor. These difficulties will likely increase the price of the house beyond what the owner is willing to pay. The specifications for the house are developed with this "give-and-take" attitude in mind.

A similar situation occurs in industry. For most new products, a company's marketing department, while representing the customer, defines the general requirements for the new product. These requirements are usually called the *market specifications* and are included in the Step One proposal along with the marketing and business aspects of the project. Many times the initial marketing specifications are impractical "pie-in-the-sky" requirements. It is important to discuss the initial market specifications and filter them down to something that reflects real market conditions before development begins. The marketing, engineering, finance, quality, and manufacturing departments will work together to help develop and review the design specifications. This will ensure that the product not only meets the customer's need, but that it can be developed and manufactured at a cost and quality level that ensures its business success.

6–2 ▶ Specification Format

To get the general idea, let us review what might be included in a set of specifications for a new electronic product. In utilizing a general specification, development engineers will look for answers to the following types of questions:

- ▶ What is the power source for the product being developed?
- ▶ What is the range in power source voltage that the product must function over?
- ▶ What ambient temperature range will the product have to function in?
- ▶ How large can the product be physically?
- ▶ What are the criteria for the appearance of the product?
- ▶ What tolerance levels should be selected for the electronic components?

In addition to answering these types of questions, the specifications should also discuss how the product is to operate. In some cases it is possible to be very specific up front in defining the product's operation. Yet, in other more complicated situations, it may be impractical to state exactly how the product will function. In these cases, the statement of function is handled in a general way.

We can categorize specifications into the following specification format:

- ▶ General description
- ▶ Performance
- ▶ Power input
- ▶ Package
- ▶ Environmental
- ▶ Operation
- ▶ Project and product cost
- ▶ Agency approvals
- ▶ Cost
- ▶ Special requirements

6–3 ▶ Specification Development

In this section we will discuss each area of the product specifications outlined in the previous section.

General Description
This will include a general write-up for the project describing the purpose of the project, the broad approach to development, and the environment of the end use and the end user.

Performance

This section of the specification deals with the quality or performance level of the project or product under development. To complete this section, it is necessary to identify the key outputs or end result of the project. Next, ideal conditions for the parameters that affect performance should be defined in order to provide a consistent basis for making the measurements. Then the acceptable range in variation of the outputs must be determined.

Example 6–1

Take the example of a digital thermometer. The primary output of the temperature indicator is the displayed temperature. The environmental parameters that affect the displayed temperature are the ambient temperature of the indicator and the quality of the power supplied to the indicator. The ideal conditions for determining this accuracy follow:

1. The thermometer is connected to a specific temperature sensor that is exposed to a temperature within the operating indication range of the indicator.

2. The thermometer indicator should be maintained at an ambient temperature of 25°C.

3. The input power is within the specification range.

The operating range of the indicator is 0°C to 400°C in this case. The acceptable accuracy of the thermometer for this application is defined as ±0.5% of range. In degrees, then, the acceptable range of the thermometer display is ±2°C (0.005 × 400°C). To check this out, the sensor is placed in a temperature within the operating range of the indicator, such as 200°C. The acceptable reading on the indicator would be between 198°C and 202°C. This is a very simple case because there is only one end result or output from the device: the indicated temperature.

In other projects, there may be many outputs with many different variables relating to each one. If we were developing a waveform generator to output a triangle wave, a sine wave and a square wave, we would list each of those outputs in our performance specifications. Each type of output would also have accuracy specifications relating to the frequency, amplitude, and overall representation of the waveform. In other words, the specifications analytically define how perfect each waveform must be.

To develop the performance aspect of a project specification, the following steps should be performed:

1. Identify all measurement aspects of all the key outputs of the project or product.

2. Identify the parameters that affect these outputs and their ideal conditions for taking the measurements.

3. Determine acceptable ranges for these parameters.

Power Input

In this section the power to be supplied to the device is specified. In most cases this problem needs to be considered. In other cases—for example, a software-only project—it may be omitted. Much of the time the power supplied is standard 115 V AC, 60 Hz power. Even in this case, it is important to note the range in amplitude and frequency that the device can expect to see, and most important, to verify that the device operates over that range. Most devices powered from standard 115 V AC use a ±10% variation in voltage level and a range in frequency of 50 Hz to 60 Hz.

In some projects a device may be capable of accepting power from both AC and DC sources. Specifications should be listed for each case. The load current or power consumption of the device, both typical and maximum values, should also be identified. If the DC source is a battery, then battery size and expected life will be important issues.

To complete the power input section of the specifications:

1. Identify the type and range of all power inputs to the device or project, making sure to include all pertinent parameters, such as frequency and amplitude.

2. Indicate the power (volt-amps) or current requirements on the power input.

Package

This part of the specification relates to the mechanical aspects of the design. The purpose is to define the general package criteria of the project without doing the design. In some cases there already exists a specific package that the project will utilize. If so, that should be stated in this section of the specifications. Other times, there may be no specific requirement for the package size; that should be stated in this section as well. Most often, however, specific criteria for the package design is necessary to successfully meet the design goals of the project. The end result of this section of the specifications will identify all the key criteria, completely defining the package design problem for the package designer. Here are the areas that should be addressed:

Mechanical size limits—The largest, smallest, or specific volume the package must conform to.

Environmental rating—The weatherproof rating for the package, how well it is sealed, and its ability to withstand corrosive atmospheres.

Shape—If important, should be specified or information included that allows the designer to determine the optimum shape.

Material—Specific material requirements for the package or any guidelines that help to determine the proper material for the package.

Human engineering aspects—It is important to list all aspects of human interaction that will be employed so that they can be considered in the design. (Are there keys that will need to be depressed? Does the device mount on a wall?) Ease of use, in every aspect of the project, should be

considered, such as unpacking, installing, using, servicing, and even re-
cycling and disposing. In the new millennium, we must increasingly ap-
preciate, preserve, and reuse our natural resources.

Environmental

This area of the specifications defines key variables in the environment that the end
result of the design process will be exposed to. These primarily include ambient
temperature, humidity, vibration, shock, and electromagnetic interference (EMI)
immunity and generation.

Ambient temperature—This is the range in temperature that the product will
be exposed to under normal operating conditions. This will be listed as
the *operating range*. It is also necessary to point out the *storage temperature range*. This is the range of temperature that the device would
be exposed to over a normal product life while not in operation.

Humidity—Humidity can have a detrimental impact on a variety of products
and, as such, is an important environmental factor. The range in humidity the product is expected to operate over, as a percent of relative humidity, should be noted here.

Vibration—The vibration levels that the project will be exposed to should be
specified. This is usually done in terms of the amplitude of the force applied, the frequency of the vibration, and the direction that the force will
be applied to the project.

Shock—The purpose of this specification is to define the one-time shock force
that the device should be able to withstand. In most cases this will turn
out to be the worst-case shock applied during shipment if the device were
to be dropped from a certain height on its side or on its corners. There
are specifications for standard tests used by shipping container manufacturers that are often specified in this section. In some applications
there may be other sources of significant shock forces that can be applied to the device. These levels should be identified to allow development
of the appropriate shock specifications.

Electromagnetic Interference (EMI) Immunity—The immunity of the design
to electrical interference of many kinds is an important requirement in
many applications. The difficulty here is the creation of a subjective statement that specifies the EMI immunity goals of the project in a way that
can be verified and tested. Designers are generally concerned with interference such as radiated electrical fields, induced magnetic fields, electrostatic discharges, and power line transients. One can try to develop a
set of requirements for each of these areas or use existing standards such
as MIL-STD-461 (limits) and 462 (test procedures) issued by the U.S. Department of Defense. This standard is very complete and stringent and
covers both immunity as well as emission. It may be a matter of simply
stating in the design specifications "meet MIL-STD-461." The actual
process of verifying and meeting these specifications is a difficult task.

Therefore, it is recommended to review the various noise standards, choose the areas most important for consideration in the subject design, and include those in the specifications. At the same time one must ensure that he or she can verify the performance by creating the environment included in the design specifications. This may mean the purchase of some sophisticated and expensive test equipment or using an outside testing laboratory.

Electromagnetic Interference (EMI) Emissions—The radiation of electronic equipment has come under increased scrutiny as the operating frequencies and the volume of equipment in operation have increased in recent years. The best source for this requirement comes from the Federal Communications Commission (FCC), which mandates two levels of requirements and testing: Class A and Class B equipment. Military Standard MIL-STD-461 also addresses this issue. An appropriate specification statement should be determined and included in the design specifications. In this case, to verify the performance, one needs to measure radiated EMI from the design accurately. This may require the purchase of equipment or the use of an outside testing laboratory.

Operation

All the operational aspects of the device will be addressed in this section. Starting with applying power to the unit, the steps and requirements for operation should be listed. All the variables that are provided for adjustment of operation should be shown and discussed. As mentioned earlier, in simpler projects it may be possible to completely define the way the device operates in the specifications. In cases that are software intensive, with many key depressions and displays, the specific operation is something that may be developed as the project design develops. However, it is important to make sure that the hardware requirements support the overall requirements for the project. When the operational requirements are implemented with software, a separate operational specification is usually generated and called *software specifications*. Software specifications will be discussed separately in section 6–4 of this chapter.

Agency Approvals

This area deals with both recognized testing agencies and recognized specifications with which the design project is required to comply. These were discussed in Chapter 3 and include agencies such as Underwriters Laboratories and CSA International. Approval by these testing agencies is increasingly important in markets for a variety of product classifications. The identification of any agency approvals required, along with the corresponding specification, is imperative before beginning the design process.

Cost Specifications

This section is one of the most important sections and often one that is overlooked. This section determines the potential for the financial success of the project. When considering all the technical details, it is easy to overlook that the cost of development and the manufacturing cost are crucial to the project's success. As such, they must be identified as part of the specifications.

Preliminary Project Cost Estimates

Save this step for last so that a reasonable estimate can be developed from the specifications completed thus far. This estimate should include both direct dollar expenditures and the total man-hours needed to complete the project.

Manufacturing Cost Goal

The manufacturing cost goal includes all manufacturing costs and is the maximum number for the design team. This number should be tied to the profitability of the product and its intended selling price as defined in the Step One project proposal. It should also be tied to the anticipated annual sales volume. It is a goal that, if met, assures meeting the intended market price for the product. If the sales volume meets or exceeds the projected numbers, the product will be a huge success. The manufacturing cost goal should be itemized to include the following categories:

- Total cost of purchased parts
- Labor cost to manufacture
- Manufacturing overhead

The projected volume at which these costs are to be achieved should be stated also.

Special Considerations

This is simply a miscellaneous category that is a good place to discuss any design criteria that do not fit in any of the other sections already covered.

6–4 ▶ Software Specifications

Many development projects today include software, either as the final product itself or for a microcomputer that is embedded in the design. The software specification is a separate document that defines how the software portion of a development project functions. It is inherently tied to the hardware specifications by the fact that the hardware must be capable of supporting the software operation. If ten keys are required to initiate various software actions, the hardware must include these keys. The location, size, and layout of the keys must fit with the overall design scheme. If the software will require a certain amount of random access memory (RAM) in order to perform the specified operations, then that amount of RAM must be provided by the hardware design.

The software specifications are also more flexible than the hardware specifications and will evolve as the project design is completed. Accordingly, the software specifications are a major problem area in many projects. The reason for this is the perception that software is changeable at any point in the project. Because of this factor, less time is spent reviewing the software requirements, and many times hardware is not provided to support the eventual software operations. The link between the hardware and software requirements must be recognized and identified in the two specifications.

At the end of Step Two, the software specifications should include an outline of the planned method of software operation. All operations that involve the end user and the hardware should be defined initially in both the hardware and software specifications. If ten keys perform operations on the product, the hardware specifications call for the need for ten keys and the relative format and requirements for the keys. The software specifications describe the functions that the keys perform. Other hardware requirements such as RAM, processor speed, inputs, and outputs must be specified as required by the software.

It is difficult to completely define the software because to do so involves completing the software design in many cases. A top-level flowchart should be developed, along with as many flowcharts of sublevel operations as possible. An excellent method for initially defining the end-user operations is to write a preliminary operator manual. This manual can be developed as the project progresses and eventually can become the final operator manual. While software may be modified with relative ease, the ramification of software changes becomes an issue later in the project. Whenever software is changed, it is necessary to verify and test it. This is a lengthy process that will be discussed further in Chapter 12. Late in the project, software changes affect operator manuals, specification sheets, advertisements, and, possibly, a long list of test procedures and processes.

The key to a good software specification is to strictly define the appropriate areas while allowing "soft" areas where there is a need for design flexibility. When developing software specifications, keep the following points in mind:

1. The software specifications must include enough requirements to completely define the hardware required to support it.

2. A preliminary operator manual must be written that describes specifically all hardware-related and end-user operations.

▶ Summary

After completing the specifications, it is important that all parties meet to discuss them. Often the process of putting together the specifications generates questions that had not been anticipated by the project's originators. Any issues should be discussed, resolved, and implemented in the final specifications. The final specifications should be initialed, dated, and distributed to all parties. Most formally controlled documents have a place for initials and dates that indicate that the document has been reviewed and approved for use. The deliverables for Step Two are shown in Figure 6–1.

During most projects, situations occur that promote some change to the specifications. This is natural simply because it is not possible to foresee everything. The difficulties that result from specification changes made during the project were discussed in Chapter 2. One of the principles of concurrent engineering is to minimize or eliminate specification changes during the project. Specification changes can be made through Step Two and possibly Step Three. But once the implementation of the project starts with Step Four, specification changes are very detrimental to the project's completion.

1. A complete set of design specifications approved by all team members

2. Software specifications

▶ **FIGURE 6–1**
Deliverables from Step Two

If specification changes must be made, they should be implemented formally in writing, noting the revision level and date. The project specification should be a formal document and controlled like any other engineering document. We will discuss the methods for controlling engineering documents in Chapter 12. With the specifications complete and approved by upper management, we are ready for Step Three: developing our solution plan—or Project Schedule—for the design problem that is now defined by the specifications. The following points are important for emphasis as you proceed through a project:

1. The project specifications are a complete definition of the design problem and, as such, they are the primary basis for measuring the successful completion of the project. The project team must keep them in the forefront of their minds.

2. Specification changes should be avoided during the project. If changes must be made to the specifications, they must be discussed and approved by all affected parties and implemented formally in the specifications.

3. In software-oriented projects, a software specification is used to define the operational aspects of the project.

Digital Thermometer Example Project

The project proposal developed in Step One was reviewed and approved by the project team and Incontrol's management. The project team developed the digital thermometer specifications that are discussed next using the process discussed in this chapter.

Digital Thermometer Specifications

General Description: To develop a digital thermometer for use with an RTD (Resistance Temperature Detector) sensor to measure and display the temperature at the sensor location. The input signal to the digital thermometer is the resistance variation of the RTD sensor. The digital thermometer will operate off of 115 V AC,

60 Hz power and will utilize a standard enclosure, available off the shelf. The thermometer will be used routinely indoors and wall mounted. The temperature will be displayed on three seven-segment red LEDs. The accuracy should be within .5°F over a maximum range of 0°F to 200°F.

Performance Specifications:

- ▸ Rated conditions:
  Ambient temperature = 25°C
  Power Voltage = 115 V AC
- ▸ Indication accuracy—± .25% of the 200°F range.

Enclosure Specifications:

- ▸ Size—Maximum size of 6″ × 4″ × 2″
- ▸ Shape—The enclosure will be a purchased component from standard enclosures available on the market. A simple rectangular volume is preferable.
- ▸ Material—Plastic preferred. Metal acceptable.
- ▸ Human engineering aspects—The following is a list key requirements:
  1. The device shall be easily installed and connected.
  2. It should be fail safe if connected incorrectly.

Environmental Specifications:

- ▸ Ambient temperature:
  Operation: 32°F to 122°F.
  Storage: –30°F to 122°F.
- ▸ Humidity—10% to 90% relative humidity, noncondensing
- ▸ Vibration—The digital thermometer shall be operable in a vibration environment with a vibration frequency from 0.3 Hz to 100 Hz with amplitudes as high as 0.2 g.
- ▸ Shock—The digital thermometer shall be capable of withstanding shock that will most likely occur during shipment and must therefore meet the requirements of appropriate ICC specifications.
- ▸ EMI immunity—The digital thermometer shall be capable of operation in an environment as follows:
  Radiated electrical fields:
  —1 V per meter from 150 kHz through 25 MHz
  —10 V per meter from 25 MHz to 1 GHz
  Induced magnetic field—20 A at 60 Hz into the enclosure
- ▸ Power line—±500 V, 50 ns duration over 360°.
- ▸ EMI emissions—This design will meet the FCC Class B specifications for emissions.

Input Power: 115 V AC, 50 Hz to 60 Hz, ±10% in amplitude. Current draw a maximum of 100 ma.

Operation: The digital thermometer has no real operational requirements other than the requirement for sensor break protection. The sensor break protection will flash the display when the sensor is out of range. The digital thermometer will simply display the temperature when sufficient power is applied. When the signal is out of range, the display will flash.

Agency Approval Requirements: UL 1092 and equivalent CSA specifications.

Special Considerations: None.

Digital Thermometer Cost Estimates:
- ▶ Project cost—$10,000
- ▶ Manufacturing cost goal:
  Purchased parts—$95
  Labor costs—$40
  Manufacturing overhead—$15
  Annual volume—1000 units

▶ Exercises

6–1 List all of the benefits that can result from the development of complete design specifications.

6–2 What are software specifications? How do they relate to the hardware in the design specifications?

6–3 Which hardware requirements are most affected by the software specifications?

6–4 List the general categories that should be included in the design specifications.

6–5 How are modifications to design specifications implemented and controlled?

6–6 Develop a list of the operational parameters that must be considered when designing a +5 V DC power supply.

6–7 Consider the design of a sine-wave generator. List the operational parameters that should be considered in this design project.

6–8 Consider the development of a "through-beam" photoelectric sensor. A through-beam photoelectric sensor consists of a transmitting light source and a photo-receiver. The transmitter and the receiver are lined up as shown in Figure 6–2. When the beam is not broken, light from the transmitter hits the receiver and turns on the output transistor from the receiver. Their purpose is to detect objects as the objects break the beam. What are the operational parameters that must be determined before starting to design a through-beam photoelectric sensor?

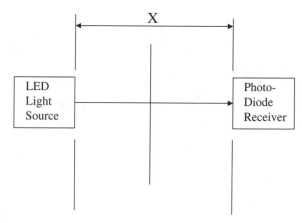

▶ **FIGURE 6–2**
Photoelectric through-beam diagram

6–9 Consider the financial aspects of the digital thermometer project. The project cost is estimated at $10,000. The total manufacturing cost of the digital thermometer, as called for in the specifications, total $150 (that is, $95 parts plus $40 labor plus $15 overhead). In order to sell each digital thermometer, consider an additional cost of $50 to cover the company's administrative overhead and selling expenses. The net selling price of each digital thermometer will be $250. Determine the following using general problem-solving skills without considering the changing value of money due to inflation:

a. What is the net profit for each digital thermometer sold?

b. If the net profit is used to pay off the project development costs, how many digital thermometers will have to be sold to recover the $10,000 investment?

c. At the projected annual volume of 1000 units per year for the product, roughly how long will it take until the $10,000 is recovered? This is the break-even point for the company.

7 Step Three: Develop a Solution Plan (Project Scheduling)

▶ Introduction

The completion of any project in a reasonable period of time requires good planning and execution. In Chapter 6 we discussed the development of specifications for larger design problems. By completing the design specifications, we have fully defined the design problem. We are now ready for Step Three of the Six Steps: developing a plan to solve the design problem. For a large design project this process is called *project scheduling*.

We have discussed previously the importance of defining the problem and how an incomplete problem definition can negatively affect the problem solution. Developing a project schedule is important in order to complete the project in an orderly way and in a reasonable period of time. If time and organized work flow were not a factor, then a good project schedule would not be required. Eventually a project team would prevail and successfully complete the project. In the real world that we live in, however, time and efficiency equal money, and we never have enough of either. Therefore, the development of a solid project schedule is *always* a critical part of the development process. This chapter discusses the following topics:

- ▶ The three phases of a project schedule
- ▶ Schedule formats—Gantt and PERT/CPM
- ▶ Project management software
- ▶ Bottleneck issues
- ▶ General procedure for project scheduling
- ▶ Scheduling Step Four—Execute (the design phase)
- ▶ Scheduling Step Five—Verify (design verification phase)
- ▶ Scheduling Step Six—Conclude (design improvements and project performance monitoring)

7–1 ▶ The Three Phases of the Project Schedule

The development of the project schedule becomes our solution plan, or Step Three, for solving the design problem. Because the project schedule is not developed until Step Three, it will include only Steps Four, Five, and Six—or the execute, verify, and conclude steps, respectively, of the Six-Step Process. For a development project Step Four, execution of the solution plan, becomes the design step, and Step Five, verify, is the design verification stage. Step Six, conclude, will be the stage of the project in which design improvements and project performance monitoring are scheduled and completed.

Before beginning the formal scheduling process, it is important to have the specifications completed. There is nothing worse than developing a detailed schedule for a project that has not been completely defined. Consider developing a schedule for the construction of a simple one-story house and then being asked to build a more complicated two-story house using the same schedule. It seems ridiculous, but in industry it happens all too often. With the specifications in hand, we can begin to develop the schedule.

7–2 ▶ Schedule Formats

As we proceed to put a project schedule down on paper or into the computer, there are two basic formats that can be used: Gantt or PERT/CPM charts.

Gantt Charts

Gantt charts are simply a variation of a bar chart. Gantt charts show a list of project tasks listed vertically in a column on a sheet of paper with their respective start and finish dates noted on a time line that goes horizontally across the bottom of the paper. Project activity can be shown as a bar or as a straight line with arrows positioned to denote the start and finish of the activity. See Figure 7–1 for an example of a Gantt chart created with a simple spreadsheet program.

Gantt charts are very simple and easy to use. Other than their simplicity, these charts are good for showing project status and slippage at all stages. This is due to the fact that actual start and finish times can be noted adjacent to the planned times, so any schedule variance can be viewed at a glance. See Figure 7–3 to see how the actual start and finish times are presented with solid lines compared with the dashed lines that signify planned activities. The primary disadvantage of a Gantt chart is that it does not show the direct interrelationships of the major tasks. In other words, Gantt charts do not show directly which major tasks must be completed before certain tasks can begin. In the project scheduling world, these are called *task contingencies*.

PERT/CPM Charts

In the late 1950s there were two similar project management methods that were developed almost simultaneously by separate groups. PERT, which stands for Program Evaluation and Review Technique, was developed by the U.S. Navy as part

Planned——x   Actual to Complete____ C

Design Stage	Weeks	1	2	3	4	5	6	7	8	9	10	11	12
Time Line													
Design Stage		v											^
Electronic Design			x										
Breadboard				x									
Mechanical Design			x										
Complete Mechanical Mockup				x									
Component Selection					x								
Cost Estimates						x							
PCB Layout							x						
Complete Preliminary Design								x					
Procure Parts							x						
Complete Prototype											x		
Test Prototype													x
Apply for Regulatory Approvals												x	

▲ **FIGURE 7–1**
Spreadsheet Gantt chart example

119

of their Polaris submarine project as a complete project management system. The Critical Path Method, CPM, developed by Dupont, focused on the critical path, which is the shortest period of time required to complete a project. Because both systems were so similar, the concepts over time have been merged together into something called simply PERT/CPM. Many books on project management discuss PERT and CPM together and most of the currently available project management software have schedule formats called PERT/CPM. The PERT/CPM system defines a project management process that includes both the schedule format and methods of project management. When discussing project schedules, the following definitions, which are defined in the PERT/CPM system, are often used:

Activity—Task in a project

Event—Point in time in a project

PERT chart—Graphical representation of the project schedule

Gantt chart—Bar chart of the project schedule

Most likely time (m)—Most likely time for completion of an activity

Optimistic time (a)—Shortest period of time for an activity

Pessimistic time (b)—Longest period of time for an activity

Expected time—Predicted time for the completion of an activity

PERT/CPM starts out with a list of major tasks and subtasks. The major task is an overall task with the subtasks being the smaller ones necessary to complete the major task. Each task has a list of contingencies associated with it. These contingencies are simply a list of the other tasks that must be completed before this task can be started. An example of a PERT/CPM schedule is shown in Figure 7–2. This is a similar schedule to the one portrayed in Figure 7–1 as a Gantt chart. See

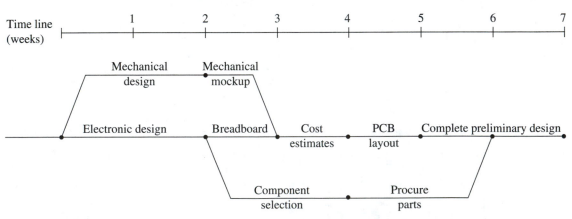

▲ **FIGURE 7–2**
PERT/CPM example

how the graphical nature of the schedule shows both the critical path and the contingencies.

PERT/CPM offers a clear advantage in showing graphically the project's interrelated aspects. The primary disadvantage is that a comparison of planned and actual and start and end times are not readily shown on the project schedule itself.

The Gantt Chart is simple and project slippage is readily shown. On relatively simple projects a spreadsheet program can be used to complete Gantt format project schedules. A list of tasks can be entered in the left-hand column's cells and a broken line is used to show a planned activity. An "X" denotes the planned completion of the activity. A "C" denotes the actual completion date. A time line is supplied as a reference to actual dates or to denote the passing of project time. See Figure 7–3 and take note of the task for completing the electronic design. This task was completed on schedule. The mechanical design in Figure 7–3 was completed one week behind schedule. The time line indicates that three weeks of project time have elapsed.

For large projects where there are detailed requirements to document resource allocation (i.e., man-hours and dollars), there are a number of different software packages designed specifically for project management. These are discussed in the following section.

7–3 ▶ Project Management Software

The process of project scheduling is basically the same whether you utilize paper, computer-drawn schedules, or project management software. Using a spreadsheet program to develop a project schedule, as discussed in the last section, is simply using the computer to draw the schedule and enter the task labels. The same result can be accomplished using a CAD software program. There are many software programs available today, however, that do much more than draw the schedule. These are software project managers and they create, monitor, and allow the management of the entire project. Project management software is currently available in prices ranging from roughly $50 to $500. These software packages are relatively easy to use once you understand some of the commonly used terminology. Specifically, project management software will perform the following project scheduling tasks:

1. Completes the schedule in Gantt chart or PERT/CPM format after information about the tasks is entered. The information needed is a description of the task, the expected duration, links (task contingencies), and resources (man-hours).

2. Allows a custom calendar to be created for the project that identifies the active project time. If project activity will take place on Saturdays, the calendar can be modified to take this into account automatically.

3. Each specific resource can be identified as well as a cost for the resource. The specific availability of the resource can also be noted. Resources are

Planned——x Actual to Complete____ C

Design Stage	Weeks	1	2	3	4	5	6	7	8	9	10	11	12
Time Line													
Electronic Design		‹-------		x C	Complete on schedule								›
Breadboard			--------	x									
Mechanical Design		‹-------		x------C	Late 1 week								
Complete Mechanical Mockup			--------	x	Incomplete due to delay in mechanical design								
Component Selection				--------	x								
Cost Estimates						x							
PCB Layout							x						
Complete Preliminary Design							--------	x					
Procure Parts							x						
Complete Prototype							--------				x		
Test Prototype											--------	x	
Apply for Regulatory Approvals										x	--------		x

▲ FIGURE 7–3
Spreadsheet Gantt chart example of actual start and stop times

typically people, and their normal hourly rate as well as their overtime rate can be entered into the resource database.

4. Once the schedule is complete, the progress of project activity is entered and compared to the initial plan. Reports on resource utilization and expenditures are available in addition to the overall project status.

The software project schedulers make the final process of putting the project schedule together very simple. They allow you to play out what-if scenarios to see how task variations affect the overall project completion date and provide for tracking resource issues, such as man-hours and project expenditures.

Example 7–1

The following example represents just one of the novel features included in project management software. On a project there is an individual we will call John who will be used to complete certain project activities. Let us say that John is available on a half-time basis only. Project management software allows the input of information about each team member into the project file. Included in this data is their time availability. In this case the 20-hour work week restriction for John and other personal issues, such as a planned vacation, are entered into the project resource file. As the software scheduler lays out the tasks in the schedule, it takes into account the amount of John's time available when calculating the linear time for the tasks he will perform. In other words, the software scheduler determines the linear time of the activity based upon the 20 hours that John is available during the work week.

Some project management software packages do not display the PERT/CPM format as it would be drawn, and some do better than others at showing the task links or *dependencies*. Most often the Gantt chart format gives the best view of the project. However, either format, Gantt or PERT/CPM, can be selected and viewed as desired. Figure 7–4 is an example of a Gantt Chart schedule generated on *Microsoft Project 2000* project management software.

The sample schedule shown utilizes resource information that has been supplied as project data. Figure 7–5 shows how the project resources are utilized on the various tasks. Figure 7–6 is a bar graph of the resource usage and indicates that the usage of the electrical engineer (Elec. Eng.) resource is over-allocated at 200% during the week of January 9, 2000. Some action must be taken to resolve this situation. Either more resources must be applied or the schedule will have to change.

Figure 7–7 shows a representation of a PERT/CPM report generated with *Minuteman* project management software.

In general, project management software is a very useful tool that can aid in the scheduling and management of projects. It is important to use the features provided as required and to learn how to utilize them properly. This type of software gives the project manager vision of the schedule from many different perspectives, but the end result will only be as good as the data provided.

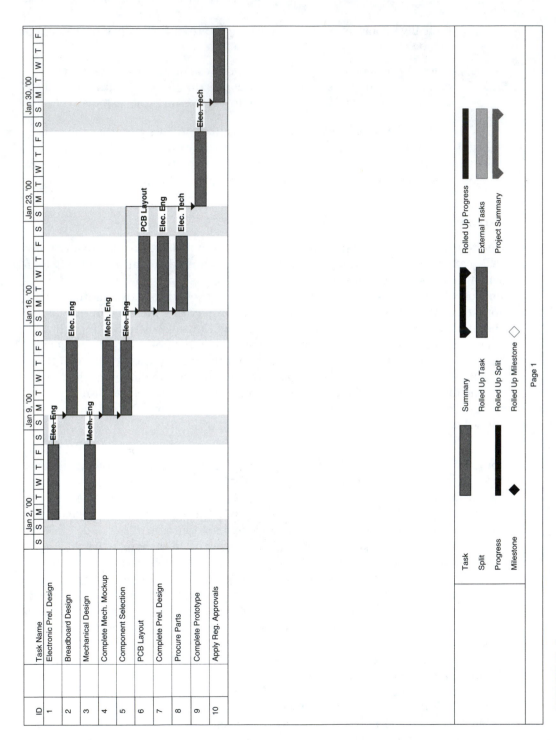

▲ FIGURE 7-4
Microsoft Project 2000 Gantt chart example

124

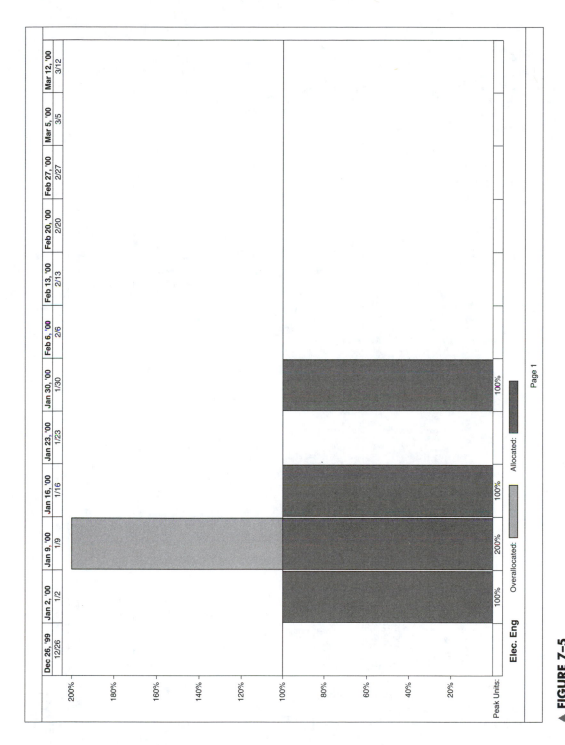

▲ **FIGURE 7-5**
Microsoft Project 2000 example of resource utilization

125

ID	Task Name	Details
1	Electronic Prel. Design	Work
	Elec. Eng	Work
2	Breadboard Design	Work
	Elec. Eng	Work
3	Mechanical Design	Work
	Mech. Eng	Work
4	Complete Mech. Mockup	Work
	Mech. Eng	Work
5	Component Selection	Work
	Elec. Eng	Work
6	PCB Layout	Work
	PCB Layout	Work
7	Complete Prel. Design	Work
	Elec. Eng	Work
8	Procure Parts	Work
	Elec. Tech	Work
9	Complete Prototype	Work
	Elec. Tech	Work
10	Apply Reg. Approvals	Work
	Elec. Eng	Work

▲ **FIGURE 7–6**
Microsoft Project 2000 example of resource usage

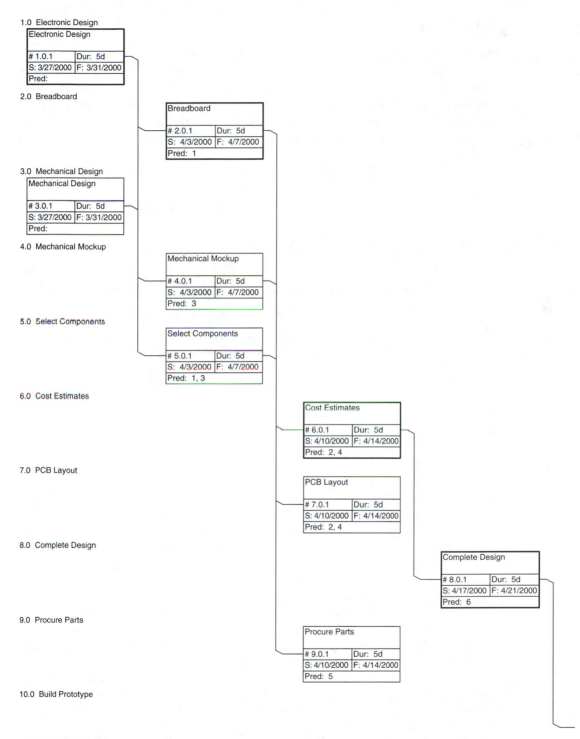

1.0 Electronic Design

Electronic Design	
# 1.0.1	Dur: 5d
S: 3/27/2000	F: 3/31/2000
Pred:	

2.0 Breadboard

Breadboard	
# 2.0.1	Dur: 5d
S: 4/3/2000	F: 4/7/2000
Pred: 1	

3.0 Mechanical Design

Mechanical Design	
# 3.0.1	Dur: 5d
S: 3/27/2000	F: 3/31/2000
Pred:	

4.0 Mechanical Mockup

Mechanical Mockup	
# 4.0.1	Dur: 5d
S: 4/3/2000	F: 4/7/2000
Pred: 3	

5.0 Select Components

Select Components	
# 5.0.1	Dur: 5d
S: 4/3/2000	F: 4/7/2000
Pred: 1, 3	

6.0 Cost Estimates

Cost Estimates	
# 6.0.1	Dur: 5d
S: 4/10/2000	F: 4/14/2000
Pred: 2, 4	

7.0 PCB Layout

PCB Layout	
# 7.0.1	Dur: 5d
S: 4/10/2000	F: 4/14/2000
Pred: 2, 4	

8.0 Complete Design

Complete Design	
# 8.0.1	Dur: 5d
S: 4/17/2000	F: 4/21/2000
Pred: 6	

9.0 Procure Parts

Procure Parts	
# 9.0.1	Dur: 5d
S: 4/10/2000	F: 4/14/2000
Pred: 5	

10.0 Build Prototype

▲ **FIGURE 7–7**
Example *Minuteman* PERT/CPM report

7–4 ▶ Bottleneck Issues

The first step in the development of a project schedule should be the identification of what can be called "bottleneck issues." Bottleneck issues are those issues most likely to be major obstacles in achieving the project's goals. By nature, these issues will consume the largest amount of a project's linear time (straight-line project time, beginning to end), as well as the total man-hours. To identify and address these critical bottleneck issues, consider the following questions:

1. Are there design areas in the project that involve completely new areas of technology for the development team?

2. Do new processes and equipment need to be developed to complete the project or product?

3. Will the design involve the use of custom components that will require tooling of some kind and have long lead times?

4. Will the design utilize standard components that have long lead times?

5. How easily and accurately can the project be prototyped?

6. Will the project involve some outside agency or another company where there might be a lack of control for getting things done in a timely manner?

Each one of these questions will be explained separately with an example:

1. Are there design areas in the project that involve completely new areas of technology for the development team?

Example 7–2

The strip chart recorder example discussed in Chapter 4 is a good example of a bottleneck issue. The product, if you recall, was a 32-channel strip chart recorder that printed graphical and alphanumeric data in six colors. A linear strip chart recorder mechanism, however, was new to the company as well as to everyone on the design team. The recorders they had worked on and manufactured for many years were circular chart recorders with significantly different requirements and mechanisms. The six-color printing and printing speed requirements represented leading-edge technology for strip chart recorders at the time, and the design team was tackling them as their first-ever strip chart recorder design. The project team had identified this early on as a bottleneck issue and had allotted much time in the schedule for the mechanism design. The design of the print mechanism became a constant problem, delaying the project many times. After the product was introduced, its lack of reliability was the ultimate cause of failure of the product in the marketplace. In retrospect, a separate research project to develop a strip chart recorder mechanism should have been completed before beginning the overall project.

As luck would have it, the very next project for the same design team included a similar situation as the one described here. On this later project, the lessons learned from the strip chart recorder project were applied. A separate research project was initiated to develop a prototype of a design area where the design team had little experience. Completing this step before starting the overall development project allowed for the successful completion of this project.

2. Do new processes and equipment need to be developed to complete the project or product?

Example 7–3

Let us take the example of a medical electronics company that is developing a new product that, because of its intended small size, must be developed using surface-mount components instead of through-hole circuit components. The company has experience manufacturing their own circuit boards, but only with through-hole components. In this case a key bottleneck issue is how the company will complete the manufacturing of the new product's circuit boards. The boards could be assembled by an outside company or completed internally. The latter would involve the development of a surface-mount technology (SMT) manufacturing process. For either solution, this key bottleneck issue must be addressed aggressively. Action items to resolve the issue should commence early in the project. In this particular case, the relatively long lead times of surface-mount components could represent another bottleneck issue.

3. Will the design involve the use of custom components that will require tooling of some kind and have long lead times?

Example 7–4

A common example of this bottleneck issue is the need to tool the enclosure or other mechanical parts included in the project. Most often this involves the tooling for plastic molded parts. Tooling is the process of creating a mold for a part that is to be molded from some material. The mechanical design process involves the initial design of the plastic part, prototyping, and evaluation. When this process is complete, the tooling can be started. The cutting of the tool usually takes 16 weeks, so it will be roughly four months before an actual part can be molded from the tool. This is a significant amount of project time and must be planned for very carefully.

4. Will the design utilize standard components that have long lead times?

The most common examples of standard components with long lead times are electronic components. The lead times of electronic components vary significantly and should be monitored on every design project. Long lead times usually result in the manufacture of the electronic component shortly after its introduction. When initial sales exceed projections, manufacturing is unable to maintain short lead times. On most projects, microcomputer and other specialized integrated circuits have the longest lead times. As a general rule, if the component is available from only one company, treat it like it has a long lead time. The more complex and specialized a component is, the more likely it is to have a long lead time.

5. How easily and accurately can the project be prototyped?

This issue usually involves tooled plastic parts. The key aspect of this bottleneck issue is the accuracy to which the prototypes reflect the design. After the initial mechanical and electrical designs have been completed, a prototype of the design is built. The electronics can be breadboarded—or laid out—as a printed circuit board in a relatively short time. Custom-designed plastic prototype parts will have to be machined. Usually one set of each part is machined. A soft tool (mold) is formed from the machined sample. Then a small quantity of prototype parts are molded from the soft tool. Only certain plastics can be used to mold parts using a soft tool. These plastic materials usually differ from those intended for the part's design. Therefore, prototype parts molded from a soft tool do not normally reflect the mechanical characteristics of the designed part. Expensive tooling must be cut without extensive testing of the molded mechanical parts that are made from the intended design materials. Once the tooling is started, the project team must wait 16 weeks before parts made from the hard tool are available. These first-run parts are fabricated using the intended design materials and can be extensively tested. The importance of this problem varies from project to project but should always be considered.

6. Will the project involve some outside agency or another company where there might be a lack of control for getting things done in a timely manner?

Many projects require approvals from at least one recognized testing laboratory such as Underwriters Laboratories or CSA International. This process is often a bottleneck issue that must be addressed in the project schedule. When working with any outside organization, whether a recognized testing laboratory or a joint venture with another company, direct control of some portion of the project is lost. The project manager must address this issue based upon any previous experience with the subject organization and schedule activity early in the project in order to explore how to interface with that organization. Many times, applications for the approval process can and should be made early in the project before the product is ready for testing. Good communication and planning are a must.

If the answer to any of these questions is affirmative, that particular issue will have to be considered very early and prioritized in the project schedule. With the bottleneck issues identified, the next step is to develop a list of tasks that will resolve these issues. The tasks should be scheduled for completion as required by the overall project. For the situation presented for the medical electronics company in Example 7–3, the SMT manufacturing process would have to be in place by the time the first production run for the new product was planned to begin.

7-5 ▶ General Procedure for Project Scheduling

This section presents a general procedure for use in developing any phase of a project schedule. The subsections that follow discuss the details for developing schedules for the design, design verification, and design improvement and project performance steps.

General Project Scheduling Procedure

1. Break down the overall project into the next largest subprojects.

Make a list of activities to complete these subprojects. Also identify any bottleneck issues and list activities to resolve those issues. For the house building example, the overall project is, of course, to construct the house. Some of the major subprojects are the excavation, foundation, cellar, overall framing, and, enclosing the house.

2. Determine the time required to complete each major activity as well as the personnel who are planned to perform them.

Estimate the total man-hours for each activity. The development team members should develop these estimates. The team members will probably have experience with similar tasks and can make relative comparisons to their previous experiences. All estimates should be challenged for being too large or too small. Estimates provided by inexperienced team members can often be too aggressive. The best way to temper all estimates is to use the process described in the PERT/CPM discussion. This process uses optimistic, pessimistic, and realistic estimates to find the best time estimate by using the formula:

$$T_e = (a + 4m + b)/6$$

where T_e = the actual time estimate you will use
a = the most optimistic estimate
b = the most pessimistic estimate
m = the most realistic estimate

3. Determine those major subprojects and activities that must be completed in series.

This determination may be due to the natural order of things (e.g., the excavation must be complete before the foundation is poured). Otherwise, logistics or manpower issues may be the overriding factors. This is the case when there are two project activities that could be completed in parallel (proceed at the same time in the schedule) but cannot because the same individuals are needed to complete both activities. The subprojects to be performed in series will form the critical path of the project. Lay out the schedule for the activities to complete each major subproject.

4. Assemble the complete schedule.

The final step is like assembling a large jigsaw puzzle. Many times, certain tasks only fit in the schedule one way to minimize the overall schedule time. The project manager must address long-term issues at the beginning of the project and use the available manpower to the fullest. Again, the team should review the schedule as it is being developed. The team members surely have ideas and thoughts about the schedule. In the end the team must agree with the schedule. The schedule should appear to be tight but doable and have planned *slack time* built into it (i.e., extra time to accommodate unforeseen problems).

Example 7–5

This is an example of how slack time is planned into a development project using the procurement of printed circuit boards. Once a circuit is designed, prototyped, and tested, and the artwork is completed, it is ready to be placed on order. The normal lead time for a printed circuit board is three weeks, which is allotted in the schedule after the board layout is planned for completion. However, a three-day delivery for printed circuit boards is available at a premium price. The premium price is included in the cost estimates for the development project. If development of the printed circuit board falls behind schedule, the project team has the ability and the budget to *buy* more than two weeks of schedule time. This is called *planning slack time* into the project schedule.

7–6 ▶ Scheduling Step Four—The Design Phase

Scheduling the design stage is difficult because the design process is inherently creative. It is important for the design team to have enough project time to allow their ideas to develop fully. On the other side, management will always be pushing for the project to be completed as soon as possible. If too much time is allotted, the project may never receive approval. Selecting the proper schedule for the design phase will be one of the first selective judgments the project manager and his or her project team will have to make.

The project manager and the project team should complete the design phase schedule using the general steps listed in section 7–5 of this chapter. This stage of the project mostly involves the design team, but all team members should be involved in its schedule. An example is the selection of a vendor to complete tooling for a new product. The purchasing, manufacturing, and quality control departments will be heavily involved with the new vendor after the project is completed and should all be involved with the vendor selection up front. It is important to address any identified bottleneck issues early in the project, even if they are not central to completing the design phase activities.

The design stage for most electronic development projects usually includes four distinct subprojects that must be completed in series. These are the preliminary design, component selection, breadboarding, and prototype stages. The preliminary design stage is where the initial design concepts are explored, tested, and developed. After the preliminary design is complete, the actual design components must be selected and procured. The components are then assembled onto a breadboard and tested. Then a prototype of the design is completed and tested by the development team. These are all part of the design stage, and each is covered in detail in separate chapters that follow.

Step Four—Design Phase Deliverables

A *deliverable* is something of substance that can be measured that results as an output from a process. In other words, deliverables are how you determine that you have completed a phase of a project. The deliverables that should result from the design phase are discussed in detail in a later chapter. The following are the general deliverables that should result from the design phase:

1. Working prototypes that reflect the design

2. Preliminary drawings and documentation that define the design

3. Set of the current design specifications

7–7 ▶ Scheduling Step Five— The Design Verification Phase

After completing the design stage schedule, the design verification stage can now be scheduled. The primary tool used to complete the design, the design specification, is the same document used to verify it. It is important that the project specifications be updated to reflect any changes that have occurred as part of the design process.

One deliverable from the design stage is a prototype unit that reflects the final design as closely as possible. A certain quantity of these may be needed and must be complete and available for the design verification process to begin. With the current design specifications in hand and prototypes available, the design verification can begin. To schedule this phase, develop a list of activities that will verify each aspect of the design and the specifications. An example of just some of these tests and activities follow:

Accuracy tests—To verify that the design can produce the accuracy required.

Ambient temperature tests—To verify that the product will function as required in the specifications over the specified temperature range.

Noise immunity tests—To verify meeting the noise immunity specifications.

Field tests—To verify the design in actual field applications.

Design review—After completion of all the tests, a design review should be conducted to review all aspects of the design with all departments: engineering, marketing, sales, customer service, field service, quality control, manufacturing, and finance.

Financial review—On every design project, money is being spent on its development and some form of financial return is planned. Therefore, the project must also be evaluated on a financial basis as part of the design verification stage. A financial review should be scheduled to discuss the cost of development as compared to the projected cost. Also, if the project is a product that will ultimately be manufactured for profit, the current estimates on manufactured cost should be compared to the design goal and must be reviewed in detail.

Verification report—Finally, a report should be put together to detail the results of the testing, the design, and financial reviews.

The report on the design verification stage evaluates, in effect, the quality of the design as it stands. It obviously has a large impact on the number of design improvements that will have to be undertaken in the design improvement stage. Having a large number of design improvements does not necessarily mean that the design team performed poorly. The number of design changes is strongly related to the innovation and technology level of the design. The newer the concept and technology, the more innovation is required, and the more likely it is there will be many ways to improve the initial design.

Design Verification Deliverables

The result of this phase of the project should include the following:

1. Summary report of all performance tests

2. Summary report of all field-test activity

3. Report of the design and financial reviews

4. List of all items that should be addressed in the design improvement stage

This list should be determined by the overall project group after a review of all the verification reports.

7–8 ▶ Scheduling Step Six—Design Improvements and Project Performance Monitoring

In some projects Step Six can be the most difficult phase to schedule. It is the phase of the project where you know the least about exactly what must be accomplished at the time you develop the schedule. The project manager must develop a schedule for this phase before even starting the design.

As discussed previously, the number of design improvements depends greatly on the type of project being undertaken. This is where the project manager must determine the experience level of the design team and the difficulty level of the project in order to develop a reasonable schedule for this stage of the project. To develop the schedule, the project manager should complete the following steps:

1. Estimate the size and scope of the design improvements to be made.

2. Determine a general task list to accomplish these design changes.

3. List activities that provide for verification of all of the design changes.

4. List activities to complete all of the deliverables noted for this phase. This includes a complete set of final design documentation for the design.

5. Develop estimates for all tasks and determine task contingencies.

6. Based upon task contingencies, lay out the critical path schedule that will complete all the required activities.

Example 7–6, Tooling Changes

When a project uses mechanical parts that must be tooled, such as molded parts, there are usually changes that must be made to the tool after the design verification phase. These changes can result from inaccuracies in dimensions as measured in the actual molded parts, or they can be design changes that will simply improve the performance of the part or make it easier to assemble. Whenever tooled parts are scheduled for a project, time should be allotted in the project schedule for changes to the tool. The amount of time allotted will depend heavily on the tooling supplier and its typical lead times. The activities that are scheduled to accommodate tooling changes follow:

1. Develop design changes to meet the desired improvements.

2. Prototype and verify the proposed changes.

3. Discuss the changes with the tool designer.

4. Modify appropriate drawings.

5. Request a quotation for price and delivery of the tooling changes, as well as any variation in piece price from the original quotes.

6. Place an order for the tooling changes.

Example 7–7, Circuit Board Changes

Most electronic projects use a printed circuit board and most of these boards will require some modification during the design improvement phase. The magnitude of the changes can vary greatly and are again dependent on how new the design concept is. In some cases the changes may be small enough that previous versions of the circuit board can be modified by the "cut-and-jumper" method and then utilized. The cut-and-jumper method involves cutting copper runs and soldering wire jumpers to modify circuit connections on a prototype circuit board. Other times, the changes are too significant and will require that the current revision of the circuit board copper be scrapped.

In any event, circuit connections that have been defined by the artwork will have to be modified and possibly other documents, such as the schematic, assembly drawing, and bill of material. The following activities are usually scheduled for implementing circuit board modifications:

1. Develop design changes to address the results of the design verification phase.

2. Schedule activities to verify the proposed changes.

3. Schedule activities to implement the changes in all drawings, procedures, and specifications.

4. Schedule the procurement of the new copper.

5. The final activity to address circuit board modifications is to verify the final revised copper by actual assembly and testing.

Example 7–8, Software Changes

Software changes seem to be a never-ending process in many projects that involve software. Software is utilized in most current products, but software changes are not as "soft" as the name might imply. Any time that software is changed even slightly, a complete verification of the software operation is usually required for products with embedded software. This is due to the difficult experience that most companies have with releasing products with poorly verified software changes.

Software changes are increasingly difficult to implement toward the end of a project. It is important to allow schedule time for developing software changes and the required reverification process as well. Also, a number of processes, procedures, and operator manuals must be modified to reflect software changes. All these activities need to be scheduled as part of the implementation of software changes.

In order to provide for the implementation of software changes, the following activities should be scheduled:

1. Determine the specific improvements to be made to the software.

2. Schedule the software development activities.

3. Schedule software verification of the areas of change as well as the overall system.

4. Schedule activities to modify all pertinent documentation to reflect the software changes.

Design Improvement and Project Performance Phase Deliverables

The design improvement and project performance stage deliverables include the following:

1. Complete set of documentation, reflecting all changes implemented in this phase, ready for formal release

2. Modified prototypes that reflect the current design

3. Final verification report that includes the results from the implemented design changes

4. List of recommended future improvements to the project that can be made as the opportunity arises.

After completion of the design improvement and project performance monitoring stage, you might think that you have reached that place called "project heaven," where the beloved project has achieved perfection and is complete. In the real world, this seldom occurs except on extremely simple projects. This stage seldom includes all of the changes that are desired, and there is always something that will

1. A complete project schedule

2. Updated design specifications

▶ **FIGURE 7–8**
Deliverables from Step Three

need to be changed at some other time. The changes implemented at this stage are those most important to the performance of the design and less likely to be achieved later. The final project evaluation report should include a list of recommended future changes that will not be completed at this time. The next time design activity is initiated on this project or product, these changes can be implemented.

▶ Summary

The Deliverables from Step Three are shown in Figure 7–8. At this point you are probably thinking that a lot of work has been done and that nothing has really been accomplished, other than gathering a lot of information and creating a pile of paper. The real value of the information and the paper will be determined by what happens next. One can have the best information, the most accurate set of specifications, and the best solution plan in the world, and it all means very little if the solution plan is not well executed. At the same time, without good information and having defined the specifications and solution plan, the project is sure to fail. At this point we at least have a chance for success. Now let us start the fun part.

Digital Thermometer Example Project

The digital thermometer specifications have been approved by the project team. To show the actual schedule development process, the digital thermometer project schedule is developed next. To start schedule development, the first step of the general process discussed in section 7–5 is utilized.

1. Break down the overall project into the next largest subprojects or activities.

For the design phase of this project, the following major activities need to be completed. Note that there is only one identified bottleneck issue for this project. The bold italicized words in the list that follows are the short form of the activity that will be placed on the schedule.

  1. Develop an ***electrical design*** for all circuitry.

  2. ***Breadboard***, test, and refine the design.

3. Develop a preliminary *mechanical design* for the package.

4. *Complete* a *mockup* of the package design and modify accordingly.

5. *Select* actual *components* to be included in the design and determine their costs.

6. Determine total *cost estimates* and compare to specifications. Make modifications in order to meet specification cost requirements.

7. *Lay out the printed circuit board*.

8. *Complete* all *preliminary design drawings*.

9. *Procure* all *parts*.

10. *Construct prototype* with actual parts.

11. *Complete prototype testing*.

12. Bottleneck issue—Determine *regulatory approval* application process requirements and timing *issues.*

2. Determine the time required to complete each major activity as well as the personnel that are planned to perform the activity.

Using the optimistic, pessimistic, and realistic time estimates and the equation shown in the general procedure on p. 131, the schedule time estimates of man-weeks for each activity were developed.

3. Determine those major subprojects and the activities that must be completed in series.

Reviewing the list of activities for the design phase shows that the following activities must be completed in series:

1. All design activity must be complete before actual components can be selected.

2. With components selected, then cost estimates can be determined and approved.

3. The circuit board can then be laid out while all the drawings are completed and parts are ordered.

4. After parts are available, the prototype can be constructed and then tested.

4. Assemble the actual schedule.

The list of activities should be combined with the final time estimates and the schedule put together based upon the serial activities noted in Step Three.

Usually the activities to be completed during the design verification stage can all be completed in parallel if manpower is available. The exception is the completion of the various reports that must wait for tests to be completed before the reports can be generated. The specific activities that will be required for the digital thermometer follow:

1. Complete performance testing
2. ***Ambient temperature testing***
3. ***Noise*** immunity and radiation ***tests***
4. ***Shock and vibration tests***
5. ***Design review***
6. ***Financial review***
7. Report development

Time estimates were developed for each of these activities. All of the testing will have to proceed in series. Testing scheduled for an outside testing agency can be scheduled in parallel.

Following is the list of activities scheduled for the design improvement stage of our example project. Note the general nature of the activities due to the fact that the specific activities are not known at this time. See Figure 7–9 for the actual schedule.

1. ***Complete design changes.***
2. ***Verify the design changes.***
3. ***Complete final documentation*** package.
4. ***Modify prototypes.***

▶ Exercises

7–1   Which three steps of the Six-Step Process are included in the project schedule?

7–2   Define the term *task contingencies*.

7–3   What does the term *PERT/CPM* mean and how did it come about?

7–4   What is the primary advantage of the PERT/CPM project schedule format?

7–5   What are the primary advantages of the Gantt chart project schedule format?

7–6   Write, in your own words, a definition for the concept described in this chapter called a project "bottleneck issue."

7–7   Develop the top-level schedule for the construction of a new home using a Gantt chart format on a computer spreadsheet program. Include the major

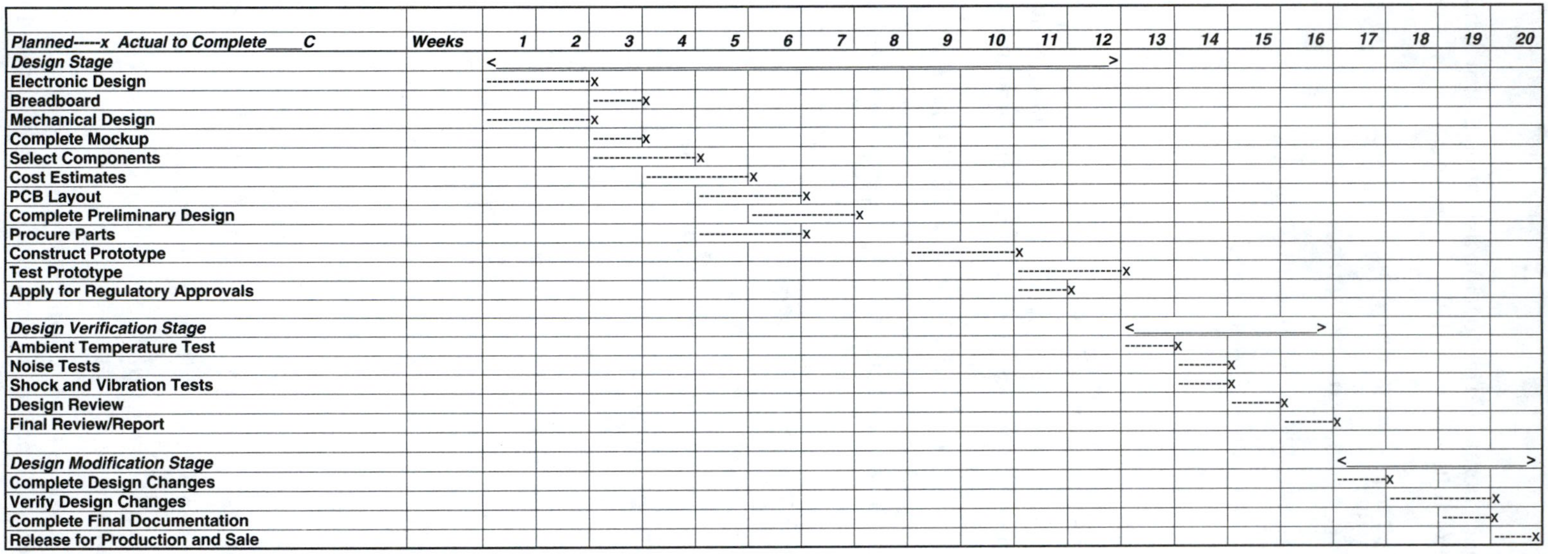

Planned——x Actual to Complete ___C	Weeks	1	2	3	4	5	6	7	8	9	10	11	12	13	14	15	16	17	18	19	20
Design Stage		∨											∧								
Electronic Design			x																		
Breadboard				x																	
Mechanical Design			x																		
Complete Mockup				x																	
Select Components					x																
Cost Estimates						x															
PCB Layout							x														
Complete Preliminary Design								x													
Procure Parts							x														
Construct Prototype											x										
Test Prototype												x									
Apply for Regulatory Approvals													x								
Design Verification Stage														∨			∧				
Ambient Temperature Test														x							
Noise Tests															x						
Shock and Vibration Tests																x					
Design Review																	x				
Final Review/Report																		x			
Design Modification Stage																		∨			∧
Complete Design Changes																		x			
Verify Design Changes																			x		
Complete Final Documentation																				x	
Release for Production and Sale																					x

▲ **FIGURE 7-9**

Digital thermometer schedule

tasks listed below in an order that you think is appropriate and schedule **tasks** in parallel wherever possible.

Site excavation

Foundation

Cellar walls and floor

Framing, including application of exterior plywood

Roofing

Windows and doors

Heating/air conditioning

Electrical

Plumbing

Interior walls

Interior floors

Bathroom and kitchen cabinets

Final site grading

Final inspection and certificate of occupancy

7–8   Repeat problem 7–7 using a PERT/CPM format schedule.

7–9   What are the advantages of using specific project management software to coordinate project activity?

8

Step Four: Execution (The Preliminary Design)

Introduction

We have completed the initial research, design specifications, and the project schedule. We have gathered information, defined the project goal, and decided how it will be accomplished. The preliminary design induces anxiety in most entry-level designers. For both simple and complicated design problems, the starting point is a blank piece of paper. Students of electronics often analyze problems for a given circuit where voltage or current levels are to be determined. But with a new design project, there is no circuit, just a blank piece of paper begging for creative thoughts and ideas to take shape. This chapter discusses how to begin the design process by addressing the following topics:

- ▶ Divide and conquer
- ▶ Preliminary design issues
- ▶ Enhancing creativity
- ▶ The initial design
- ▶ Design considerations
- ▶ Circuit simulation software
- ▶ Software development

8-1 ▶ Divide and Conquer

The best way to start the process is to break down the design problem into smaller blocks or modules, these being the next largest functional modules contained within the overall design. Look at the design problem process as if it were a gift-wrapped present that, when unwrapped, revealed another fully wrapped gift, then another, and so on. A complex project will sometimes have many sublevels.

Example 8–1

Consider the problem of designing an audio tape deck to play back and record standard audiocassette tapes. In this example the top-level design problem is broken down into the next level of subprojects. In order to define the functional requirements inside the tape deck, think about the functional requirements from the operator's point of view. The top-level operational requirements of a standard tape deck are as follows:

1. Power on/off

2. Insert and eject cassette tapes

3. Play tapes

4. Fast forward

5. Rewind

6. Pause

7. Record

8. Stop

9. Adjust record levels

10. View record levels

The operational requirements listed can be categorized into the following common functional circuits in a block diagram:

1. *Tape deck control*—This module will control turning the tape deck on and off, all tape deck movement, insertion and ejection of tapes, and selection of play and record.

2. *Audio signal playback*—Includes the audio playback head and amplifiers to the output.

3. *Audio signal record*—Includes the audio record head, the signal from the record inputs, the adjustment of the input signals, and the display of the record levels.

4. *Power supply*—All of the modules listed previously will need a variety of DC power that should be centrally developed and supplied to the various circuits.

5. *Tape transport*—The cassette tape and the complete mechanism.

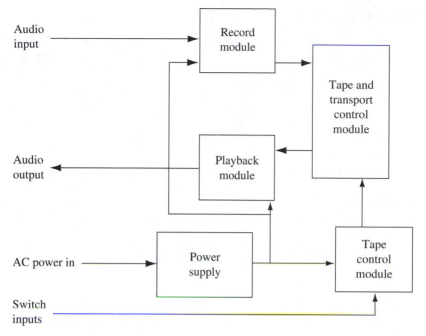

▲ **FIGURE 8-1**
Audio tape deck block diagram

The functional requirements for the audio tape deck design have been broken down into five modules that can be approached as separate designs. Figure 8–1 shows the actual block diagram.

After subdividing the project, Step One and Step Two should be reapplied to each module. The level to which this occurs depends on the complexity of the module. In some projects the modules are highly complicated circuit boards for which it is appropriate to develop a complete specification for each. For other projects the requirements can simply be listed as they are designed.

8-2 ▶ Preliminary Design Issues

After the design problem has been broken down into modules, there is more research to be done before actually designing the circuits. This is necessary because even though data was collected and the design problem was defined for the overall design problem, it has now been broken down into smaller modules. There are more details about the design of these smaller modules that must be determined. In a sense we are starting the Six Steps all over again as we attempt to solve the smaller design problem of the submodules. The research process described in Step One in Chapter 5 should be repeated for the submodule designs. If the module is complex, then Step Two should be completed for the module by developing design specifications.

Technology Selection

Another important choice to be made regarding the design project is the package technology that will be used for the electronic components. This is important for many reasons that will become obvious as we discuss them. The primary decision to be made is whether to use through-hole or surface-mount electronic package technology. Through-hole technology (see Figure 8–2) is the package technology that requires a hole in the middle of a circuit pad located on the PCB for mounting the component. With surface-mount technology (SMT) (see Figure 8–3), the component solders directly to a pad on the surface of the board. Because a hole is not required in the printed circuit board for surface-mount technology, the lead spacing can be made much smaller. This allows the reduction of component sizes as well. The decision between through-hole and surface-mount technology is an easy one because of their key differences and their impact. To make this decision, first pose the following questions:

1. Do the product specifications call for very low costs at very high volume levels?

2. Does the product require an extremely small physical size?

3. Does the company that will manufacture the circuit boards have SMT manufacturing equipment and capabilities?

If the answer to any one of these questions is *yes*, then SMT should be strongly considered. Otherwise, through-hole technology should be utilized. To get a better understanding of this question, let us review the advantages and disadvantages of SMT.

▶ **FIGURE 8–2**
Through-hole technology
*(Alfred Pasieka/Science
Photo Libary/Photo
Researchers, Inc.)*

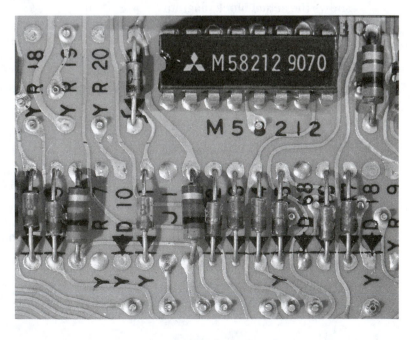

▶ **FIGURE 8–3**
Surface-mount technology

SMT advantages:

> Very small package sizes are possible.
>
> Low assembly costs are possible in high volume through automated assembly.

SMT disadvantages:

> Difficult to breadboard and prototype.
>
> Components may have longer lead times and larger minimum buy quantities.
>
> Large investment in equipment and know-how.
>
> Not easily repaired.

The criteria for this decision have been presented. Unless one of the unique advantages of SMT is required (small size or low costs at high volume) or the company already possesses SMT capabilities, then it will not be cost effective to make the investment required for SMT. In other words, if SMT capabilities are not in place and the unique advantages of SMT are not required, then SMT components should not be used in the design. This situation will change when SMT components cost significantly less than through-hole components and or SMT equipment prices continue to decline.

There are many times when SMT is utilized, and all the components needed for the product are not available in SMT packages. This causes a difficult problem because both technologies must then be utilized. This detracts from one of the key SMT advantages: automated assembly. In these cases the SMT components are installed first with an automated pick-and-place machine. Then through-hole components are inserted manually as a secondary operation. SMT packages are unavailable when the power level is too high to be dissipated in an SMT package or when the volume level is too low for an SMT package to be developed. When using both through-hole and SMT package types, it is advisable to keep the application of the "mixed" technologies as localized as possible. If possible, try to keep all the SMT components on one circuit board. When both technologies are used on the same board, place all through-hole components on one side and SMT parts on the other side.

Manufacturing Cost Budget

At this point it is a good idea to consider the manufacturing cost of the design to assure that the design process begins with this in mind. A manufacturing cost budget should be developed for each of the functional modules. This is done by dividing the total manufacturing cost goal for the product among each module, packaging and enclosure costs, and total assembly costs. As the design develops, this cost budget should be used as a tool to measure the performance of the design in meeting the cost goals. The total cost of each module will be determined by adding up the total cost of the parts and components within that functional module.

8–3 ▶ Enhancing Creativity

Creativity is a critical aspect of many careers, and it is rarely studied in school—or in the industrial world—as a subject in itself. Creative thinking, especially in engineering, is a most useful process. And for engineers, creative thinking is enhanced when both information gathering and problem definition occur before starting the creative process. We have completed both of these steps thus far.

Being creative involves the use of the subconscious: that part of our minds that is a mystery to us because we cannot directly access it. Think of the mind's conscious and subconscious parts as being like a computer operating in a multitasking environment, where the background operating system controlling all of the basic functions compares to our subconscious mind and the specific applications programs represent the conscious mind. The applications programs do not appear to directly access the operating system software. Nevertheless, the operating system is there operating and controlling all critical operations. The operating system in this case actually does a lot of work for the applications programs that are running and is an integral part of their results.

Compare your subconscious mind to the operating system software in your computer. It is a powerful and necessary part of your mind, and it can do much good work for you. It differs from software in that it has tremendous creative power. Some of its best work is in the area of producing ideas, novel, off-the-wall ideas that can solve problems in unique ways. However, the subconscious mind has some of the same limiting factors as a computer operating system. It has many application programs running at the same time (i.e., making sure you are breathing, speaking, and that your heart is beating, to name a few), and it has a hard time being creative during busy periods or after a busy day. The best way to put your creative subconscious mind to work is to feed it with all the information and a problem definition, then assign it the task of coming up with some ideas. Relax and let your subconscious mind perform its magic. Most of us have been troubled by a problem, then "sleep on it," only to wake up with a solution right on the tip of the tongue.

Many experts believe that the subconscious mind is our most valuable creative tool. Studies of extremely creative people indicate that their unique capabilities result from their natural skill combined with an extraordinary way of working. They set problems aside and let them incubate in the subconscious; then suddenly a solution appears. When planning daily activities, schedule the review of particularly difficult problems toward the end of the day. Tell your subconscious, almost in jest, to figure out a solution. Then relax, sleep on it, and collect the ideas in the morning. The next time you are faced with a significant problem, try giving your subconscious mind a special assignment while you go take a nap. It has been said that the best work flows forth effortlessly. This is true especially if the preliminary groundwork is done to set the stage.

After your subconscious mind has had time to review the problem, let your mind ramble as you write down the fresh ideas that come. It is important not to be too critical at this point. Suppress the urge to edit (i.e., discipline) your flow of ideas. Now spread out the different ideas, let your mind focus on one idea, and see what else your mind comes up with that is an offshoot of the original idea. Let your mind ramble again like this for each of the differing root ideas.

The end result should be a list of different and unique ideas. Many of them will be somewhat ridiculous, but others will surprise you. At this point you should evaluate the list of ideas with only a semi-critical outlook and attempt to combine the positive aspects of one with another. If things do not fit together, try turning one idea inside out, upside down, or invert it. Play with the ideas until they either work for the situation or you are convinced that they will not work. It helps to involve more than one person as soon as you can articulate your ideas. A novel idea has a better chance of acceptance and success if it is a team effort.

After evaluation, if an acceptable idea has not been developed, take note of what has been learned in the process and go back through the cycle again. Figure 8–4 shows a list of the steps to enhance creativity.

Example 8–2

This is an example of brainstorming experienced at a seminar on creativity. With 20 people in one room, they were asked how many different facts about a very average pencil could be listed on the blackboard. The group estimated that the number of facts about the pencil would be between 20 and 40. Then they began listing

1. Gather information.

2. Define the problem.

3. Relax, let your subconscious mind
   work on the problem.

4. Collect the ideas by listing all of them.

5. Categorize the initial ideas.

6. Try combining or modifying the initial list of ideas
   to create more ideas.

7. Edit the list of ideas by excluding the ones that
   obviously do not apply to the problem.

▶ **FIGURE 8–4**
Creativity steps

facts, one by one, as they went in order around the room six times. They developed a list of 114 facts about an average yellow pencil before being unable to continue. The process is revealing as a unique idea would reveal a number of other undisclosed facts about the pencil. Then another unique fact would feed the process a little longer, until, finally, the brainstormers were no longer able to think of any more. This shows the idea power that is possible by combining the minds of a number of people.

8–4 ▶ The Initial Design

We are now ready to proceed with the electronic design. The design should proceed in a logical sequence with the input section first. The power supply designs should be completed last, if possible. Many times the designs of the different functional modules will proceed in parallel with different designers assigned to each module. The input and output of each of the functional blocks must be known in order to begin the design. One way to begin is to sketch circuits on scrap paper. As the design takes on a finished look, make a neater sketch on paper with a light grid background.

Example 8–3

Continuing Example 8–1, let us proceed to complete the preliminary design of the tape deck control module of the audio tape deck. As shown in Figure 8–5, the inputs to the tape deck control module are the stop, play, record, fast forward, and rewind buttons. The necessary outputs to the tape transport are digital signals that represent the following actions: record and play (1 = record, 0 = play), rewind motor, fast-forward motor, and fast and slow speed pulley (1 = fast, 0 = slow).

This is a typical logic design problem, where the input and output relationship needed can be shown on a truth table. The truth table for the input and output relationship is also shown in Figure 8–5. The truth table shows that there are four specific input codes that will result in a specific output code. For all other input codes the output code will be all zeros. Using logic circuit design methods, the circuit shown in Figure 8–6 was developed.

8–5 ▶ Design Considerations

The effects of ambient temperature and electromagnetic interference (EMI) are important considerations when designing any electronic device. These issues should be addressed at the very beginning of the design process. This section will develop overall guidelines to consider each of these issues.

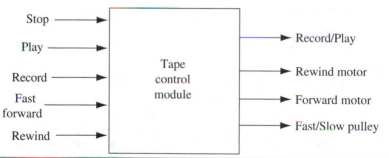

Stop	Play	Record	Fast Forward	Rewind	Record/Play	Rewind	Forward	Fast/Slow
1	x	x	x	x	0	0	0	0
0	0	0	0	1	0	1	0	1
0	0	0	1	0	0	0	1	1
0	1	0	0	0	0	0	1	0
0	1	1	0	0	1	0	1	0

▲ **FIGURE 8–5**
Example 8–3 tape control module block diagram and truth table

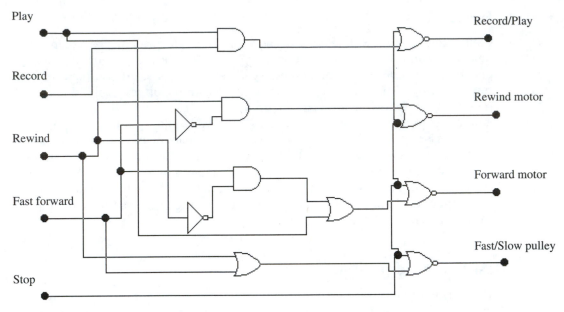

▲ FIGURE 8–6
Example 8–3 circuit solution

Ambient Temperature

The effects of ambient temperature on a circuit can either change the value of an analog signal or it can cause a component to fail due to excessive ambient temperature levels. When considering circuit reliability, the higher the ambient temperature, the shorter the operating life of the circuit. The effect of ambient temperature on a component can be determined from its data sheet.

To determine the change in analog signal levels that result from ambient temperature changes, review the temperature drift values in the specifications for each component in the circuit. The sum total of these effects can be determined analytically from the equations for the circuit in question. To meet the requirements for the design, it may be necessary to select different types or high-grade components that are less sensitive to ambient temperature. Another approach is to control the ambient temperature of the sensitive circuit.

To consider the effect of ambient temperature on the reliability of electronic circuits and components, first determine the maximum temperature ratings for all of the components. The most critical are those that will get the hottest, the ones that dissipate the most power. Voltage regulators, amplifiers, and power switching circuits usually require the most power. These component packages usually accommodate heat sinks. Heat sinks are metal components designed to mount on electronic component packages for the purpose of dissipating heat. Make sure that these components have the proper heat sink to disperse the power they are dissipating. Next determine the rise in temperature inside the enclosure for the electronic device being designed. Mechanical engineers have more experience with this and can determine this analytically because of their background in thermodynamics.

The temperature rise can be determined experimentally by placing the amount of power (use an appropriate number of miniature light bulbs) that the circuit will generate in a simulated enclosure fabricated from the design material intended. Measure the rise in ambient temperature that occurs when the circuit goes from a no-power state to full power. Once the rise in ambient temperature is determined, add that number to the ambient temperature range that the device will be exposed to (see the project specifications). The total that results is the range in the internal case temperature to which each component will be exposed. Verify that each component's maximum temperature of operation is at least 20% less than the maximum temperature allowed. Components that exceed this value can possibly be replaced with those that have a higher maximum operating temperature. Other resolutions involve reducing the power consumption of the electronic device or revisiting the design of the enclosure (its size and material) to reduce the expected internal ambient temperature rise.

EMI Immunity

For EMI to be a problem, there must be a source for the noise, a circuit that is sensitive to it, and a means for coupling or connecting the noise to the sensitive circuit. The primary sources for EMI are the AC power grid, radio signals, microprocessor-based equipment, electronic switching circuits, inductive switch-gear (relays and contactors), ignition systems, and arc welders. EMI can enter an electronic instrument through external wire connections (conductive EMI) or it can present itself as an electromagnetic signal (radiated EMI). EMI can also be generated within the electronic device that can affect its own operation.

The result of EMI noise presented to electronic circuits is realized either as an analog signal that fluctuates incorrectly or as a circuit that ceases to operate properly in some way. This effect can last for a moment or continuously until the circuit is reset or powered down. One example is a digital display that fluctuates momentarily due to a noise spike on the AC power line. Another is a microprocessor-based circuit that *locks up* (does not respond to key depressions) due to noise induced from a cell phone transmission.

Conductive EMI is electrical noise that is introduced to the electronic circuit by a circuit connection to an external device. The noise either already exists on the conductor, is picked up by passing through an electromagnetic field, or results from sharing the same power supply or ground with another conductor. The AC power grid contains many noise signals that result from its generation and the frequent switching of large inductive loads. Whenever changing current flows through a wire, a magnetic field is created around the conductor. When a current-carrying conductor is exposed to another electrical or magnetic field, current can be induced in the conductor from this field. When this current is unplanned and unwanted, it is called noise or EMI that is magnetically coupled onto the conductor.

When two circuits share the same ground or power supply, the current flowing through one circuit affects the voltage supplied to the other. In this case it is

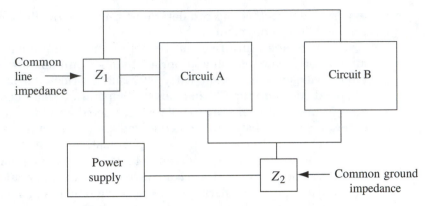

▶ **FIGURE 8-7**
Common ground and
power supply circuits

important to realize that each conductor represents a resistance to the circuit (see Figure 8–7). If the resistance value is made small (by using a larger conductor), the amount of noise voltage induced from one circuit to the other is minimized. This method of inducing noise signals onto conductors can affect external conductors and is a major source for generating noise internal to an electronic device. Radiated EMI occurs when an electronic circuit is exposed to an electromagnetic field directly and some noise current is induced in the circuit.

The methods for promoting EMI immunity in electronic devices combine the following concepts:

1. Minimize the amount of conducted EMI that can enter the electronic circuit.

2. Minimize the amount of radiated EMI the electronic circuit can pick up.

3. Minimize the amount of internal EMI created in the circuit.

4. Minimize the impact of any EMI noise that is presented to the circuit.

These are accomplished by implementing the following techniques:

1. *Elimination of the noise source*—If possible, the simplest and best approach is to eliminate the source of the noise completely. This can be done in a number of ways. If the source of the noise is an electromechanical contactor that is switching an inductive load, then the use of a transient suppressor such as a metal oxide varistor or MOV, or an RC network placed across the contacts, can eliminate the source of the noise.

2. *Shielding*—The concept of shielding can be applied to conductors or complete electronic devices. It involves placing a conductive material around a conductor or device to pick up noise that would otherwise be induced in them. The shield is grounded to drain off the induced noise signal with minimal impact to the conductor or device. Shielding works well in minimizing the effects of electrical fields over a wide range of frequencies. Shielding has a lesser effect on reducing noise generated by magnetic fields, especially at

higher frequencies, because the current flowing in the shield can be induced on the shielded conductor unless the shield is made from a magnetic material. Shielded cable is available in a variety of configurations to apply the shielding solution to conductors. It is important to connect the shield to the zero signal reference for the signal being shielded. To shield circuits or devices, the entire device should be surrounded by the conductive shield and connected to the zero signal reference. Openings or gaps in the shield should be minimized.

3. *Grounding*—Grounding is the establishment of a zero signal point in the circuit. The basis for good grounding is to separate grounds for circuits where the possibility of noise voltages affecting either circuit exist. The creation of ground loops should also be avoided. Ground loops occur when a circuit is grounded in two places that are actually at different potentials. The result is a noise voltage that is the difference in the two ground potentials. Ground loops are also susceptible to noise from magnetic fields (see Figure 8–8).

    Good grounding practices involve keeping the grounds from different types of signals separate. The best example is a circuit that includes both analog and digital circuitry. In this case the analog circuit grounds and analog power supplies should be kept separate from digital circuit grounds and digital power supplies. In the case of long ground conductors, avoid the establishment of ground loops. For low-frequency signals, a single-point ground can be used, eliminating the ground loop that results from multiple ground connections.

4. *Impedance matching*—This approach is based upon matching the source impedance with the load impedance to create a balanced situation in a circuit. When this is done, induced noise voltages in the circuit become common mode noise as opposed to differential mode noise. The term *common mode noise* means that the noise is common to both the positive and negative sides of a signal. *Differential mode noise* is present on only one side of the signal. Common mode noise is easily removed from a signal by using a differential amplifier that amplifies only the differential signal and rejects the common mode signal.

5. *Noise filters*—To eliminate EMI from a circuit, noise filters can be used to attenuate noise from a signal. These are usually designed with capacitors

▶ **FIGURE 8–8**
Ground loop examples

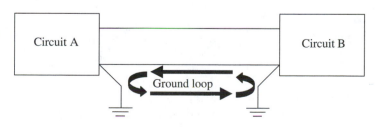

but could be any combination of R, L, and C circuits. The values of the components are selected to attenuate the expected noise-voltage frequencies and pass the signal frequencies.

6. *Physical orientation*—When conductive or radiated noise sources exist, try to keep them as far away as possible from sensitive circuits. On circuit boards locate a switching power supply away from any analog signal amplifiers. In cables and on connectors, locate signals that can interact away from each other and place ground levels in between them. In the case of magnetic fields, place circuit runs so that they are perpendicular with the magnetic field.

7. *Circuit isolation*—Transformers are used to isolate circuits for safety reasons, but there is also a benefit for eliminating noise voltages. When a ground loop exists between two circuits, a noise voltage is created between the two circuits if the grounds are at different potentials. Connecting an isolation transformer between the two circuits can break the ground loop (see Figure 8–9).

8. *Ferrite beads*—These are essentially a magnetic shield that serves to attenuate high frequencies without affecting DC or low frequencies. They are installed by passing a conductor through the ferrite beads. The result is a series RL circuit that serves as a high-frequency filter.

9. *Minimize the EMI effect*—There is always some level of EMI that will be presented to a sensitive circuit. In microprocessor circuits the result of EMI often causes the processor out of the normal program loop. It eventually locks up, unresponsive to any keyboard commands. To prevent this, a watchdog timer can be used to make this invisible to the user. A watchdog timer is a device that is reset at the beginning of every main program loop. The timer value is set at a time that is slightly greater than the maximum time to execute the main program loop. If the timer counts down before being reset at the beginning of the next program loop, then the processor must be out of the program loop. The watchdog timer invokes a reset and the processor will begin again at the beginning of the main loop. One other detail is that the RAM that contains parameters being used by the program must be preserved for this approach to work well. If the RAM is corrupted by EMI, then operation will not commence properly.

▶ **FIGURE 8–9**

Isolation transformer example

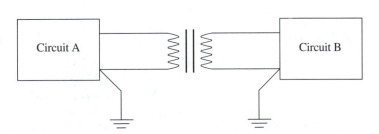

Generated EMI Levels

It is also important to minimize the EMI generated by an electronic device. The FCC publishes minimum acceptable limits of conducted and radiated EMI generated by electronic devices. European Committee approval directives also have limits for generated EMI. Basically, all of the techniques mentioned previously can be used to minimize the generated EMI from an electronic circuit.

8–6 ▶ Circuit Simulation Software

Circuit simulation software has become increasingly effective and easy to use over the last 15 years. Circuit simulators allow the circuit designer to draw a circuit schematic and simulate circuit operation on a personal computer. The simulation allows the examination of any voltage or current in the circuit as a function of time. Circuit simulators provide a very quick and easy way to evaluate a particular design concept without using expensive test equipment or having to procure parts and assemble a breadboard.

The first widely accepted circuit simulator was SPICE2, which was developed at the University of California, Berkeley, in the mid-1970s as a modification of its original SPICE program. SPICE stands for *Simulation Program with Integrated Circuit Emphasis.* SPICE, SPICE2, and the most current Berkley version, SPICE3F5 (commonly called BSPICE), are the recognized standard for analog circuit simulation and were developed with public funds. Therefore the software is in the public domain and available to U.S. citizens. XSPICE is a custom version of SPICE made for the U.S. Air Force that includes special modeling subsystems. PSPICE is a commercial version of SPICE developed by the MicroSim Corporation to operate on personal computers. PSPICE was followed by Electronics Workbench and many other commercially available software circuit simulators that can be run on personal computers. The key competitive differences between these simulators are as follows:

1. The number of device models available—in other words, how many of the available electronic devices have circuit simulation models

2. The allowable circuit complexity—this is usually realized as a limit to the number of components, connections, or circuit nodes

3. The types of analysis available

4. Functional complexity and ease of use

5. Ability to output schematic to circuit board layout software

The importance of these differences depends on your perspective. Electronics students may be interested in just the basic analysis functions but in industry it is important to have a widely functional software circuit simulator that interfaces directly with circuit board layout programs.

Circuit simulators function by having a computer model for all electrical and electronic components. The circuit designer keys in the schematic diagram for the circuit to be analyzed using the available component models and specifying any parameter values. If a model does not exist for a particular device, one has to be generated. All currently available circuit simulators allow the schematic to be drawn on the screen, whereas on older SPICE simulators the schematic was defined by specifying the devices to be connected between various circuit nodes. The schematic must be complete in every respect, including all power supplies and ground connections. A wide range of DC and AC power supplies are available as well as input signal sources. The circuit on the screen is almost the equivalent of a software breadboard. The designer then selects the voltages and currents to be analyzed. Some simulators simply plot a graph voltage/current over time while others allow the connection of a software-driven DVM or oscilloscope to a particular point in the circuit to display the waveform as it would be seen on a real oscilloscope. The circuit simulation described thus far can be classified as basic DC and/or AC circuit analysis, nothing more or less than could be accomplished on a circuit breadboard. However, it can be completed more quickly and without any physical components or test equipment. The following discussion covers the many types of analysis provided by state-of-the-art circuit simulators that provide functions and features that are often impossible to perform on a breadboard.

Transient Analysis

Transient analysis is where the circuit simulator starts to present a significant functional advantage over breadboards. Transient analysis is the determination of instantaneous changes that occur at a circuit node after power-up or some other starting point. The analytical treatment of transient analysis often involves rigorous mathematical operations and the definition of boundary conditions. On a laboratory breadboard transient conditions can be difficult to simulate and usually require a digital scope, storage scope, or some other data-recording device. Circuit simulators accomplish this task with relative ease and control, allowing the precise variation in initial conditions and analysis start and stop times. The transient analysis function provides accurate plots of voltage and or current over the specified time period.

Fourier Analysis

Fourier series analysis is another key circuit simulation tool. Fourier theory says that any non-sinusoidal periodic function can be described by a DC component with some number of sine and cosine functions. With this type of analysis, it is possible to determine the sine and cosine components that make up a complex waveform that exists at any particular circuit node. This information provides the circuit designer with the harmonic frequencies present in a signal as well as their relative amplitude. This can help to filter out unwanted signals, because the circuit designer can determine their frequency and their possible source. The total harmonic distortion (THD) can also be calculated with the Fourier analysis function in many circuit simulators.

Noise Analysis

Circuit simulators can also simulate the various types of noise generated by components: thermal noise, shot noise, and flicker noise. Thermal noise is noise caused by the temperature and its induced affect on the interaction of electrons and ions in a conductor. Shot noise results from the discrete nature of electrons (there can be one electron or two electrons flowing in a circuit, but not 1.5 electrons) flowing in a semiconductor and is the most significant cause of transistor noise. Flicker noise is a noise present in BJTs and FETs at low frequencies. When noise analysis is utilized on circuit simulators, the total noise present at a particular node resulting from these three types of noise is calculated and recorded.

Distortion Analysis

Distortion results when an electronic device such as an amplifier fails to duplicate an input waveform correctly. The causes of distortion are non-linear gain of a circuit or relative phase variations. Distortion caused by nonlinear gain is called harmonic distortion, while phase induced distortion is known as intermodulation distortion. Both types of distortion can be determined for a circuit a plotted as a function of frequency for a particular circuit node.

DC Sweep Analysis

As discussed in Chapter 6 in the development of specifications, variations in the DC power supply value are always an important consideration affecting circuit accuracy. Many circuit simulators provide a DC sweep analysis function that provides analysis of selected voltages or currents, as one or two DC supply values are varied. When selecting this type of analysis, the circuit designer specifies the particular DC supplies to be varied and the circuit node being analyzed, as well as the beginning and ending supply values and the increment steps. The results will indicate the effect of the DC supply variations on the voltage/current of the particular node.

Sensitivity Analysis

Sensitivity analysis serves to determine the component variations that will most affect the accurate function of a circuit. DC sensitivity analysis includes the variation of all component values, one at a time, in order to determine which component has the greatest impact on a critical circuit voltage value. On the other hand, AC sensitivity analysis varies the value of just one component, providing the critical variations of the component along with its impact on the circuit.

Parameter Sweep Analysis

The sensitivity analysis just described is usable for determining which component affects the circuit accuracy the most. Parameter sweep analysis provides for the variation of any component parameter value over a range, and in increments specified by the user. Semiconductor components will have a number of parameter values that can be varied as compared to passive components that will have few.

Temperature Sweep Analysis

In Chapter 2 we discussed the importance of considering ambient temperature effects on circuit performance. Temperature sweep analysis provides a helpful tool in the determination of ambient temperature sensitivity at very early stages in the

design process. During this process, circuit operation of selected nodes is recorded for different ambient temperatures. The parameter value for all components
that vary with temperature are changed accordingly and the impact on the circuit
function is plotted.

Transfer Function Analysis

A transfer function mathematically describes the operation performed on an input
signal by a functional circuit block to its output. Circuit simulators can analyze and determine the transfer function for a particular circuit. To do so the inputs and outputs
of the circuit are specified to the transfer function analysis feature whereas the transfer function, input and output impedance of the circuit is analyzed and determined.

Worst-Case Analysis

Worst case analysis is an extremely useful design tool. During the design process,
it is often desirable to know the maximum and or minimum voltage for a particular circuit node. Worst-case analysis can provide this by making sensitivity analysis runs for each component and then plotting the maximum and minimum values
found over the course of the sensitivity runs. This is critical information when determining accuracy specifications and selecting component tolerances.

Monte Carlo Analysis

Monte Carlo analysis involves the statistical probability of the variation in the parameter value of a circuit component. In other words, it uses the probability distribution function of a parameter value change to determine its value and the
ultimate effect of the circuit node. Each parameter value is selected at random
over the range specified using the selected probability distribution function (usually Guassian).

   Current software circuit simulators offer powerful design simulation tools,
and there are many from which to choose. There are a number of circuit simulators currently available that offer a wide range of features. There are three distinct
levels of performance and application that are apparent: academic, medium performance, and professional.

Academic-Oriented Circuit Simulators

These circuit simulators usually are low-priced (around $200) and are marketed pri-
marily to high schools and colleges that offer technical and electronic programs.
They represent a lower-level analysis function but are a very beneficial educational
tool, primarily due to the ease with which circuits can be built and analyzed. While
they may interface with other circuit board layout software, they have no or minimal circuit board layout capabilities themselves.

Medium Performance Circuit Simulators

These usually offer medium- to high-level circuit simulation and analysis. This often
includes mixed signal simulation (analog and digital circuits combined) and programmable logic design. The suppliers usually have functionally limited student

versions available for under $200, but the fully functional versions cost between $800 and $1000. There may be additional circuit board layout software that can interface directly with these circuit board simulators offered by the same supplier (at an additional price of between $1000 to $2000), or other circuit board layout software can be use directly. The circuit board layout software for the medium-level circuit simulators usually does not match the performance of the professional circuit board layout packages.

Professional Performance Circuit Simulators

These circuit simulators are available at the highest cost, usually around $2000 to $4000. They provide optimum circuit simulation, including mixed signal and programmable logic design, combined with professional circuit board layout capabilities. Circuit board layout features include multi-layer circuit boards, surface mount components, powerful auto-routing, and 3-D renderings of the final board.

As with all software, circuit simulators are rapidly changing as new features are offered and new suppliers enter the market. There is a significant learning curve required to take advantage of the key features on new products and software versions. This is similar to what occurred with mechanical CAD software and printed circuit board layout software. The market is still developing, but what is desired most is a circuit simulator that is accepted and used widely in industry, that functions directly with professionally accepted circuit board layout software, and that is available as a limited-function student version at a reasonable cost.

8–7 ▶ Software Design

The design of software programs has been an integral part of development projects since the expanded application of microprocessors and microcomputers in many products and systems. Software design is completed in a process similar to hardware design. The software specifications are the definition of the design problem for the software developers. The overall software requirements are broken down into smaller subprojects that are often called subroutines. Creativity is also a critical aspect of the software development process. The overall challenge for the software designer is to write programs that meet the following general objectives:

1. To develop programs that meet the functional objectives stated in the software specifications

2. To develop programs that operate reliably

3. To develop programs that take up as little memory space as possible

4. To develop programs that operate as fast as possible

In meeting these objectives, the software designer is emulating the tasks of the hardware designer as he or she tries to meet the hardware's design requirements with a minimal amount of circuitry that operates fast and uses little power. The primary differences between hardware and software design are the information and tools needed to perform them. The software world is one of computers, programming languages, methods of structuring data, and encoding and decoding information. Software programmers use a different language to communicate their ideas. For example, programmers use what is called a flowchart to lay out and design programs. A flowchart is very similar to the block diagram used by the hardware designer.

There are a couple of issues unique to software development that should be considered at the beginning of the preliminary design phase. As software is developed, there is the possibility that many portions of it may be used later in other products or systems. If this is a possibility, then the software in question should be written so that it can be separated and easily used for other applications. Another issue is the software's ability to operate in any particular memory area. This is called software portability. The portability requirements of the software should be defined in the specifications. As each stage of the development project is discussed, the special issues that relate to the software design should be discussed at that time as well.

The preliminary design of the software should be an activity than can be completed in parallel with the other design activities. The steps for the completion of the preliminary software design are discussed next.

Breakdown of the Problem

As with the hardware design, the overall software design problem must be broken down into logical, next-level subprojects that can be addressed separately. It is important to break down the requirements efficiently so that required functions are not duplicated in other modules. The process is very similar to the one described for hardware design. The modules can be assigned to and completed by different individuals on the project team.

Developing Flowcharts

Flowcharts are developed using the same process employed to create the block diagram for the hardware design. Flowcharts have a different shape for each functional block that symbolize the operation of that block. Figure 8–10 shows a summary of typical flowchart designations. Flowcharts are necessary for the development of efficient and concise software.

Example 8–4

Let us examine the overall software requirements for the design of a microprocessor-based alarm clock. The clock must keep time, compare the current time to one of two alarm settings, and provide for setting the time and the two alarm settings. The inputs to the microprocessor will be the square wave that the clock will use and the various switches that will invoke the set time, set alarm, fast, slow, and engage alarm functions. The outputs from the microprocessor are the displayed time (on four seven-segment LEDs) and the alarm output. A flowchart for how the software might be divided is shown in Figure 8–11.

The clock operation and clock parameter settings routines can then be broken down and developed separately.

▶ **FIGURE 8–10**
Flowchart symbols

Flowchart symbol	Symbol meaning
⟶	Direction of data flow
◯	User interaction
▭	An operation
⬭	An operation where the data is transformed
◇	A decision

Initial Code Development

After the flowcharts are complete, the code development can begin. The software can be written in assembly code for a particular microcomputer or in a higher-level language. Most software developed at this time is done with high-level languages. High-level programming languages offer a significant advantage over

▶ **FIGURE 8–11**
Alarm clock top-level flowchart

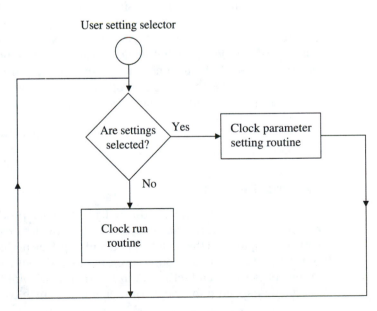

developing software with assembly code. High-level languages allow programming to occur much more quickly, but they result in software that takes up more memory. The benefit of programming speed far outweighs the additional program memory. In cases in which code execution time and space are critical, the software should be developed in assembly code.

The initial software should be developed to operate as planned on the flowchart. The program statements should be keyed in and compiled (high-level languages) or assembled (assembly code), and all errors should be corrected.

Simulation

The operation of the software can usually be simulated if there are not many real-time requirements in the system. Real-time requirements are those that must occur at an actual time in order to be valid. In the example alarm clock discussed previously, the clock input that represents a one-second count is a real-time event. This clock input must be read instantly as presented to the microprocessor; otherwise the clock will lose time. Without some hardware to generate the one-second clock input, it is hard to simulate the software operation completely in real time. However, the general operation could be verified by simply counting up in time, verifying the proper operation of all aspects of the software as key times are encountered, such as alarm settings and the like.

The various subroutines should be combined together and checked for errors. The resulting software should be simulated and tested to whatever degree possible.

8–8 ▶ The Mechanical and Industrial Design

As the preliminary electrical design is being worked on, the mechanical design is progressing concurrently in a similar way. The electronic aspects of the project are the focus of our discussion, but the student should also be aware of the other aspects of the development project. The project schedule will usually have the electrical and mechanical designs proceeding in parallel. The mechanical design process resembles the electronic process in many ways. The mechanical designer develops sketches and drawings of mechanical components and can simulate these designs on advanced CAD systems. In some projects the overall appearance of the product will require the expertise of what is called an industrial designer before the mechanical design can begin.

Industrial Designers

The industrial design involves the look and feel of a product. The industrial designer defines the appearance of the product and how it is used. The industrial designer can be compared to the architect of a building. The architect defines the appearance of the building and its layout to promote its efficient use, while the civil design engineer puts the architect's plan into a realizable building design. So too the mechanical designer implements the scheme put forth by the industrial designer.

One of the key decisions made on a project is whether to employ the services of an industrial designer. Only the largest companies retain industrial designers on their staffs so the average- and small-sized firms utilize private consulting industrial designers. The decision to use an industrial designer always requires additional investment by the company. Many times the size and appearance of a product are already strongly defined by the application or by the appearance of previous products. In these cases the industrial design is often left up to the mechanical engineer. If there is a need for a unique appearance, however, or there are significant human engineering aspects to the design, then the use of an industrial designer is worth the investment in the eventual success of the product. Industrial designers have unique ways of presenting subtle and different images and are thoroughly trained in human engineering principles. For example, the industrial designer will be aware of the current average size of the human hand in every respect.

▶ Summary

Having completed the preliminary design, let us note what has been accomplished and the steps that achieved those accomplishments. The following steps can be used as a guideline for completing the preliminary design in any development project:

1. Divide the design problem into smaller modules.
   Complete a block diagram.

2. Define the design problem relating to these modules.
   Develop submodule specifications.

3. Do research on areas relating to each design problem.
   Gather more information.

4. Develop a cost budget.
   Develop cost goals for each submodule.

5. Apply creative thinking!
   Use your subconscious mind. Let your ideas flow.

6. Complete the preliminary design.
   Complete a preliminary design schematic.

7. Perform software simulation.
   Simulate the preliminary design.

The blank piece of paper we started with at the beginning of the chapter can now have many schematic symbols with their associated connecting circuit traces. The result of the effort put forth so far should be a solid preliminary design. The next steps involve the selection of the physical components, their procurement, and breadboarding the design. A design exists but it is a paper design that is just on the verge of becoming a physical reality.

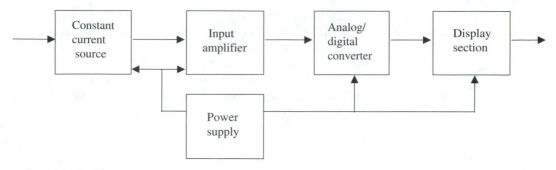

▲ FIGURE 8–12
Digital thermometer block diagram

Digital Thermometer Example Project

The preliminary design for the digital thermometer is completed next. The digital thermometer is broken down into modules as shown in Figure 8–12.

The five modules shown in Figure 8–12 represent five subdesign projects or modules that must be completed. We have successfully broken down a complicated problem into a set of smaller problems. Now the Six Steps are applied to each of these smaller design problems.

Sensor constant current source: The RTD temperature sensor varies in resistance as temperature changes. By injecting a constant current through the RTD, a voltage signal that is proportional to the temperature is developed.

Input section: The input section has to accept and amplify the input voltage signal created by the RTD sensor and the constant current course. The output signal from this section needs to be at an appropriate level and range for input to the analog-to-digital converter. The overall gain and the proper output level must be determined.

Analog-to-digital (A/D) converter: This section samples the input signal and converts it to a digital number. The circuit that accomplishes this is an analog-to-digital (A/D) converter. The bit resolution of the A/D converter and its sample rate must be determined and the A/D converter will have to be selected.

Display section: The display section takes the digital number developed for the temperature, converts it to binary coded decimal information and then to seven-segment data format, and sends it to the individual seven-segment LEDs to display the measured temperature.

Power supply: The power supply converts the AC power source called for in the specifications to supply the DC power supply needs of the circuits. The design of this module should be left until the other modules have been designed on a preliminary basis. This way the overall current and voltage requirements of the power supply can be more accurately identified.

These are the results of the research conducted before starting the design of the digital thermometer submodules:

Sensor Constant Current Source

RTDs come in a variety of nominal resistance values (their resistance at 0°C) and materials. The digital thermometer will use a standard 100 Ω platinum RTD with a temperature coefficient of 0.00392 Ω/Ω/°C. The actual resistance and temperature table was obtained for this RTD. The resistance change per °C is roughly .22 Ω/°C over the 0°F to 200°F range.

Other research indicates that the amount of current passing through the RTD should be less than 1 ma to preclude self-heating of the sensor. Self-heating would cause erroneously high temperature indications by increasing the temperature in the sensor area higher than it would be otherwise.

Input Amplifier

Past experience dictates that the input amplifier should be a differential amplifier to eliminate the effect of any common mode noise. The input impedance of the input amplifier should be as high as possible. The gain required will be determined in the next stage of the design.

Analog-to-Digital Conversion

The range of the digital thermometer is from 0°F to 200°F. The bit resolution of an A/D converter equals 2 to the Nth power, where N equals the number of digital bits output from the A/D converter. The 200°F range, if converted by a 10-bit A/D converter, would give a resolution of 200/1024 = 0.1953°F. A 12-bit A/D converter yields a resolution of 200/4096 = 0.0488°F. The accuracy specification of ±.25% of range can be calculated by multiplying 0.0025 by 200°F = 0.5°F. The accuracy requirements, in °F, are ±0.5°F. A 10-bit A/D converter will meet the accuracy specifications if there are no other major sources of error. The 12-bit A/D converter would be exceptional. A decision on which bit resolution to use will be determined later when the cost and complexity issues of the available A/D converters will be better understood.

The other factor affecting the A/D converter selection is the speed of conversion. This is dependent on the sample rate required by the application. The sample rate determines how often the A/D converter will have to sample the signal and convert it to a digital number. The conversion speed of the A/D converter must be faster that the sample rate in order to keep up with the system requirements. Because temperature is a very slow changing process, the sample rate need only be twice a second. This corresponds to conversion times of 300 ms to support this sample rate. Subsequently, integrating-type A/D converters can be utilized. These are the slowest and least expensive A/D converters available.

Display Section

The specifications call for the digital display to be 0.5″ high with seven-segment red LEDs. Four will be required to indicate the degrees 0.0 to 200.0 in tenths of a °F. ICs are available to convert binary data over to seven-segment data with the LED drivers included.

Power Supply

The power supply will be a standard linear type of supply. A switching-type power supply would be considered if efficiency or small size was more of a concern. In addition, 5 V at roughly 250 ma will be needed to power the LEDs and all the digital circuitry while ±5 V at 25 ma will be needed for the analog amplifier and constant current source sections. An AC ripple factor of 10% will be used as a design goal for each supply.

Cost Budget

The following cost budget was developed for the digital thermometer:

> Current source—$5
>
> Input section—$10
>
> Analog to digital converter—$15
>
> Display section—$20
>
> Power supply—$15
>
> Enclosure and packaging—$20
>
> Total assembly costs—$40

The total budget is the digital thermometer manufacturing cost goal of $125.

Constant Current Source Design

The RTD sensor varies in resistance with temperature according to the table shown in Figure 8–13. From the table we can see that the constant current source will have to supply 1 ma to a resistance that can vary from 92.9 Ω at 0°F to 136.7 Ω at 200°F. That is the design goal of the constant current source.

▶ **FIGURE 8–13**
RTD table showing resistance vs. temperature for 100 Ω platinum RTD

Temperature	Resistance
0°F	92.9 Ω
20°F	97.3 Ω
40°F	101.8 Ω
60°F	106.2 Ω
80°F	110.6 Ω
100°F	115.0 Ω
120°F	119.3 Ω
140°F	123.7 Ω
160°F	128.0 Ω
180°F	132.4 Ω
200°F	136.7 Ω

There are basically two kinds of current sources: those with ungrounded loads or those with grounded loads. Because the first stage of amplification is expected to be a differential amplifier, we can use the ungrounded variety of constant current source, since this is a much simpler and cheaper circuit. The differential amplifier will subtract the common mode signal resulting from the ungrounded RTD from the overall RTD voltage signal.

The ungrounded constant current source is shown in Figure 8–14. A voltage reference is needed as an input to the constant current source circuit to provide a stable input. The value of the constant current varies with the input voltage, so it is important that this input remains relatively constant. A 2.5 V reference has been chosen because this type is readily available. With a 2.5 V reference input, the constant current output is required to be 1 ma. We will assume that no current actually flows into the op amp negative input terminal. That means that the current flowing into the input resistor, R_1, will flow in the feedback path that is through the RTD. The value of the constant current will be equal to V_{ref}/R_1 or 2.5 V/R_i = 1 ma. Solving for the value of R_1, we calculate 2500 Ω. We will select the actual physical resistor later, but the closest available standard value is 2.49 kΩ.

We must make one design check in order to verify the operation of this circuit over the range in resistance variation of the RTD. We must verify that the voltage at the output terminal of the op amp required to maintain the 1 ma current is not greater than the saturation voltage for the op amp. The saturation voltage of an op amp is roughly 2 V less that the power supply voltages attached to the V+ and V– terminals. If the op amp is powered with ±5 V, the saturation voltages are roughly ±3 V.

The largest output required from the op amp to maintain the 1 ma current equals the sum of R_1 and the maximum RTD resistance multiplied by the 1 ma current. The maximum RTD resistance is 137 Ω + 2490 Ω = 2627 Ω × 0.001 amps = 2.63 V. This is significantly less than the 3 V saturation voltage, so the design checks out.

Finally, the value of compensating resistor R_2 must be determined. The nominal value for R_2 is the parallel combination of the input resistor R_1 and the nominal RTD resistance. Since the RTD resistance will vary with temperature, the best solution we can possibly provide is to assume the RTD is midway between the 0°F

▶ **FIGURE 8–14**

Constant current source circuit

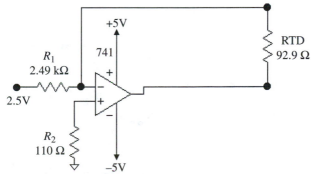

to 200°F range. At 100°F the RTD resistance is 115 Ω. R_2 should equal the parallel combination of 115 Ω and 2.49 kΩ. This calculates out to 109.92. The closest 1% value is 110 Ω.

Input Section Design

The resulting output of the RTD and the constant current source circuit is a voltage that is equal to the RTD resistance times 1 ma riding on top of a common mode voltage of roughly 2.58 to 2.62 V. The common mode voltage is the op amp output voltage of the constant current source discussed earlier. When the RTD is at 0°F (92.89 Ω), the resulting signal will be 92.9 Ω × 0.001 A = 0.0929 V on top of 2.58 V. When the RTD is at 200°F (136.7 Ω), the resulting signal will be 136.7 Ω × 0.001 A = 0.1367 V on top of 2.62 V. The design goals of the input section are to subtract the common mode signal and to amplify the difference signal by an amount that directly corresponds to the A/D circuit. The required gain of the input section depends upon the voltage range of the A/D converter.

After reviewing the A/D converters in more detail, a decision is made to use a digital multimeter type of A/D converter that has an input range of 0 to 2 V. Therefore the input section must have an output of 0 V to 2 V over the RTD input signal range of 0°F (92.9 mV) to 200°F (136.7 mV). The net RTD input signal range is 136.7 mV – 92.9 mV, which equals 43.8 mV. The overall gain of the input section will need to equal 2 V (the range of the A/D input) divided by the 43.8 mV range of the RTD, which equals 45.66.

The first stage of the input section will be a differential amplifier as shown in Figure 8–15. The differential amplifier will strip off the common mode signal that results from the ungrounded constant current source. In order to keep the input impedance of the differential amplifier high and the value of the feedback resistor low, the differential amplifier will have a gain of 1. The 45.66 gain factor required will be generated by the op amp stage that will follow. To provide the unity gain, 100 kΩ is selected as the value for all the resistors shown.

▶ **FIGURE 8–15**
Differential amplifier circuit

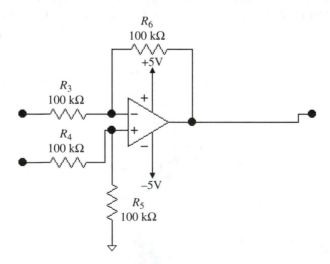

The output of the differential amplifier is equal to the RTD resistance multiplied by the 1 ma constant current source referenced to circuit ground. The common voltage has been subtracted from the net RTD signal. The result of this subtraction and the fact that the difference signal is input to the negative input terminal of the differential amplifier cause the output of the differential amplifier to be negative. Next the RTD signal must be offset so that the 0°F voltage, which at this stage is equal to 92.9 mV, is equal to 0 V. The RTD signal must also be amplified by a gain factor of 45.66. A linear combination circuit like the one shown in Figure 8–16 will be utilized to accomplish all of this. This circuit will invert the output signal from the differential amplifier, subtract off 92.9 millivolts, and amplify the resulting signal by a factor of 45.66.

The output of the linear combination circuit is the following:

$$V_o = -[(V_{RTD} \times R_F/R_9) + (V_Z \times R_F/R_8)]$$

To provide a gain of 45.66, the total feedback resistance is selected to be 456.6 kΩ ($R_{11} + R_G$) and R_9 will be 10 kΩ. The equation for V_0 reduces to

$$V_0 = -45.66(V_{RTD} + V_Z)$$

V_Z is generated by the +5 V supply and a voltage divider circuit that includes the zero potentiometer. The zero potentiometer will be adjusted when the circuit is initially calibrated so that V_Z is equal to V_{RTD} at the 0°F RTD resistance. The output of the linear combination circuit equals the amount of RTD resistance greater than the 0°F resistance multiplied by the 1 ma current source times the gain of 45.66 of the linear combination circuit.

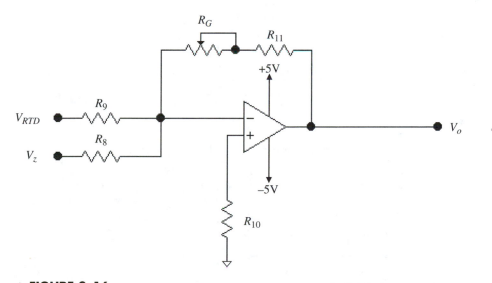

▲ **FIGURE 8–16**
Linear combination circuit

In order to subtract the negative RTD voltage at 0°F from V_{RTD} voltage, V_Z must be positive. V_Z is generated with +5 V and a voltage divider circuit. The nominal value for V_Z will equal the RTD voltage at 0°F, which is 92.9 mV. Potentiometer R_Z is selected to be 1 kΩ, ten times smaller than the 10 kΩ input resistor to preclude any loading effects. The value of R_Z that will provide the 92.9 mV offset signal ideally will be the midpoint of the potentiometer adjustment or 500 Ω.

$$V_Z = +5 \ V[500Ω/(500 + R_7)] = -0.0929 \ V$$

Solving for R_7 yields 26,410 Ω. The closest standard 1% value is 26.1 kΩ. This is the value selected for R_7.

The final consideration for the linear combination amplifier is the development of an adjustable gain to provide the nominal gain of 45.66 required. The feedback path includes a potentiometer R_G that will make the gain adjustable (see Figure 8–17). Potentiometer R_G will be adjusted to provide exactly 2 V output to the A/D circuit when the RTD is at 200°F. R_Z of the linear combination circuit will be adjusted to provide 0 V output when the RTD is at 0°F. To provide the adjustable gain of 45.66, the value of R_{11} plus the potentiometer setting of R_G must total 456.6 kΩ with the input resistors equal to 10 kΩ. The nominal value of R_G is equal to 456,600 Ω – 430,000 Ω = 26,000 Ω. This is close to the 25,000 Ω center position of the potentiometer.

Compensating resistor R_{10} should equal the parallel combination of all the resistors between the op amp negative terminal and ground. There are three resistive paths to consider: 456.6 kΩ, 10 kΩ, and 10.5 kΩ (the nominal value for R_Z + 10,000 Ω). The parallel combination of these three resistors calculates out to 5,065 Ω. The closest 1% value is 5.11 kΩ.

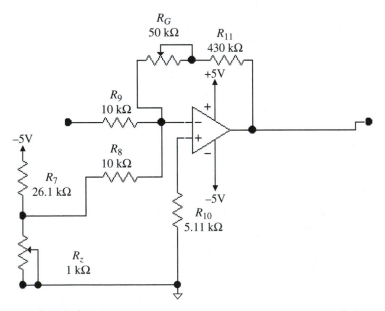

▲ **FIGURE 8–17**
Gain adjustment circuit

A/D Converter

The input signal presented to the A/D converter will be 0 V to 2 V and will represent the temperature seen by the RTD from 0°F to 200°F. The A/D converter will sample the voltage at its input and convert the analog voltage to a digital number. The resulting digital number will depend on the number of bits available on the particular A/D converter chosen. As discussed earlier, either 10 or 12 bits will be adequate.

The output of the A/D circuit should be the digital signals that will light up the proper digits of the four seven-segment LEDs with the temperature in °F. In order to accomplish this, the binary number computed by the A/D converter in 10 or 12 bits will have to be converted over to binary coded decimal (BCD) for the four digits to be displayed. The BCD data will then be converted to seven-segment format in order to light up the proper segments on the LED display. Figure 8–18 shows a visual explanation of this process.

The slow sample rate of two samples per second will allow the use of the slower-integrating type of A/D converters. It will also allow us to consider the use of the digital multimeter type of A/D converter that already has the binary-to-seven-segment decoder and drivers built into the integrated circuit (IC). Much of the circuitry shown in block diagrams in Figure 8–18 is already incorporated into the digital multimeter integrated circuits. These ICs were designed for use in digital multimeters.

The particular IC chosen for the A/D converter module is the 7117 device offered by a number of manufacturers. It is always desirable, although not always possible, to have more than one source for all components. The 7117 is a 3½-digit, 11-bit (2000 counts) A/D converter with LED decoder drivers already built in. The half digit means that the most significant digit can only be a zero or one. The resulting output display from this A/D converter would be 0 to 199.9, which is acceptable.

The display output from the 7117 is *multiplexed*. This means that the data for each digit is presented to the same seven-line output port that connects to all four display digits in parallel. The multiplexer output is a four-digit selector, which enables the appropriate digit when the data for that digit is available on the output port. The sequence of operation for the 7117 follows:

Sample input voltage

Perform A/D conversion

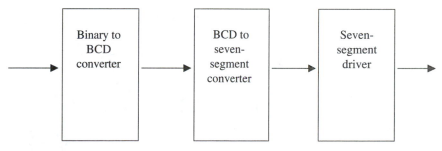

▲ **FIGURE 8–18**
Binary/BCD/seven-segment conversion

Binary data is converted to $3\frac{1}{2}$-digit, seven-segment data for each display

Data for least significant digit is output to output port with digit 1 line high

Data for digit 2 is output to output port with digit 2 line high

Data for digit 3 is output to output port with digit 3 line high

Data for digit 4 is output to output port with digit 4 line high

The 7117 continuously strobes and updates the display data. It all happens so quickly that the display appears continuous to our eyes.

The 7117 requires ±5 V DC power that has already been planned for and it also has included its own internal reference for the A/D converter. Many A/D converters require an external reference to determine the analog range of operation for conversion (see Figure 8–19).

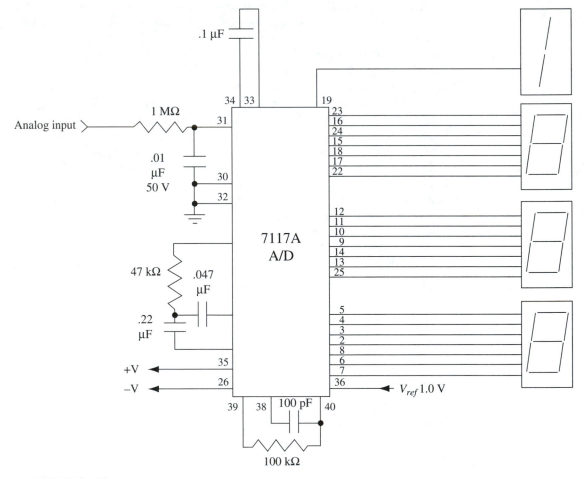

▲ **FIGURE 8–19**
7117 A/D converter with display driver

Digital Display Circuit

Since the A/D converter circuit will drive the seven-segment LEDs directly, all that is required for this circuit is the dropping resistors for each LED segment of the display selected by the A/D multiplexer. The LED displays will consist of 0.5″-high, red seven-segment displays that are available in either common cathode or common anode configurations. *Common anode* means simply that all the anodes of the seven LEDs that comprise the display are connected together. In order to turn on a common anode segment, the individual cathodes must be connected to ground through a current limiting resistor. On the other hand, the common cathode display is lit up when the individual anode is driven by +5 V through a current limiting resistor.

Power Supply Design

The power supply section will supply the DC voltage to all of the circuits included within the digital thermometer. Positive 5 V @ 250 ma is needed to power all the digital circuitry and ± 5 V @ 25 ma are the voltages required by the analog circuits. Two positive 5 V supplies will be developed because it is desirable to keep the analog and digital circuits separate. The large amount of switching that takes place in the digital circuits will induce small noise voltages in the analog circuits if they share the same power supply.

The input to the power supply circuit will be nominal 115 V AC at 60 Hz. The specifications state that the input AC may vary in amplitude by as much as ±10%. This means that the input AC voltage may vary by as much as ±11.5 V or from 103.5 V AC to 126.5 V AC. The line frequency can vary from 50 Hz to 60 Hz. The design of the power supply will have to take into account these variations. On the low end of the AC voltage input (103.5 V AC at 50 Hz), the design must ensure that there is enough voltage fed to the voltage regulators to maintain regulation at the output DC supply voltages. On the high end (126.5 V AC at 60 Hz), the voltage drop across the regulators must not exceed the rating of the regulators at the current rating of the supply.

The block diagram shown in Figure 8–20 provides an outline for the design. The input AC voltage is input to a transformer to step down the 115 V AC to a smaller AC voltage. A center-tapped transformer is selected to generate the plus-and-minus voltage supplies. The lower AC voltage is then rectified by a full-wave rectifier circuit that will result in fluctuating DC voltages as the output of the rectifier circuit. The rectifier output is connected to a filter circuit that will minimize the fluctuation or ripple before connection to the IC regulator, which will accept a higher DC voltage and regulate it down to some lower DC voltage.

Starting at the output of the power supply, each regulator will be regulating at ±5 V. Each regulator will need about 3 V to regulate with, so the inputs to the regulators should be at least 8 V when the input AC is at the low line condition. The output of each filter circuit must be 8 V DC.

Positive 5 V Digital Supply Filter Circuit

This filter circuit will have to supply 250 ma to the digital voltage regulator. A ripple factor of 10% will be the design criteria for the input of the regulator as design

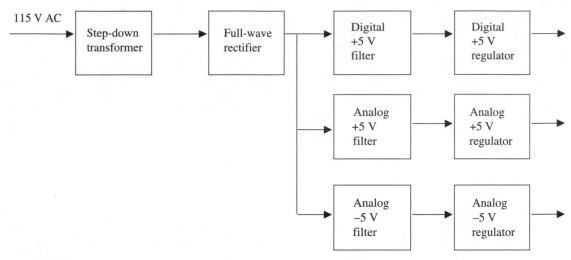

▲ **FIGURE 8–20**
Power supply block diagram

criteria. The output of the regulator will also be filtered to reduce the net output ripple even further.

$$\text{Ripple Factor} \rightarrow r = V_{rpp}/V_{DC} \qquad (8\text{–}1)$$

where V_{rpp} = the amount of ripple
V_{DC} = the DC output voltage from the filter

$$V_{rpp} \cong \left(\frac{1}{fR_LC}\right)V_{RECT} \qquad (8\text{–}2)$$

$$V_{DC} = \left(1 - \frac{1}{2fR_LC}\right)V_{RECT} \qquad (8\text{–}3)$$

where f = frequency of the rectifier output, which is twice
the frequency of the input AC for a full-wave bridge
R_L = the load resistance
C = the capacitor filter value
V_{RECT} = the peak value output from the rectifier

Starting with the ripple factor of 10%, the amount of ripple acceptable can be calculated by substituting into equation 8–1:

$$V_{rpp} = r \times V_{DC} = 0.10 \times 8 = 0.80 \text{ V ripple}$$

From equation 8–2:

$$V_{rpp} = (1/120R_LC)V_{RECT} = 0.80 \text{ V}$$

For the +5 V supply

$$R_L = 8 \text{ V}/0.250 \text{ A} = 32 \ \Omega$$

V_{RECT} will be one-half the ripple value above V_{DC} or

$$8 \text{ V DC} + 0.40 = 8.40 \text{ V DC}$$

and

$$C = 8.4/(0.8 \times 120 \times 32) = 2700 \text{ } \mu F$$

From equation 8–3:

$$V_{DC} = (1 - [1/(240)(32)C]) \text{ } V_{RECT}$$

Substituting $C = 2700 \text{ } \mu F$:

$$V_{DC} = [1 - (1/240 \times 32 \times 2700 \text{ } \mu F)] \text{ } 8.4 = 7.99 \text{ V}$$

± 5V Analog Supply Filter Circuit

These two filter circuits will have to supply 25 ma to the ±5 V analog voltage regulators. Again, a ripple factor of 10% will be used at the input of the regulator as a design criteria. The output of each regulator will also be filtered to reduce the net output ripple even further. Starting with the ripple factor of 10%, the amount of ripple acceptable can be calculated by substituting into equation 8–1:

$$V_{rpp} = r \times V_{DC} = 0.10 \times 8 = 0.80 \text{ V ripple}$$

From equation 8–2:

$$V_{rpp} = (\frac{1}{120}R_L C)V_{RECT} = 0.80 \text{ V}$$

For the each ±5 V supply

$$R_L = 8 \text{ V}/.025 \text{ A} = 320 \text{ } \Omega$$

V_{RECT} will be one-half the ripple value above V_{DC} or

$$8 \text{ V DC} + 0.4 = 8.40 \text{ V DC}$$

and

$$C = 8.40/(0.80 \times 120 \times 320) = 273 \text{ } \mu F$$

From equation 8–3:

$$V_{DC} = [1 - (1/(240)(320)C)] \text{ } V_{RECT}$$

Substituting $C = 273 \text{ } \mu F$:

$$V_{DC} = [1 - (1/240 \times 320 \times 273 \text{ } \mu F)] \text{ } 8.4 = 8.39 \text{ V}$$

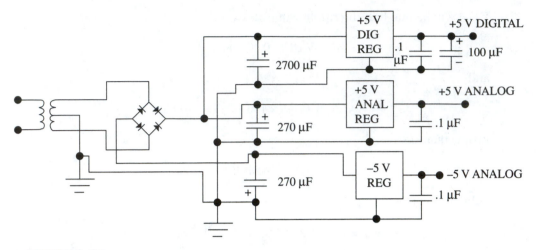

▲ **FIGURE 8–21**
Power supply schematic

Step-Down Transformer Design

The final detail required for completion of the design of the power supply is the determination of the parameters for the step-down transformer: input voltage and current and output voltage and current. The circuit schematic for the transformer and the full-wave bridge rectifier is shown in Figure 8–21. The previous calculations for the filter sections indicate that a V_{RECT} of ±8.2 V is the required output from the full-wave bridge rectifier. There is a voltage drop of about 1.4 V across the diodes in the full-wave bridge circuit. The net peak DC voltage output from the transformer secondary will be dropped by 1.4 V. The net output from the secondary should be 16.4 V (2 × 8.2) plus 1.4 V or a total of 17.8 V under low line conditions. The low line condition was previously calculated to be 103.5 V. The turns ratio of the transformer should be 0.172. At the nominal AC voltage of 115 V AC, the secondary output will be 115 V AC × 0.172 = 19.78 V. The design requirements for the step-down transformer are 19.78 V center tapped at 275 ma. The complete power supply circuit is shown in Figure 8–21.

There are two areas of the digital thermometer project that are worthwhile simulating in software before breadboarding: the input section and the power supply. These are shown in the two examples that follow.

Example 8–5

In this example, we will review the simulation of the constant current source and the entire input section of the digital thermometer on *Electronics Workbench*. To accomplish this, the entire circuit for each was input into *Electronics Workbench* with all power sources connected. The resulting circuit is shown in Figure 8–22. The real advantage of this simulation is that the actual RTD table resistance values can be keyed into the RTD resistance and the resulting output of the input section can

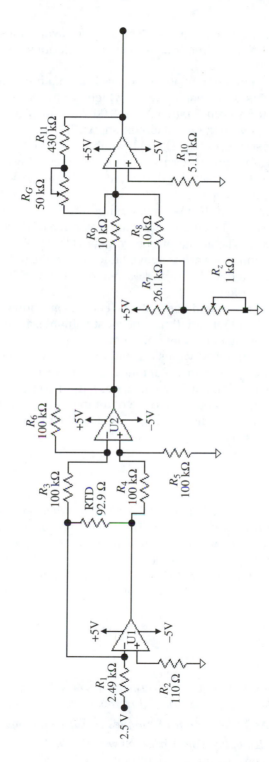

▲ **FIGURE 8-22**
Input section simulation

be read by the simulated multimeter. The impact of variations in the constant current source can be seen directly as well as the variations of any other component or situation.

The first step in the simulation process is to make sure that the circuit as designed performs its basic requirements, which are to convert the RTD resistance variations from 0°F to 200°F into a 0 V to 2.00 V output signal for input to the A/D converter. After powering up the circuit shown in Figure 8–22, the RTD was set equal to the 0°F setting of 92.9 Ω and trimming resistor R_Z was adjusted to make the output voltage as close to zero as possible. The resulting output voltage was actually –15 mV due to the fact that the potentiometer in the software simulator was variable in minimum steps of 1% of its value. Next the RTD resistance value was changed to equal that of the maximum temperature 200°F (135.7 Ω) and the gain adjustment R_G was adjusted until the output voltage of the input section was approximately 2.00 V. The actual output voltage was 2.022 V, again because of the minimum potentiometer variation available. The adjusted value of each potentiometer that resulted in the proper output voltages verified the design calculations performed earlier.

Because the potentiometers will correct for variations in the nominal values for all other static circuit variations, the most valuable simulation analysis will be the variation in parameter values with temperature. *Electronics Workbench* has a built-in analysis option that varies the parameters of the components over a temperature range that you specify. Running this analysis over a range of 0°C to 50°C, one can observe only a 20 mV change in the output voltage over this temperature range. Since every 10 mV equates to 1°F of temperature indication, this means that the output temperature, when displayed at an ambient temperature of 25°F, varies ±1°F from 0°F to 50°F.

Example 8–6

The power supply design completed for the digital thermometer in this section was input into *Electronics Workbench*. The circuit was tested to determine the net output voltage from each power supply along with the amount of ripple for variations in the amplitude and frequency of the input AC voltage. During high AC line conditions, the effect on the input regulator ICs was verified and determined to be satisfactory. Variations in load and the impact on the regulator outputs were studied as well.

▶ ## References

Gamache, R. D. & R. L. Kuhn. 1989. *The Creativity Infusion: How Managers Can Start and Sustain Creativity and Innovation.* New York: Harper & Row.

Ott, H. W. 1976. *Noise Reduction Techniques in Electronic Systems.* New York: Wiley.

Pinker, S. 1997. *How the Mind Works.* New York: W. W. Norton & Co.

Sommerville, I. 1995. *Software Engineering.* 5th ed. Reading, MA: Addison-Wesley.

▶ Exercises

8–1   You are working with a design team on a new product and are assigned to design the power supply. The requirements for the power supply are as follows:

Input AC: 115 V AC ± 10%, 50–60 Hz

Outputs:

+5 V DC at 100 ma

±12 V DC at 50 ma

Ripple: 10% ripple factor on each supply output

Break down this design problem into subdesign problems and show a block diagram for the entire power supply system.

8–2   Review the last two op amp stages of the digital thermometer project. It is decided to utilize a different A/D converter that has an input range of 0 V to 5 V. Modify the circuit so that the output signal from the input section has a range of 0 V to 5 V instead of 0 V to 2.5 V.

8–3   List the two steps that should be completed first before trying to utilize creative thinking to solve a problem.

8–4   What is the key advantage of utilizing surface-mount technology?

8–5   Surface-mount technology is always cheaper to manufacture than through-hole technology—true or false?

What are the issues that determine which is cheaper?

8–6   What are the key disadvantages of utilizing surface-mount technology?

8–7   Which are the two most important environmental design considerations discussed in this chapter?

8–8   List the general concepts for maximizing EMI noise immunity.

8–9   In your own words, define what is meant by the term *ground loop*.

8–10 List the general objectives for the development of a software program.

8–11 What is an industrial designer? What is the result of his or her work, and on which types of projects should he or she be utilized?

8–12 List the advantages of using software simulators in general when compared to the breadboard alternative.

8–13 List the shortcomings of software simulators in general when compared to the breadboard alternative.

8–14 To make the process of software simulation a more efficient part of the design process, what consideration is most important regarding the resulting schematic capture file?

8–15 You are the project manager on a development project working on a new line of laptop computers. The project has been broken down into the following modules:

Power supply

Processor and motherboard with RAM

I/O circuit board

CD-ROM drive

Floppy drive

LCD display

Keypad

Enclosure and all mechanical parts

Final assembly

The total manufacturing cost goal for the project is $500. Develop an initial cost budget for the project that will meet the target manufacturing cost. Use your best estimates of how to divide the amount of the project goal. Does the $500 goal seem reasonable? Would a $250 goal be reasonable?

9

Step Four: Execution (Component Selection)

▶ Introduction

In Chapter 8 we completed a discussion of the preliminary design process. At this point the design is on paper and some sections of the circuit have been computer simulated. The design includes a variety of active and passive components. Before we start to prototype the circuit, we must select and procure the components for use in constructing the breadboard and prototype. Whether you are completing a two- or four-year electronics program, you have seen many resistors, capacitors, and inductors used in circuits. You should possess a solid understanding of their function in a circuit. The discussion that follows is an attempt to summarize the important aspects of selecting passive components and to discuss the many different types available. More importantly, the advantages and disadvantages of each type are reviewed. A method for selecting the right type of component for a specific application is presented. The selection of active components is also discussed in a general way. Appendix A includes much reference information on both passive and active components. The specific topics of this chapter are as follows:

- ▶ Resistors
- ▶ Variable resistors
- ▶ Capacitors
- ▶ Inductors
- ▶ Transformers
- ▶ Switches and relays
- ▶ Connectors
- ▶ Selecting active components

9–1 ▶ Resistors

Once we have determined the nominal design value for a resistor in a circuit schematic drawing, we must convert that information into an actual physical resistor that we can connect in the circuit. To select a resistor, the value, the type (material and construction), tolerance, power rating, and temperature coefficient must be defined. The available types of fixed resistors are described in the following list:

Molded carbon (carbon composition) resistors—These are a common older type of resistor formed by molding carbon, insulating filler, and a resin binder. They are low cost and feature 5% and 10% value tolerances, a wide variety of power ratings, and a negative temperature coefficient of anywhere from –200 ppm to –500 ppm/°C. Example 9–1 describes a simple way to handle coefficients stated in parts per million (ppm).

Carbon film resistors—These have replaced molded carbon resistors as the most widely used fixed resistor. They are constructed by depositing a resistive carbon film material onto a ceramic rod. They offer smaller size and more stability than the molded carbon resistor and are still low cost. Carbon film resistors come in 1%, 5%, and 10% tolerances with negative temperature coefficients of –200 ppm to –500 ppm/°C and a variety of power ratings.

Metal film resistors—These resistors utilize the same construction method described for the carbon film resistors but metal film (nickel chromium) is deposited on the ceramic rod instead of the carbon film. Metal film resistors come in 1%, 5%, and 10% tolerances and a wide variety of power ratings and feature a positive temperature coefficient of less than +100 ppm/°C.

Metal oxide resistors—Tin oxide is deposited on a glass rod to form these resistors. The stability is very high. Metal oxide resistors come in 1% and 2% tolerances and power ratings of 1/5 W and 1/2 W and feature a positive temperature coefficient of less than +60 ppm/°C.

Wire-wound resistors—These are formed with resistive wire wrapped around an insulating rod. Wire-wound resistors typically have both higher power ratings, and very low resistance values are available. A wide range of values and power ratings are available at medium to high cost. Their temperature coefficient is positive and less than 100 ppm/°C.

Example 9–1

The temperature coefficients of many types of components are given in a format called *parts per million* (ppm). This ppm format requires some thought when first trying to apply it. A simple way to use any coefficient given in ppm form follows:

1. Divide the ppm temperature coefficient value by the number 1,000,000. This will be the decimal value of the temperature coefficient.

2. Multiply the decimal ppm value calculated in step 1 of this example by the nominal value of the component.

A practical example would be to determine the resistance variation of a 10,000 Ω carbon composition resistor with a temperature coefficient of -200 ppm/°C from 25°C to 55°C. Follow the steps listed:

1. Divide

$$\frac{-200 \text{ ppm/°C}}{1,000,000} = -0.0002 \text{ /°C}$$

2. Multiply

$$-0.0002/°C \times 10,000 \ \Omega = -2 \ \Omega/°C$$

3. Multiply

$$-2 \ \Omega/°C \times 30°C \text{ (increase in temperature from 25°C to 55°C)}$$
$$= -60 \ \Omega \text{ change}$$

Since the temperature coefficient is negative, the 10,000 Ω resistance decreases by 60 Ω to a value of 9,940 Ω.

Resistor Selection

The criteria for selecting the actual resistor for use in a circuit depends upon the nominal resistance value, the level of stability, the tolerance required, the temperature variation that is acceptable, the power level, and cost criteria. The following process is suggested to select all fixed resistors:

1. Determine which types of resistors have the value and power rating required.

2. Determine the minimum temperature coefficient that is acceptable for the application.

3. Chose the lowest cost and most available of the resistor types that will meet the requirements of the circuit.

Figure 9–1 shows the resistor comparison chart to help make comparisons.

Thick Film Networks

These networks are fabricated with the same technology as metal film and carbon film resistors but are incorporated in a package that allows for multiple resistors of the same value. The packages include dual-in-line and single-in-line as

Resistor Type	Range of Values	Tolerance	Range of Temperature Coefficient	Power Rating	Cost Factor x = .02	Primary Advantage
Carbon composition	All standard 5% values	+/- 5% to +/- 10%	−200 to −500 ppm	1/8, 1/4, 1/2, 1, 2 W	1x	Cost and size
Carbon film	All standard 5% values	+/- 5% to +/- 10%	−200 to −500 ppm	1/8, 1/4, 1/2 W	1x	Cost and size
Metal film	All standard 1% values	+/- 1%	+100 ppm	1/4 W	2x	Tolerance and temperature coefficient
Precision metal oxide	All standard 1% values	+/- 1% to +/- 2%	+60 ppm	1/4, 1/2 W	4x	Temperature coefficient
Power metal oxide	All standard 5% values	+/-5%	+100 ppm	1/2, 3 W	15x	High power and temperature coefficient
Precision wire-wound	All standard 1% values	+/- 1%	+100 ppm	1, 2, 3 W	100x	High power and tolerance
Power wire-wound	All standard 5% values	+/- 5% to +/- 10%	+200 ppm	1 W on up	75x	High power

▲ **FIGURE 9–1**
Resistor comparison chart

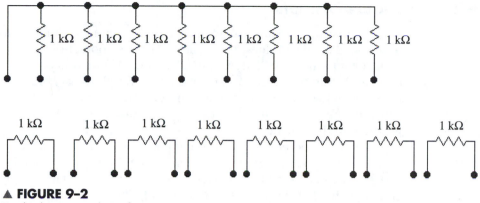

▲ **FIGURE 9–2**
Dual-in-line network configurations

well as SMT packages. For example, you can purchase eight 1000 Ω resistors in one dual-in-line package, and the resistors can all be connected on one side to the same pin or kept completely separate. Figure 9–2 shows a schematic diagram of these two configurations. These are handy in situations in which eight of the same value resistors are needed, for example, as pull-up resistors on an 8-bit data bus.

Power Resistors

You can typically classify power resistors as anything over 1 W. These are typically wire-wound or metal film resistors that have the same characteristics as other wire-wound or metal film resistors except that the packages are physically larger and designed to dissipate heat. Figure 9–3 shows examples of different SMT power resistor packages.

▲ **FIGURE 9–3**
Power resistor packages *(Courtesy sof Vishay)*

9–2 ▶ Variable Resistors

Variable resistors are also called *potentiometers* and are simply resistors that can be adjusted to many values. These can be full-sized potentiometers or smaller trimming potentiometers called *trimpots*. They come in a variety of different packages that can be circuit-board or panel mounted. Variable resistors are available that are adjustable with circular or linear motion. The circular adjustment types can be single or multi-turn. The multi-turn devices provide a more precise angular adjustment of the resistance value.

Inside the potentiometer is either a wire-wound or a continuous film type of resistive element. The wire-wound type is similar to wire-wound fixed resistors and is fabricated by winding fine resistance wire around an insulating bobbin. Each end of the bobbin is connected to a terminal. The resistance between the two end terminals is the maximum resistance of the variable resistor. A third terminal, called the *wiper arm*, is moved along the surface of the resistive windings to achieve the variable resistance value. Wire-wound potentiometers perform much like their fixed resistor counterparts. They are stable and have a good temperature coefficient but are plagued by resolution issues inherent in their construction. As the wiper arm moves across the resistance wire wrapped around the insulating bobbin, the resistance measured will increase in steps as the wiper goes from contact with one "turn" to the next "turn." The number of turns (the number of times the wire is wrapped around the bobbin) will determine the resolution (the smallest resistor value change) of the variable resistor.

Continuous-style potentiometers are fabricated with a thick film composition consisting of metal film on a ceramic substrate. These variable resistors use materials such as cermet, carbon composition, carbon film, and metal film for the thick film compositions. Because they are constructed of a continuous length of conductive material, these potentiometers greatly improve on the resolution problems experienced with wire-wound variable resistors. However, they are limited to smaller power ratings.

Example 9–2, Wiring Potentiometers

In the op amp circuit in Figure 9–4, determine the proper connection of the potentiometer, so the gain of the circuit increases with a clockwise adjustment of the variable resistor. The gain of the circuit increases as the value of the potentiometer increases.

It is important to understand the correct wiring of potentiometers so that they will function properly, which means providing the proper adjustment for the desired direction of adjustment. Potentiometers are three-terminal devices that have their total resistance between end terminals 1 and 3 and a variable resistance between terminal 2 (the wiper) and either of the other terminals. The direction of rotation (or the linear direction for linear adjustment types) that increases the resistance between terminals 2 and 1 and 2 and 3 is the critical point.

Notice that the circuit shown has the second terminal connected to one of the end terminals. This is commonly done when the desired effect is to function simply

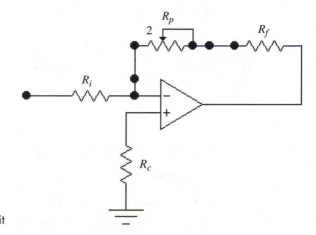

▶ **FIGURE 9–4**

Op amp adjustment circuit

as a variable resistor. In this case there are two common connections to the potentiometer. When a voltage divider function is required, all three connections to the potentiometer are separate.

To solve this problem, we must determine the end terminal (pin 1 or pin 3 of the potentiometer) to which the wiper arm (pin 2 of the potentiometer) should be connected. The solution will come from the answer to one question: When the potentiometer is turned clockwise, between which set of terminals does the resistance increase? This question can be answered by taking resistance measurements on a sample potentiometer or by remembering that for all variable resistors, the resistance between wiper terminal 2 and terminal 1 increases as the potentiometer is turned clockwise. To resolve our example problem, what is known about the problem can be listed:

1. We desire the gain to increase when the potentiometer is adjusted clockwise.

2. The gain increases when the potentiometer resistance increases.

To resolve the problem, determine the two terminals where the resistance increases when the potentiometer is adjusted clockwise. The answer is between terminals 1 and 2. To which end terminal should wiper terminal 2 be connected? The answer is terminal 3. Connecting terminals 2 and 3 together shorts terminal 2 to the terminal 3 end. Since the resistance increases between terminal 2 and terminal 1, when the potentiometer is turned clockwise, the desired performance will result.

9–3 ▶ Capacitors

Capacitors are probably the second most utilized passive electronic component, and there are many variations from which to choose. When you are selecting capacitors, your goal will be to find a capacitor with the following:

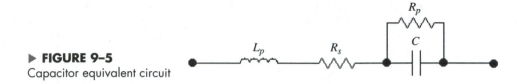

▶ **FIGURE 9–5**
Capacitor equivalent circuit

1. The proper capacitance value

2. An acceptable tolerance rating

3. The proper working voltage

4. A temperature coefficient that is acceptable for the application

5. In some cases insulation resistance, quality factor, and dielectric absorption will also be important.

6. Smallest size and cost available

The physical model for a capacitor is shown in Figure 9–5. Inductance, L_P, and series resistance, R_S, are shown in series with a parallel capacitance, C, and resistance, R_P. The series resistance is a result of the resistance of the leads, plates, and any contact points. The resistance in parallel with the capacitance represents the leakage resistance that occurs through the insulation material around the plates. The capacitance shown is the true capacitance of the capacitor.

Insulation Resistance

The insulation resistance limits the capacitor's ability to completely block DC current as, theoretically, it should. That is why it is also called *DC leakage current*. The insulation resistance represents the ability of a capacitor to hold a charge for a period of time. It is an important parameter for capacitors used in integrator, sample hold, and peak detector circuits, which must hold a charge for a long period of time.

Equivalent Series Resistance, Dissipation Factor, and Quality Factor

The equivalent circuit shown in Figure 9–5 can be converted to an *equivalent series resistance* (ESR) and a capacitance as shown in Figure 9–6. The ESR is calculated by calculating the equivalent series impedance for the parallel RC network at a given frequency and adding the resistance component to the series resistance, R_S, shown previously. L_P is usually negligible. The dissipation factor (DF) is a measure of ESR/X_C. As such it is a measure of the AC loss of the capacitor. It is a unitless number that is expressed as a percent. The lower the DF number, the less loss is dissipated by the capacitor at a certain frequency. The quality factor or Q is simply $1/DF$.

▶ **FIGURE 9–6**
ESR circuit

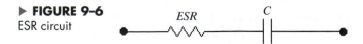

It is the inverse of the dissipation factor. The DF and Q factors are important when precision operation at a certain frequency is required and when sampling a signal. Resonant frequency, precision filters, and sample and hold circuits are a couple examples of circuits that require high Q and low DF.

Dielectric Absorption (DA)

Dielectric absorption is the phenomenon that allows a capacitor, which is quickly discharged and open circuited, to recover some of the charge that was discharged. In this case the charge is actually absorbed by the dielectric. In applications such as sample and hold circuits, this recovered charge will add to the signal the next time it is sampled, causing an error. Dielectric absorption (DA) is an important consideration in sampling, timing, and high-speed switching circuits. The characteristic is expressed as a ratio and given as a percent. The higher the percent of DA rating for a capacitor, the greater the amount of DA effect.

Capacitor Types

To meet the circuit's requirements for capacitors, they will be selected from many types of construction that utilize many different dielectric materials. The key is to find the desired performance characteristics in the smallest package at the least cost.

Ceramic Capacitors

These are widely used because they have a wide range of available values, and they are relatively small and inexpensive. The high dielectric constants available with the ceramic materials used in these capacitors result in large capacitance values for their size. The most common types of construction available for ceramic capacitors are the disc, multilayer, and chip variations. Figure 9–7 shows a typical disc type ceramic capacitor. The performance characteristics have the following ranges:

▶ **FIGURE 9–7**
Ceramic capacitors

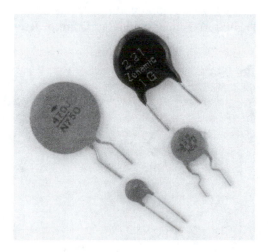

Capacitance range: 1 pF to 1 μF

Working voltage: 25 V to 30 kV

Tolerance: ±5% to as high as +50% and −20%

Temperature coefficient: ±15% over temperature range

Relative size: Small

Relative cost: Inexpensive

Typical use: As medium- to high-frequency bypass, coupling, and filter capacitors

Mica Capacitors

Mica capacitors (see Figure 9–8) offer superior performance when compared to the best quality ceramic capacitors when used at frequencies above 200 MHz. They are available in small capacitance values with tight tolerances, high working voltages, and stable temperature characteristics. Their performance characteristics have the following ranges:

Capacitance range: 2.2 pF to .01 μF

Working voltage: 50 V to 5 kV

Tolerance: ±0.5% to as high as ±20%

Temperature coefficient: ±200 ppm/°C

Dissipation factor: 0.02% to .1%

Relative size: Medium to large

Relative cost: Medium

Plastic Film Capacitors

This is a category of many different plastic film dielectric materials that have similar performance characteristics. Plastic film capacitors (see Figure 9–9) are characterized by high working voltages, more stable temperature characteristics, low dielectric absorption, and a higher Q factor. These premiums result in their larger

▶ **FIGURE 9–8**
Mica capacitors

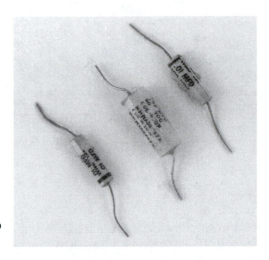

▶ **FIGURE 9–9**
Film capacitors

size and higher cost. The most common types of dielectric used to make plastic film capacitors are polyester, polypropylene, polystyrene, polyethylene, polycarbonate, teflon, and mylar. The following performance characteristics are the broad range of specifications for all film capacitors:

Capacitance range: 20 pF to 500 μF

Working voltage: 30 V to 10 kV

Tolerance: ±1% to as high as ±20%

Temperature coefficient: ±2.5% to ±10% over temperature range

Relative size: Medium to large

Relative cost: Medium to high

Typical use: In low- to medium-frequency ranges where better than average capacitance tolerances, stability, temperature coefficients, and Q factors are required

Next we will look at the individual characteristics that are unique to the different types of dielectric materials used in plastic film capacitors.

Polyester   These are considered as medium performance plastic film capacitors. They are designed for mounting on printed circuit boards and are available in values from 0.001 μF to 2.2 μF. Polyester capacitors feature low inherent inductance, tolerances of ±5%, ±10%, and ±20%, and a temperature coefficient ±10% change over the rated temperature range. Their size tends to be larger than other plastic film capacitors and they are available at medium cost. They are typically used as coupling capacitors.

Polypropylene   These capacitors (see Figure 9–10) feature very low dielectric absorption (0.001% to 0.02%) in a wide range of values (10 pF to 0.1 μF) and they are inexpensive. Their disadvantages are large case size, an inability to withstand

▶ **FIGURE 9–10**
Polypropylene capacitors

high temperatures, and high inductance. For lower temperature applications, these would be a good choice for sample and hold circuits.

Polystyrene Polystyrene capacitors offer the highest performance of the plastic film capacitors. They are available in ranges from 10 pF to 0.1 μF at tolerances from ±1% to ±10%. Their stability is excellent with a temperature coefficient of around ±1%. The dissipation factor is low, which means the Q factor is high. They also feature a very low dielectric absorption factor. Polystyrene capacitors tend to be smaller than other plastic film capacitors and more expensive. They are typically used in filter networks, tuned circuits, and other precision charging circuits in the low- to medium-frequency range.

Polyethylene These are specialized capacitors designed to suppress transients and noise from the input power connections to many industrial and commercial products. They are available in values from 0.001 μF to 1 μF and temperature coefficients of ±10%. The primary applications are as filter capacitors placed across the primary AC voltage lines.

Polycarbonate These offer performance almost as good as polystyrene capacitors at a smaller size. If a high-quality capacitor is required in the smallest size, a polycarbonate capacitor may be a good choice. They are available in ranges from 10 pF to 0.1 μF and temperature variations on the order of –2.5%, +1% over the useable temperature range. Their cost is in the medium to high range. Applications include filter networks, tuned circuits, and other precision charging circuits in the low- to medium-frequency range, where small size is an overriding requirement.

Mylar Mylar capacitors are the general purpose capacitors of the plastic film types. They are available in ranges from 0.001 μF to 0.22 μF and working voltages up to 100 V DC. Their tolerances and temperature coefficients are on the order of ±10%. They are available in small to medium sizes and low to medium cost.

Electrolytic Capacitors

Electrolytic capacitors are designed to achieve high capacitance values in as small a size as possible. The resulting design tradeoffs give electrolytics a lower range of working voltage, and they are polarized. *Polarization* means that one plate is made positive and the other negative. This polarization is due to the fact that the plates are made of different materials, unlike standard capacitor types. Reversing the polarity of the voltage applied to electrolytic capacitors will destroy them if the voltage is large enough. There are two dielectrics used to make electrolytic capacitors: aluminum and tantalum. Aluminum electrolytics are more common, physically larger, and have a wider range of values. The general range of capacitance values for electrolytic capacitors values goes from 0.1 μF to 5700 μF. Working voltages go from 10 V DC to 500 V DC, depending on the value and type. Tolerances are typically –10% to +50% and temperature coefficients are not usually listed. Their common use is as power supply rectifier filters, so the actual value of capacitance seldom requires high accuracy.

Aluminum Electrolytics   These feature a cylindrical package with either axial or radial leads (see Figure 9–11). The radial lead is usually preferred for circuit board mounting. There are variations in marking the polarized leads. Some manufacturers will mark the plus lead accordingly. Many others mark the negative lead with a large band with a negative sign embedded in it. There are various grades of aluminum electrolytics available that primarily determine the expected operational life. Their selection will involve using the lowest working voltage and smallest package size combined with the desired package style, grade, and capacitance value.

Tantalum Electrolytics   These electrolytics are typically smaller and restricted to a narrower range of capacitance and working voltage values. Capacitance values range from 0.1 μF to 100 μF and voltage ratings from 10 V DC to 35 V DC. There are two different packages: cylindrical with axial leads and teardrop shape as shown in Figure 9–12. The teardrop shape is typically used in circuit board applications.

Capacitor Selection

With all the different types of capacitors to choose from and all the parameters to consider, capacitor selection may seem overwhelming. Yet, with a little experience, the more typical selections become routine. Most of the time the value and

▶ **FIGURE 9–11**
Electrolytic capacitors

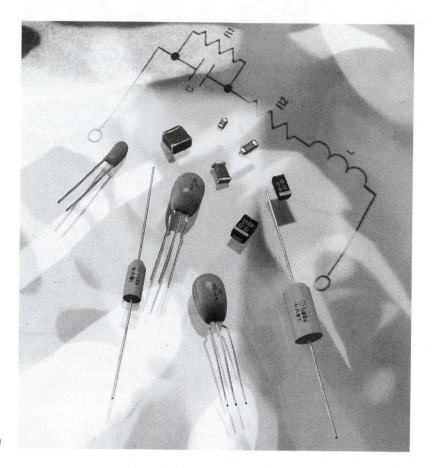

▶ **FIGURE 9–12**
Tantalum capacitors
(Courtesy of Vishay)

function required will result in the use of ceramic disc or aluminum electrolytic capacitors. The nontrivial applications will require the most effort.

To help in selecting capacitors, use the following steps in conjunction with Figure 9–13:

1. Determine which capacitor types have the value required.

2. Determine the capacitor types that have a working voltage sufficient for the application.

3. Determine those capacitor types that will function over the frequency range of the application.

4. From the application, determine the capacitor types that have an acceptable capacitance tolerance combined with the temperature coefficient.

5. Review any special aspects of the application, such as operation or function at specific frequencies (high Q and low DF), data conversion at low frequencies (insulation resistance), or accurate sampling, timing circuits, and

high-speed switching circuits (low DA). Review the insulation resistance, dissipation and quality factor, and dielectric absorption as required by the application and select the capacitor type that will meet these needs.

6. Review the sizes available of all the remaining capacitor types that meet all the criteria discussed so far.

7. Determine the cost and availability of all the capacitors reviewed in step 6 above. Make the decision based upon the combination of size, cost, and the other key factors defined in steps 1 through 5 above.

Back-to-Back Electrolytics

When large capacitance values and small size are required for voltages that vary both positive and negative, the polarized plates of electrolytic capacitors are a significant limitation. To get around this, it is possible to place two polarized capacitors in series, back-to-back, with the negative plates connected together. Doing this achieves the effect of one nonpolarized capacitor equal to the series equivalent of the two polarized capacitors. Remember that two capacitors in series act like two resistors in parallel when determining the equivalent capacitance.

There are some limitations of this practice that relate to tantalum capacitors. These limitations regard the equivalent capacitance value of the back-to-back capacitors and the fact that as the signal level across the capacitor increases, the equivalent capacitance will vary from the value calculated. Also, many manufacturers recommend against using "wet anode" tantalums capacitors in the back-to-back configuration.

Feed-Through Capacitors

These are special capacitors for suppressing unwanted signal noise in the form of radio frequency interference (RFI). The purpose of feed-through capacitors is to shunt unwanted noise to ground at the entry or exit of a grounded metal enclosure or cavity. One plate of the feed-through capacitor makes contact with the feed-through conductor and the other capacitor plate is connected to both sides of the enclosure or cavity wall. These are intended for high-frequency applications and the range of capacitance values usually used is between 0.01 μF and 2 μF.

Example 9–3

On many circuit schematics, you will notice the placement of two bypass filter capacitors in parallel with values such as 100 μF and 0.1 μF. The 100 μF capacitor is usually an aluminum electrolytic type while the 0.1 μF value is usually a ceramic disc capacitor. The purpose of both of these capacitors is essentially the same: to bypass any frequency higher than DC, removing them from the power supply output. Why then are both capacitors required? Why can't the 100 μF capacitor filter perform this function alone? In analyzing this situation we will use the equation $X_C = 1/(2\pi f C)$ for capacitive reactance. The larger the value of the

Capacitor Dielectric	Range of Values	Tolerance	Temperature Coefficient	Range of Working Voltage
Ceramic	1 pF to 1 µF	+/-5% to +/- 20%	+/- 15%	25 to 30 kV
Mica	2.2 pF to 0.01 µF	+/-0.5% to +/- 20%	200 ppm/°C	50 to 50 kV
Plastic film types	20 pF to 2.2 µF	+/- 1% to +/- 20%	+/- 2.5% to +/- 10%	30 to 10 kV
Polyester	0.001 µF to 2.2 µF	+/- 5% to +/- 20%	+/- 10%	30 to 10 kV
Polycarbonate	0.001 µF to 2.2 µF	+/- 1% to +/- 10%	+/- 10%	30 to 10 kV
Polystyrene	10 pF to 0.1 µF	+/- 1% to +/- 10%	+/- 10%	30 to 10 kV
Polypropylene	10 pF to 0.1 µF	+/- 1% to +/- 10%	+/- 10%	30 to 10 kV
Polyethylene	0.001 µF to 1 µF	+/- 1% to +/- 10%	+/- 10%	30 to 10 kV
Mylar	0.001 µF to 2.2 µF	+/- 1% to +/- 10%	+/- 10%	30 to 10 kV
Aluminum electrolytic	0.1 µf to 10,000 µF	−10% to +50%	NA	10 to 500 V DC
Tantalum electrolytic	0.1 µF to 100 µF	−10% to +50%	NA	10 to 35 V DC

▲ **FIGURE 9–13**
Capacitor selection table

capacitance, C, the smaller X_C is at a given frequency, f, which is what is desired: a low impedance value that will bypass the frequency component to ground. The 100 µF capacitor presents a much lower (100 times lower) capacitive reactance at any frequency when compared to the 0.1 µF capacitor. This reinforces the question: Why is the 0.1 µF capacitor necessary? The answer comes from the equivalent model of the capacitor discussed earlier in this section. The equivalent series resistance of the aluminum electrolytic capacitor and the

Frequency Range	Dielectric Absorption	Quality Factor	Size	Cost	Primary Advantage
High frequency	High	Low	Small	Low	Size and cost
High frequency	High	High	Medium	Low	Tolerance and temperature coefficient
Low to medium	Low	High	Medium to large	Medium to high	High quality factor and low dielectric absorption
Low to medium	Low	High	Medium to large	Medium to high	High quality factor, tolerance
Low to medium	Low	High	Medium to large	Medium to high	High quality factor, tolerance
Low to medium	0.001 to 0.02 %	High	Medium to large	Medium to high	Low dielectric absorption
Low to medium	0.001 to 0.02 %	High	Large	Medium to high	Low dielectric absorption
Low frequency	Low	High	Medium to large	Medium to high	High quality factor, tolerance
Low to medium	Low	High	Medium	Medium	General purpose plastic
Low frequency	High	NA	Small capacitance to size ratio	Low	Large capacitance for small size, polarized
Low frequency	High	NA	Small	Low	Large capacitance for small size, polarized

▲ **FIGURE 9–13**

Continued

ceramic capacitor also change with frequency. This limits their effectiveness as capacitors as frequency varies. The aluminum electrolytic is not effective above the range of 10,000 Hz. The ceramic capacitor is effective over a frequency range of roughly 1000 to 1 MH. So neither capacitor by itself can perform the complete needs of the circuit: bypassing all frequencies to ground. Both capacitors together function as a tag team, one taking over where the other one leaves off as the frequency increases.

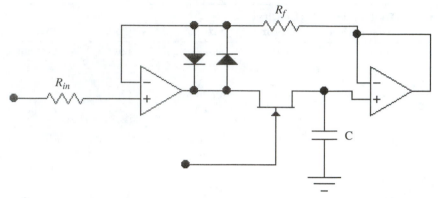

▲ **FIGURE 9–14**
Sample-and-hold circuit

Example 9–4

In this example we will select a capacitor for use in a sample-and-hold circuit (see Figure 9–14). The circuit requires a value of 0.01 μF and a 20 V rating. The temperature range for the circuit is 0°C to 85°C. A sample-and-hold circuit is often used with an A/D converter to sample the signal at one point in time and save it for the A/D converter to process. In sample-and-hold circuits a capacitor is connected to the input signal for a brief period of time so that the capacitor will charge up to the signal value. At the specified sample time, the input signal is disconnected and the A/D converts the sampled signal.

In this case the dielectric absorption of the selected capacitor, as discussed earlier in this section, is critical. The correct capacitor will have a minimal DA factor. The capacitor types with the lowest DA factor are polystyrene, polypropylene, and teflon. Polystyrene will not be used because of its temperature limitations. Either polypropylene or teflon can be used for this application. Teflon has a little higher DA factor and will be higher in price. Polypropylene will have a lower DA factor and a larger size and will be cheaper.

Variable Capacitors

As with resistors, variable and trimming capacitors are available to provide for adjustable capacitance values. There are two basic types: air-variable and trimmer capacitors (see Figure 9–15).

Air-variable capacitors are used to tune resonance circuits like those utilized in the front end of a radio receiver. Air-variable capacitors use an interleaved set of metal plates, with air as the dielectric. Capacitance variations of 1 pF to 200 pF are available. Trimmer capacitors are fabricated with mica, ceramic, and glass dielectrics and are used for fine-tuning a capacitance value. The key features of the variable capacitor types are:

▶ **FIGURE 9–15**
Variable capacitor
(Frank Labua/
Pearson Education/
PH College)

Mica—Good stability and low temperature coefficient; good capacitance to size ratio and can handle moderate shock and vibration; low inductance and cost

Ceramic—High Q factor with predictable temperature coefficient; good capacitance to size ratio with low inductance; not usable in high shock and vibration environments; limited to 180° rotation

Glass—Possesses a high voltage capability and can be environmentally sealed; can handle many adjustments with smooth and nearly linear capacitance changes with rotation

9–4 ▶ Inductors

Inductors are the least-used passive electronic components because of their relative size and cost. Usually, the desired effect of the inductance can be achieved with a cheaper and smaller capacitive circuit. This is shown by the fact that we can construct both high- and low-pass filters from either resistor-capacitor circuits or resistor-inductive circuits. There are times when inductors are simply best suited and are the only practical way to provide the desired effect. Inductors are used in filter and tuned circuits, as current limiters, and as ripple filters in power supplies.

The three principal types of inductors have air, iron, or ferrite cores. The core is the material that the conductor coil is wrapped around. The conductive coil is covered with an insulating layer, usually a varnish, to prevent conduction between the adjacent coils. Magnetic core inductors offer a higher range of inductance values. There are also fixed and variable inductors. Figure 9–16 shows the schematic symbols that correspond to the core materials and fixed and variable symbols.

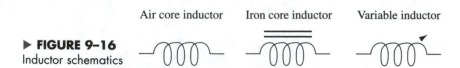

Air core inductor   Iron core inductor   Variable inductor

▶ **FIGURE 9–16**
Inductor schematics

There are four specific applications of inductors, and each has different performance requirements.

Ripple Reduction

This application involves smoothing the ripple on the output of DC power supplies (see Figure 9–17). The critical parameters in this application are the minimum inductance value, DC current, working voltage, and maximum DC resistance.

Swinging Inductor

Swinging inductors (see Figure 9–18) are applied to the AC input of many power supplies to reduce the ripple input to the supply. The critical parameters in this application are both the minimum and maximum inductance values, DC current, working voltage, and maximum DC resistance.

▶ **FIGURE 9–17**
Ripple reduction

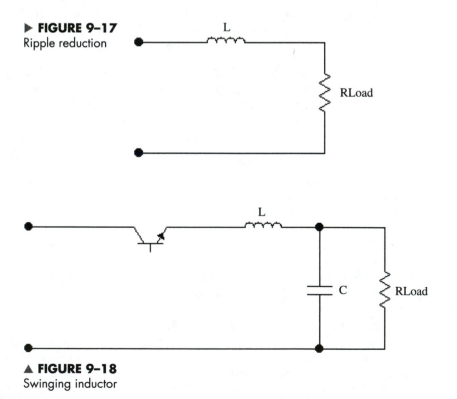

▲ **FIGURE 9–18**
Swinging inductor

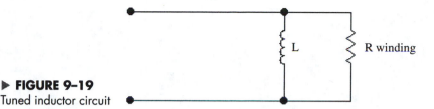

▶ **FIGURE 9–19**
Tuned inductor circuit

Current Limiting

As current limiters, the value of inductance and the tolerance becomes important along with the maximum AC current and maximum AC voltage drop for which the inductor is rated.

Tuned or Timing Circuits

These applications (see Figure 9–19) require tighter tolerances of inductance values and consideration of the inductor Q value, current, and DC resistance. Inductor Q value is a ratio of the resistance of the inductor to the inductive reactance (R/X_L) at a specific frequency.

Air Core Inductors

Air core inductors are available from a few tenths of a microhenry to several hundred microhenries. These are usually used at high frequencies in the range of 100 kHz to 1 GHz. Very small air core inductors can be fabricated by a few loops of wire or loops on a printed circuit board.

Iron Core Inductors

Iron core inductors are available from 0.1 to 120 mh with DC resistances ranging from a few to several hundred ohms. Maximum DC currents are on the order of 100 ma with Q values of about 50.

Variable Inductors

Adjustable magnetic core inductors are available that vary inductance by moving the magnetic core in or out of the coil with screw adjustments. Inductors come in a variety of sizes and shapes. Figure 9–20 shows some of the inductor packages available.

9–5 ▶ Transformers

A transformer consists of two separate coils that are wrapped around a closed magnetic circuit. They are used to step down, step up, isolate, and impedance match AC circuits. The ratio of the number of turns on the secondary to the primary

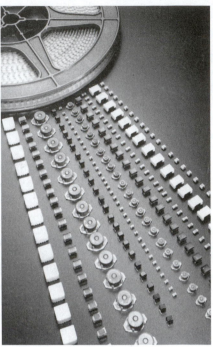

▲ **FIGURE 9–20**
Coilcraft inductor family *(Courtesy of Coilcraft)*

turns is called the *turns ratio, n*. The voltage output from the secondary, V_{sec}, is equal to the primary voltage, V_{Pri}, divided by the turns ratio. Conversely, the secondary current, I_{sec}, equals the turns ratio multiplied by the primary current, I_{pri}.

$$V_{Sec} = V_{pri} \times n$$
$$I_{sec} = I_{pri} / n$$

There are a wide variety of transformer types based on primary and secondary voltages and currents. There are also many variations of primary and secondary windings. In either winding there may be a center tap that is a connection point in the middle of the winding. Center taps are used to create full-wave rectifiers with two diodes or combination plus-minus power supplies. In other cases a winding may actually be two equal windings that can be connected in parallel or series.

Example 9–5

In this example a power supply is designed for use as an industrial product that will operate off of 115 V AC or 230 V AC, depending on which is available to the user. To accomplish this, the product will have to be altered to change the turns ratio of the transformer used. This can be achieved by using a transformer that has two separate and equal windings on the primary side (see Figure 9–21). If 230 V AC operation is desired, then the two primary windings are wired in series. For 115 V AC

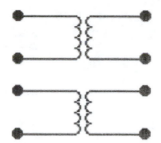

▶ **FIGURE 9–21**
Transformer example schematic

operation, the windings are wired in parallel, which reduces the turns ratio and increases the current capability of the primary. The desired secondary voltage is 12 V. If the input voltage is 230 V AC, then the required turns ratio is 230/12 = 19.16. For 115 V AC, the turns ratio should be 115/12 = 9.58. Since a half a turn cannot be fabricated, a primary with two separate windings that are equal and 10 times the number of secondary turns will be the best solution. If the secondary has 10 turns, then each of the primary windings should have 100 turns. The turns ratio will vary from 20, when the two windings are wired in series, to 10 with the parallel connection.

The resulting option to the user will be a choice of how to wire the two primaries, in series or parallel. An even easier option is to include a switch in the power supply design that changes the series and parallel wiring of the primaries with the flip of that switch.

The primary applications for transformers are discussed next, followed by the criteria for selecting transformers.

Step-Down Transformers

This is the most common application of transformers and involves the step down of an AC power supply voltage to a smaller voltage AC voltage for use as a low voltage AC or DC power supply. In this case the turns ratio is less than one, meaning that the primary has more turns than the secondary.

Step-Up Transformers

These are utilized whenever it is necessary to increase the voltage level from the AC voltage level available. The best example of this is the development of 230 V AC when only 115 V AC is available. There are many other applications where the AC voltage must be increased slightly. Increasing 208 V AC to 230 V AC is another example. Step-up transformers have fewer turns in the primary than the secondary.

Isolation Transformers

These are used solely to isolate one AC voltage from another AC voltage. Transformers inherently accomplish this because there is no electrical connection between the primary and the secondary. The energy is transferred magnetically. The output voltage and current should be the same as the input voltage and current. The

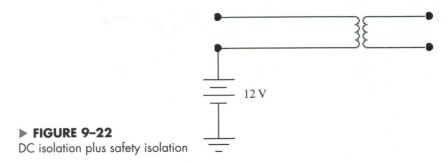

▶ **FIGURE 9–22**
DC isolation plus safety isolation

reason for their use is either to eliminate DC noise levels from a signal or to isolate the secondary circuit for safety reasons (see Figure 9–22). In selecting these transformers, the primary issue is finding a transformer that will handle the required power level.

Impedance Matching

Impedance matching involves the matching of a source impedance to the load impedance. This is desirable because maximum power is transferred only when the load impedance equals the source impedance. A transformer can achieve this because of the way the windings reflect the impedance value through the windings. The most common example is the typical 75 Ω source resistance of an antenna lead that is connected to the 300 Ω input resistance of a television receiver. An impedance matching transformer can be used to match these two resistances or, in other words, make the 300 Ω television load appear to the antenna as a 75 Ω load.

The selection of impedance matching transformers is most dependent on the ideal turns ratio. The turns ratio can be determined by the formula

$$n = \sqrt{R_{Load} / R_{Primary}}$$

where $R_{primary}$ equals the value of the source resistance and R_{load} is the actual load resistance. The proper power rating must also be determined by multiplying the maximum secondary output voltage times the output current. This will be the volt-amp requirement for the transformer.

9–6 ▶ Switches and Relays

Switches are mechanical devices that switch electrical circuits. The two basic types of switches are *maintained* and *momentary*. Maintained switches maintain the switch position after they are engaged until they are mechanically repositioned. Momentary switches are spring loaded and revert back to their normal state after being released. Switch contacts are defined as common (C), normally open (NO), or normally closed (NC). The normal state of the contacts are the mechanical "off" state of the switch. Switches are classified by the terms *poles* and *throws*. Poles are merely the numbers of sets of contacts included in a switch. One set of contacts is needed for control of one circuit. If two or more circuits must be controlled, then additional sets of isolated contacts will be needed. *Throws* is derived from the number of positions for old-fashioned "knife" switches. The term has been carried over to today's switches. A switch that makes or breaks only two points in a circuit is said to have a single-throw contact. A switch that makes or breaks one common contact to either a normally closed or a normally open contact is called a double-throw contact. To achieve more than two throws, some sort of rotary switch is required that will connect one common contact with potentially many other contacts. Switch selection is determined by mechanical size and style, contact arrangement, and the voltage and current rating of the contacts. Figure 9–23 shows a variety of switches.

Relays

Relays are electromechanical or solid-state devices that can best be defined as voltage-controlled switches. When a voltage is applied to the input, the output contacts change states. Electromechanical relays have been used in control circuits for many years. They continue to be utilized, in spite of the electronic alternative, because of

▲ FIGURE 9–23
Switch assortment *(Courtesy of C&K Components, Inc., Watertown, MA)*

their ease of use, versatility, contact ratings, and relatively low cost. They consist of a coil that, when energized, pulls in an armature, which changes the common contact—or switch position. These relays feature both AC and DC coils and contact arrangements up to four poles in single- or double-throw variations. The poles and throws specification for electromechanical relays is the same as described for switches. Devices called *contactors* are simply large electromechanical relays typically used in motor starter circuits. There are many different packages available for panel mounting, plug-in modules, or direct solder in circuit board variations. Selection of these relays involves the mechanical size and configuration, the coil voltage, the contacts arrangement (number of poles and number of throws), and the contact voltage and current ratings.

Solid-state relays are the electronic equivalent to electromechanical relays. They consist of an input circuit to which an input voltage is applied to change the state of the output contacts. The output contacts are an electronic switch that is either a transistor for switching DC circuits or a triac for AC applications. The input and output circuits are isolated by using opto-isolators. Opto-isolators use an LED and a photodiode to transmit a light signal that turns on the output when the input voltage is applied. Solid-state relays offer a significant improvement in switching life over electromechanical relays. Input voltages are limited to 5 V DC to 15 V DC and only one contact arrangement is available: single pole/single throw (SPST). Care must also be taken to derate the current rating of solid-state relays when they are used at higher temperatures. To select solid-state relays, the input voltage, output voltage being switched (AC or DC), and the voltage and current ratings of the contact are the key parameters.

Example 9–6

Determine the number of poles and throws for the switches and relays shown in Figure 9–24.*

9–7 ▶ Connectors

Connectors are an important aspect of any electronic design. Connectors involve any device that serves to make an electrical connection from one circuit element to another. There are generally three types of connector situations that occur in electronic design:

---

*Answers: A. DPST B. SPDT C. SPST D. 3PDT

▶ **FIGURE 9–24**
Poles/throws examples

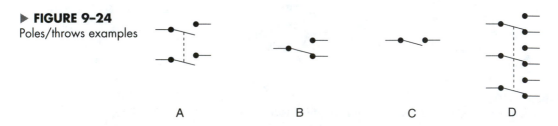

| A | B | C | D |

Wire-to-wire

Wire-to-PCB (printed circuit board)

PCB-to-PCB

The key aspects that determine the applicability of a type of connector to a specific application are as follows:

1. Application type (wire-to-wire, wire-to-PCB, PCB-to-PCB)
   Wire-to-wire could be flat cable, coaxial cable, or conventional wire.

2. Mounting (none, panel, printed circuit board)

3. Voltage across contacts

4. Current through contacts

5. Size

6. Number of contacts and spacing

7. Termination type (solder or crimp)

8. Environmental aspects (temperature, humidity, seal)

9. Contact resistance, which depends on the contact material
   Typical contact materials include beryllium copper, phosphor bronze, spring brass, and low-leaded brass. In higher-quality connectors, selective gold or silver plating is used on the specific contact areas. Low-cost connectors utilize an electrotin plate.

10. Keyed or not keyed
    Keying prevents incorrect connection.

11. Insertion force and pin alignment tolerance

12. Reliability (number of disconnects over life, design life)

13. Cost and ease of assembly

The most common types of connectors will be discussed next, along with their typical applications and advantages.

Printed Circuit Board Edge Connectors

Printed circuit board edge connectors are very cost effective because they form a one-piece connector that either connects a printed circuit board to some wires or to another printed circuit board. They rely on the printed circuit board to substitute for the male portion of the connector. Nevertheless, when utilizing printed circuit board edge connectors, care must be taken to ensure that the design and quality of the printed circuit board will provide a secure connection. They are

▶ **FIGURE 9–25**
Printed circuit board edge
connector examples
(Courtesy of Vishay)

available in both crimp and solder terminations. Typical spacing of printed circuit board edge conductors are 0.156″ centers or 0.1″. Figure 9–25 shows examples of edge connectors.

Flat Cable Connectors

Flat cable connectors are designed to work with standard flat cable and form a termination that pierces and crimps the flat cable wire. This allows a very quick and low-cost connector assembly, which is its primary advantage. Flat cable connectors come in many configurations that include male pin and female receptacles as well as printed circuit board edge connectors. The flat cable connectors are connected to the flat cable by sandwiching the flat cable between the two sections of the connector and carefully pressing them together to pierce the insulation and form a crimp contact between the connector and the wire. The pressure must be applied in a consistent manner and is usually performed with a vice. Flat cable and the corresponding connectors come with many contact arrangements, which can be up to 64 contacts wide. Figure 9–26 shows examples of flat cable connectors.

▶ **FIGURE 9–26**
Flat cable connectors *(Reprinted
with permission of Tyco
Electronics Corp.)*

► **FIGURE 9–27**
D-type connectors
*(Reprinted with permission
of Tyco Electronics Corp.)*

D-Type Connectors

Connecting data input and output lines between two pieces of equipment is usually accomplished with some sort of D-type connectors. D-type connectors are terminated in a variety of solder lugs, including direct mounting to printed circuit boards. Standard D-type connectors include contact numbers of 9, 15, 25, and 37. Figure 9–27 shows examples of D-type connectors.

Coaxial Connectors

Coaxial connectors make connection on one axis and are usually restricted to connecting two wires: a signal and ground. These connectors (see Figure 9–28) are typically used for radio frequency (RF) and audio applications. The most popular RF coaxial connector is the standard BNC connector. However, there are many other popular varieties, such as SMA and SMB. Audio-type connectors are usually called phone plugs, phonojacks, and mini- and micro-plugs. The phone plugs are usually 1/4″ plugs and receptacles. Phono jacks are 3.18 mm in size while the mini-plug is 3.58 mm and the microplug 2.46 mm.

► **FIGURE 9–28**
RF connectors *(Reprinted with permission of
Tyco Electronics Corp.)*

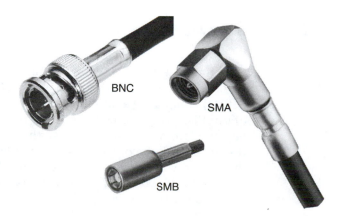

▶ **FIGURE 9-29**
Circular connectors *(Reprinted with permission of Tyco Electronics Corp.)*

Circular Connectors

These are usually multi-pin connectors of higher reliability. Many military-type connectors are of the circular type. But military connectors will be discussed separately in the next section. A good example of a typical circular connector application is the keyboard connector on most personal computers. Figure 9–29 shows examples of circular connectors.

Military Connectors

Military connectors possess extremely high reliability and can withstand extreme environments such as moisture, ambient temperature, vibration, shock, and EMI. There are many different configurations and types and these are used whenever the requirements of the application justify their high cost.

Zero Insertion Force Connectors

Zero insertion force (ZIF) connectors are used whenever there are a high number of contacts—and the number of disconnects is also very high (see Figure 9–30). The primary purpose is to provide superior contact and preclude any damage to the connection point on either side caused by insertion. These connectors mate without any insertion force. The contact force is applied separately after insertion. The most typical application is EPROM programmers.

9-8 ▶ Electronic Displays

Most electronic products and projects include electronic displays of some type, even if only a single indicator is utilized to show that power is on. Light emitting diodes (LEDs), liquid crystal displays (LCDs), and vacuum fluorescent displays (VFDs) comprise the most commonly used technologies today and are the focus of this discussion. Cathode ray tubes (CRTs), an old technology still very much in demand, are not covered because they are typically applied as a complete functioning module (a computer monitor, television, etc.) and a detailed discussion of their inner workings is beyond the scope of this book. In this section we will review the theory of operation, specifications, and the various configurations of LEDs, LCDs, and VFDs.

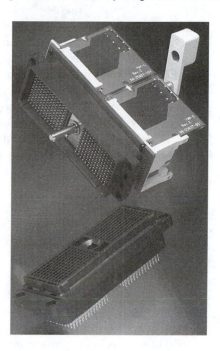

▶ **FIGURE 9–30**
ZIF connectors *(Reprinted with permission of Tyco Electronics Corp.)*

All electronic displays can be broken down into segments that are bars, dots or small pixels. Information is displayed by energizing the particular segments to show the desired image.

All three types of electronic displays can be driven in static or multiplexed mode. In static mode there are permanent connections made between the segment driver and the individual segments of a display character that are energized continually, lighting up the segment. In multiplexed mode there is a common segment driver that is connected to many different character displays. For example, the bottom segment of four seven-segment LEDs would be connected together. The data that determines which segment of which characters are to be displayed is sent to all characters at the same time, but only the selected character (the one with data being sent at that time) is enabled. In the multiplexed mode each character is strobed with the appropriate data, one at a time. After all of the characters have been strobed, the process is repeated again and again. In order for multiplexed operation to perform well the strobe rate must be faster than the eye's capability of detecting any display flicker. The primary benefits of multiplexed displays are that the drive circuitry and the connections between drivers and the display are both minimized.

Electronic displays can be further categorized as character/digit displays or graphic displays. Character/digit displays use a variety of shapes called *segments* to display a specific set of characters. Character/digit displays are identified by the number of segments that comprise one character, the overall character height, and the quantity of characters. For example, a .56″, four-digit, seven-segment display with decimal point includes four seven-segment digits that are .56″ high, each with a decimal point in one display module. Sometimes the most significant digit in a seven-segment display can only display either a "1" or be off. In this case the most significant digit is called a *half digit*. This is because the addition of this digit essentially doubles the range of the succeeding less-significant digits. For example, a three-digit seven-segment display can display a number up to 999. Adding a half digit (a fourth digit capable of displaying only a "1") increases the range of the display to 1999. This display would be called a 3-1/2-digit seven-segment display.

Graphic displays include a number of dots or pixels, organized in rows and columns that can create a variety of images by selecting which dots or pixels are energized. The number and size of the pixels determine the quality of the displayed image. Graphic displays with large dots or pixels (called *dot-matrix displays*) are designed to display a very large character set and are not intended to display detailed graphic information. A large number of small pixels can display detailed pictures. These are the types of displays commonly used for computer monitors.

Light Emitting Diodes (LEDs)

In a standard silicon power diode the .7 V drop that occurs across the semiconductor p-n junction is converted into thermal energy in the form of heat. The form of energy given off by the diode is determined by the energy level difference between electrons in the p material and those in the n material. This energy level difference is called the *energy bandwidth*. By modifying the p-n materials that comprise a diode, an energy bandwidth can be developed that produces an electromagnetic wave. If the wavelength of the resulting electromagnetic waveform is in the visible spectrum, our eyes see this energy as light. LEDs have been developed that radiate red, yellow, amber, green, blue, and white light in the visible spectrum as well as some (ultraviolet is the most common) that operate outside this spectrum.

The elements most commonly used to manufacture LEDs are indium, gallium, arsenic, aluminum, nitride, and phosphorous. The gallium nitride or indium gallium nitride combinations represent the latest developments in LED technology that have allowed the development of bright blue LEDs. A combination of gallium aluminum arsenic is used most often for red LEDs while gallium arsenic phosphate results in amber and yellow light LEDs. Green LEDs are fabricated by combining gallium and phosphorous.

LEDs can be packaged individually to function as indicators or can be combined in a variety of combinations to show bar graphs or display numbers (7-segment format) and alphanumeric information (14-segment or dot matrix configurations). Figure 9–31 shows the various types of multiple LED display configurations. LEDs work well in interior or other relatively low background light conditions. When used outdoors under bright conditions, the output light is readily washed out.

LED Specifications

When specifying an LED for use in a design, it is important to consider both the electrical and the light specifications. The electrical specifications are similar to those for a rectifier diode and include the operating voltage, peak inverse voltage, and the

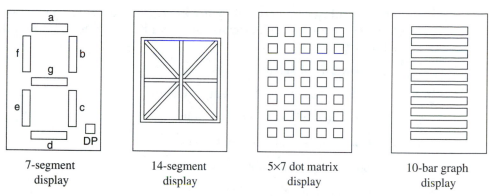

| 7-segment display | 14-segment display | 5×7 dot matrix display | 10-bar graph display |

▲ FIGURE 9–31
LED display configurations

operating continuous forward current. The voltage drop across a forward biased LED is approximately 1.4 V, much greater than the .7 V experienced with a silicon diode.

The reverse breakdown voltage for LEDs is much less than the value for standard diodes. Therefore, LEDs can be damaged by applying a relatively low reverse voltage across them. Twenty milliamps is the typical forward current value used to qualify the LED's intensity.

The light output of an LED is specified by its color, spectral width, and intensity. The color of the LED light is indicated by the center wavelength of the light produced. A red LED, for example, will output light with a wavelength of approximately 650 nanometers. The light output from an LED is not pure 650 nm light but includes a range of wavelengths around the center wavelength of 650 nm. Spectral width is a measure of how wide a band of wavelengths are generated by a particular LED. LEDs in general have spectral widths on the order of 30 nm. This means that an LED that outputs 650 nm red light with a spectral width of 30 nm actually outputs a range of wavelengths that goes from about 635 nm to 665 nm. The intensity of the LED is specified in terms of brightness using the units of millicandellas of foot lamberts at a specific forward current value.

The epoxy used to package the p-n junction that forms the LED can either be neutral (transparent and clear) or a color such that transmits the light produced by the LED. The epoxy can also be formed into a lens to diffuse (disperse) the light. A dispersed light pattern produces overall less brightness and less clarity with the benefit of a wide viewing angle. An epoxy that does not diffuse the output light results in higher brightness with a crisp image and a narrow viewing angle. The combination of lenses and colors results in four categories of LEDs in regard to their appearance and output light pattern: colored transparent, colored and diffused, non-colored diffused, and non-colored and clear. The shape of the finished individual LED package can be either round or rectangular, and both through-hole and SMT packages are available. There are a variety of mounting angles available as well.

LEDs can be driven by powering them on continually, or they can be strobed at a frequency high enough to appear constant to our eyes. LEDs will light up approximately 60 ns after they are energized and have operating lives greater than 40,000 hours. Summarizing, the key LED specifications and their units are as follows:

Forward voltage drop, volts

Continuous forward current, milliamps

Reverse breakdown voltage, volts

Color, wavelength in nanometers

Spectral width, nanometers

Output light pattern, viewing angle

Average brightness, millicandellas

Speed

Life

LED Indicators

Individual LEDs are often used as status indicators on a variety of products and systems designed for use in interior or otherwise low background light conditions. Individual LEDs are very simple to apply, requiring just a low-voltage DC power supply, a current limiting resistor, and some device for switching the power connection on and off. They are available in round or rectangular shapes and a variety of sizes.

LED Bar Graphs

LED bar graphs are used to simulate analog indicators or provide a cluster of individual status indicators. They consist of a number of individual rectangular LEDs packaged together in one IC. Seven-, eight-, and ten-segment bar graph LEDs are the most common packages available. Package connection variations include completely separate connections for each LED and common anode or common cathode versions as well. Additional circuitry is needed to determine which LEDs are to be lit and to provide the proper drive current (usually 20 mA). The layout of a ten-bar LED bar graph is shown in Figure 9–31.

Seven-Segment LEDs

When seven-segment displays were initially developed they became the popular replacement for "nixie tube" displays in the 1970s. The availability of these displays combined with microprocessors resulted in the first hand held calculators and digital wrist watches. Today, LEDs are seldom used in these applications because of their relatively high current requirements per LED segment.

Seven-segment LED displays utilize seven long, rectangular LEDs to develop a pattern that can create any one-digit decimal number from 0 through 9. Each digit is created by energizing some combination of the seven segments. Each segment is labeled a, b, c, d, e, f, and g as shown in Figure 9–31. Most seven-segment displays actually possess an eighth segment in the form of a decimal point labeled simply, "DP." Seven-segment displays are available in various heights (.3″, .56″, 1″, and so on) and either individually, in pairs, or other digit quantities. Connections are provided for each segment and the LEDs are wired internally to either a common anode or common cathode connection. To drive common anode LEDs, $+V_{CC}$ is connected to the common anode. Pulling the cathodes low through a current limiting resistor energizes individual segments. Common cathode displays are driven by connecting ground to the common cathode connection. Each segment is energized by applying $+V_{CC}$ to the anode through a current limiting resistor.

Standard logic devices have been developed to drive seven-segment LEDs that accept various types of codes and convert these to drive the seven segments correctly. Most common is the binary-coded decimal to seven-segment decoder driver, where four binary-coded decimal inputs are converted directly to drive a seven-segment display to indicate the encoded decimal number.

Seven-segment LEDs were designed to indicate digits only, and this fact posed a significant limitation to the application of LEDs in microprocessor-based products in which it was desirable to display messages. Designers will sometimes use a crude set of characters, a mixture of uppercase and lowercase letters (i.e., A, b, C, d, E, F, etc.) to display messages on seven-segment LEDs in these situations. The next section discusses 14-segment LEDs, which offer a much better alternative.

14-Segment LEDs

Another option, 14-segment displays, provide the product designer with a complete set of well-defined numbers and letters for displaying alphanumeric information. These LED displays have the same operating parameters as seven-segment displays, with seven additional segments to show the various types of characters. The layout for the 14-segment display is shown in Figure 9–31. ASCII character codes are usually used to drive 14-segment displays through the appropriate decoder driver circuits. 14-segments can display many additional characters, such as +, >, <, /, and so on, but they cannot present the complete set of ASCII characters.

Dot Matrix LEDs

Dot matrix-format LED displays show the complete ASCII character set. There are two dot matrix formats, the 3×5 and 5×7 arrays. The 5×7 array is the most popular and is capable of displaying all ASCII characters. They possess 35 individual LEDs and develop characters just as shown on dot matrix printers. The layout of a dot matrix LED is shown in Figure 9–31.

Liquid Crystal Displays (LCDs)

Liquid crystal displays are probably the most popular devices used to display information on electronic products today. This is because they are generally smaller and lighter and use less power than other popular alternative display devices, such as LEDs, VFDs and CRTs.

Liquid crystal displays are unique because they do not generate light as do all other electronic displays. They function by either reflecting or transmitting light that is incident to the liquid crystal assembly.

Liquid crystals are materials that exist in a state between solids and liquids, at room temperature. You may recall from high school chemistry that when matter is a solid all of the molecules point in the same direction. On the other hand, the molecules that make up liquids are free to move about the liquid and can point in any direction. Because liquid crystals are between solids and liquids, they can be manipulated to have the structure of either. When liquid crystals act like a solid, they do not let light pass through, but when acting like a liquid, they will let light pass. The liquid crystals used for electronic displays exist in what is called the *nematic* phase, one of many possible phases for liquid crystals.

One of the ways to manipulate the characteristics (acting as a solid or a liquid) of liquid crystal materials is to pass electric current through them. One of the most popular types of liquid crystals used for display purposes are called *twisted nematics*. Twisted nematics liquid crystals exist naturally in a twisted configuration. When twisted together, the liquid crystals will act as a liquid, allowing light to pass. When electric current is passed through twisted nematic liquid crystals, they unravel and act like a solid, blocking the transmission of light.

LCDs actually consist of six layers. Starting from the bottom, these layers include a mirror, polarizing film, liquid crystal material with electrodes layers on both sides, and, on top, another layer of polarizing film. The electrode connected to the bottom of the liquid crystal layer is called the *back plane*. The back plane is connected to all of the segments included within the LCD. The top electrode is connected separately to each segment. When the LCD segment is not energized, light passes through the top, directly through the twisted liquid crystal materials, and is reflected back by the mirror on the bottom layer. When the LCD is energized, the liquid crystals untwist and block the light so that this segment of the display appears darker than the surrounding liquid crystal materials. LCDs simply reflect or absorb reflected light to show a particular image. These types of LCDs are called simply *reflective*. As with the LED, segments can be shaped in any form, but most often segments are in the shape of bars, dots, or pixels.

LCDs can also be used in low light conditions by backlighting the display. In this case a light source, usually an LED, is placed behind the LCD. This type of LCD is called *backlit* or *transreflective* because it either reflects light incident to the surface of the display or it transfers light through the display from the backlight. Light passes through the segments in the display, but it is not energized; it appears simply as reflected light entering the top of the display. Liquid crystals can be arranged in 7-segment formats to display numbers, 14-segment formats to show alphanumeric characters, and dot matrix/pixel formats with resolutions that challenge CRT displays.

The speed at which so-called *passive* LCDs change state is relatively slow compared to other display technologies. Passive LCDs are especially slow at turning off. Temperature significantly affects the operation of liquid crystals. Cool temperatures impede the liquid crystal materials from changing states when the electrical current is either applied or removed. Therefore, passive LCDs have a low temperature limit that is much higher than other types of electronic displays.

▶ **FIGURE 9–32**

LCD segment driver

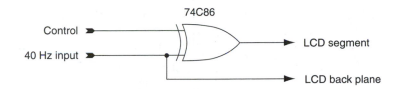

Applying an AC voltage between the back plane and the segment's top electrode energizes an LCD segment. Connecting out-of-phase 40 Hz square waves to the back plane and the segment is the same as applying an AC voltage. This is accomplished by connecting a 40 Hz square wave through an Exclusive-Or gate, as shown in Figure 9–32.

The Exclusive-Or gate produces a high output when the inputs are not the same. When the control input to the circuit shown in Figure 9–32 is low, the output of the Exclusive-Or gate is the same as the input 40 Hz square wave; the LCD segment is not energized. When the control signal is brought high, the output of the Exclusive-Or gate is inverted, which means it is 180 degrees out of phase with the input 40 Hz square wave. This creates a voltage of +5 or –5 V (effectively a 5 V AC signal) between the segment and the back plane, energizing the segment.

CMOS drivers are typically used instead of TTL devices to drive LCDs because of the lower power requirements and because the low output for CMOS is closer to 0 V. The low output from TTL devices can be as high as .4 V. This adds a DC component to the AC waveform that reduces the life of the LCD. A seven-segment LCD can be driven as shown in Figure 9–33 by connecting the outputs of a CMOS BCD-to-seven-segment decoder/driver (74HC4511) through a series of Exclusive-Or gates that are also connected to a 40 Hz square wave. Decoder/drivers are available that are designed to drive LCDs directly. The 74HC4543 is an example. It includes the BCD-to-seven-segment decoder/driver, the Exclusive-Or gates, and a 40 Hz square wave generator, all on one IC.

Seven-segment LCDs are laid out in the same format as seven-segment LEDs. They are available in a variety of character heights and quantities, with or without backlighting. These are commonly used in digital clocks, watches, calculators, and digital multimeters.

14-segment LCDs are used in products in which alphanumeric characters must be displayed. These are usually battery-powered devices that offer the user a selection of a variety of operational modes or features that are best explained with prompted messages.

Dot-matrix LCDs are used to show a much larger set of characters and possibly limited graphics or symbols. The quality of the images is determined by the number and the size of the dots. This is specified as the *dot pitch* (width × height). A typical dot matrix LCD might include 128 columns × 64 rows of dots with a dot pitch of .52 × .52. New technology has resulted in the development of active matrix LCDs, often called *TFT LCDs*. This type of LCD includes a thin film transistor (TFT) that is assembled on a piece of glass that is part of the LCD. The TFT provides an active switch on-board to switch the LCD pixels on and off, significantly improving the speed and temperature performance of LCDs.

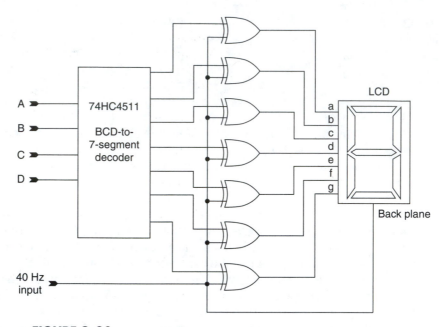

▲ FIGURE 9–33
LCD seven-segment decoder driver

Color LCDs are accomplished by utilizing three LCD subpixels that possess red, green, and blue filters. Colors are created for each pixel by varying the level of each of these primary-color filters. Sophisticated and specific VLSI circuits are needed on board to drive these types of displays because of the complex information and the number of interconnections that are needed.

The general specifications to be considered when applying LCDs are as follows:

Power supply voltage

Drive current

Segment "on" speed

Segment "off" speed

Viewing angle

Reflective or backlit (transreflective)

Ambient temperature range

Vacuum Fluorescent Displays (VFDs)

Vacuum fluorescent displays offer the highest brightness levels of all electronic displays. VFDs function in a manner very similar to the vacuum tubes of yesteryear. Each display segment includes a common control grid sandwiched between a common cathode filament and an individual segment anode. Electrons are emitted from the cathode filament toward the control grid. If the control grid is positively charged, the electrons are attracted and their motion is further amplified toward the anode. When the control grid is negatively charged, the electrons are repelled. When the anode is connected to a positive voltage, the electron is attracted to that particular segment of the display. In order for the segment to light, both the control grid and the anode must be connected to a positive voltage. Light is emitted from the segment when the electrons strike the phosphor coating deposited on the anode.

In order to power a VFD, first a filament voltage must be applied to the cathode filament. In most applications, the cathode filament is continually powered whenever power is supplied to the operating device. In order to prevent a variation in brightness across all of the characters included in the display, it is best to use a center-tapped transformer to power the filament (see Figure 9–34). The center-tap is then used as a common reference for the anode driver. The level of the filament voltage will have a large impact on the display brightness and longevity, so it is best to supply a filament voltage that is within the manufacturer's recommendations.

With the filament voltage in place, supplying a positive DC voltage to both the control grid and the anode will light up the VFD segment. The anode drive voltage must be well filtered because any ripple that is coincident with the grid scan frequency can cause the display to flicker.

Most VFDs are available with much of the circuitry included to develop the various power supply voltages and drive signals necessary. Some are microprocessor-based and completely control all display operations after receiving the characters to be displayed through parallel or serial communications. VFDs with fewer characters/segments are usually controlled externally but provide the actual drive circuitry and power supply circuits on board. VFDs come in 7- and 14-segment and dot matrix formats similar to LEDs and LCDs. An example of a common variety of 14-segment VFD is a 1-line-by-16-character, 14-segment display. The characters are 9 mm high and include 64 character fonts. Power supply requirements are simply 12 V DC at 150 mA. Data to be displayed is transmitted serially to the display.

▶ **FIGURE 9–34**
VFD cathode filament
voltage connection

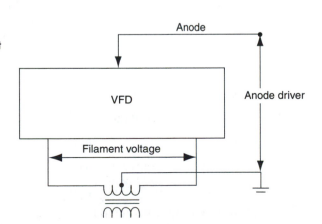

A higher-performance VFD display is one that includes 2 lines of 20 characters in a dot matrix format to display an extended ASCII character set. Each character is 9 mm high. In this case the VFD is usually microprocessor-controlled and designed for serial communications up to baud rates of 19,200. Parallel I/O communications are also available. Power requirements are 12 V DC at 260 mA.

The most advanced VFDs are called *dot-graphic displays*. An example dot-graphic display might include 112×16 dots that are .33 mm wide and .575 mm high. Each dot is controlled individually by the on-board microprocessor so that any graphical image or character can be displayed. The power supply requirements are 5 V DC at 260 mA, and both serial and parallel communications are supported.

Summarizing, the key specifications for applying VFDs are as follows:

Input power supply voltage

Segment drive current

Color, wavelength in nanometers

Spectral width, nanometers

Output light pattern, viewing angle

Average brightness, millicandellas

Speed

9–9 ▶ Selecting Active Components

Selecting active electronic components is a broad area that is well covered in most electronic courses and texts. This section is a general review of active components, discussing their critical parameters and summarizing their selection.

The specifications for most active components start out with a section called "Absolute Maximum Ratings," where the maximum ratings for all parameters are listed. It is imperative to review every aspect of these ratings with the intended circuit application to ensure that these ratings are not exceeded. Exceeding these ratings usually means the component will malfunction and may sustain permanent damage. There should be at least a 10% safety factor between the maximum value of a parameter in a circuit and the absolute maximum value listed in the specifications. The key parameters of concern are usually:

Maximum power supply voltage

Maximum input voltage

Maximum differential input voltage

Maximum output current (source)

Maximum output current (sink)

Maximum operating temperature

Maximum voltage on any pin

Beyond the maximum parameter ratings, the determination of which active component fulfills the functional requirements of a circuit comes down to the speed, accuracy, and power consumption of the component.

9–10 ▶ Developing a Parts List

At this point a parts list (also called a *bill of material* or BOM) should be developed. A parts list should include the following:

1. The company part number and the general description of the assembly for which the part lists was developed.

2. For each component included in the assembly, the following are included:
   a. The suggested manufacturer of the part and manufacturer's part number
   b. The company's designated part number
   c. A description of the part with all key parameters
   d. The component ID number that is referenced on the assembly drawing
   e. The quantity of the part that will be used in the subject assembly
   f. For development purposes and cost calculations, the cost of the part per unit in different quantities should be included, and the total cost for that component per assembly should be calculated as well. The quantities chosen will depend on the planned annual volume for the product.

It is useful to use a spreadsheet program to complete the parts list because component cost information can be included on the spreadsheet, tabulated, and compared to the project cost goals. Figure 9–35 shows an example of a parts list.

▶ Summary

In this chapter we have discussed the process of selecting components for use in the preliminary designs for most kinds of projects. By completing this task, we have taken an important step toward the physical realization of the circuit. In the next chapter we discuss the procurement of parts and breadboarding. From this point on, the amount of time spent in the development lab will increase until the end of the project, when the final project reports and documentation are finalized. This is a positive aspect of the design engineering position and the project development cycle. The activities and challenges change as you progress through the project, making every day of the design process unique. Remember when the paper was blank at the beginning of the project? It now includes a complete schematic of a design that has been simulated. The design also includes a parts list, where all of the parts are fully specified. The project is ready to move into the development laboratory. The next chapter discusses the "hands on" skills and the tools that will be required to complete the design breadboard.

Item #	Description	Company Part Number	Manufacturer	Manufacturer Part Number	Component ID	Quantity
1	2.49 k ohm, 1% Metal Film, 1/4w Resistor	10000000	Res. Inc	2.49kX	R1,R13	2
2	110 ohm, 1% Metal Film, 1/4 w Resistor	10000001	Res. Inc	110X	R2	1
3	100 k ohm, 1% Metal Film, 1/4w Resistor	10000002	Res. Inc	100kX	R3--R6,R15	5
4	26.1 k ohm, 1% Metal Film, 1/4w Resistor	10000003	Res. Inc	26.1kX	R7	1
5	10 k ohm,10 turn Trimpot	10000011	TP Inc.	102tp1	RR	1
6	.1 uF, 50 V, Ceramic Capacitor	10000012	Cap. Inc.	P1101	C1--C4	4
7	.01 uF, 50 V, Ceramic Capacitor	10000013	Cap. Inc.	P1102	C5	1
8	5 V.1 A Negative Voltage Regulator	10000022	IC Inc.	LM340LAZ-5.0	U3	1
9	5 V 1 A Voltage Regulator	10000023	IC Inc.	LM340T-5.0	U4	1
10	Quad Operational Amplifier	10000024	IC Inc.	LM324P	U5	1
11	Voltage Reference- 2.5 Volts	10000025	IC Inc.	LM4040CZ-2.5	U6	1
12	115/20VAC,.3A Center Tapped Transformer	10000026	IC Inc.	MT2111	T1	1
13	Seven Segment .56" Red LED Display	10000027	IC Inc.	67-1463	D1--D4	4
14	Printed Circuit Board	10000028	PCB Inc.	NA	NA	1
15	Terminal Block -5 Position	10000029	Conn. Inc.	W5500	J1	1

▲ **FIGURE 9–35**
Parts list example

224

Digital Thermometer Example Project

Selecting Components for the Digital Thermometer

In this section we use the information presented in this chapter to select the passive and active components that are used to complete the design of the example digital thermometer project.

Resistor Selection

There are two different areas in the digital thermometer design where resistors are used, and their selection must be considered: the input and display sections. The input section is more critical, because any variation in resistance will cause a corresponding change in the input to the A/D converter and the resulting displayed temperature. The design of the input section is shown in Figure 9–36. In our discussion of the design in Chapter 8, remember that the two trimpots provided in the final stage of the input section will compensate for the variation in nominal resistance values. The two trimpots will allow the calibration of the circuit. The primary concern, when selecting the resistors, involves their temperature coefficients. However, there are areas of the circuit where the nominal resistance values are of concern. The initial differential amplifier stage is a prime example. The purpose of the differential amplifier is simply to subtract the common mode voltage and amplify the difference signal. The degree to which the differential amplifier will accomplish this depends on the matching of the four resistor values. So the relative matching of these resistors is important.

The resistors selected for use in the input section are 1%, metal film, 1/4 W resistors. These provide a temperature coefficient of ±100 ppm and tolerances of 1%. At approximately four cents each, metal film resistors are a good choice for this part of the design. In the display circuit, there is no real concern for tolerance or the temperature coefficient of the resistors, so 5%, carbon film, 1/4 W resistors are selected for this portion of the circuit.

There are two variable resistors or trimpots, that must be selected also. In this case the turns resolution is a concern as well as the temperature coefficient. If the turns resolution is not fine enough, then the trimpot may not be able to adjust the voltage to the exact desired value: 0.00 V for 0°F and 2.00 V for 200°F. A 15-turn cermet trimpot, with a ±10% overall resistor tolerance and ±100 ppm temperature coefficient, is selected for trimpots R_Z and R_G of the input section.

Capacitor Selection

The capacitors required in the digital thermometer are all power supply filter types of applications. In power supply filter capacitor applications, the actual capacitor value is not very critical. Neither is the temperature coefficient or overall stability of the capacitor. Aluminum electrolytic capacitors are used for the power supply capacitors above 1 µF, and 0.1 µF ceramic disc capacitors are used in almost all other cases. Many times both electrolytic and ceramic disc types will be used for the reasons explained in Example 9–3.

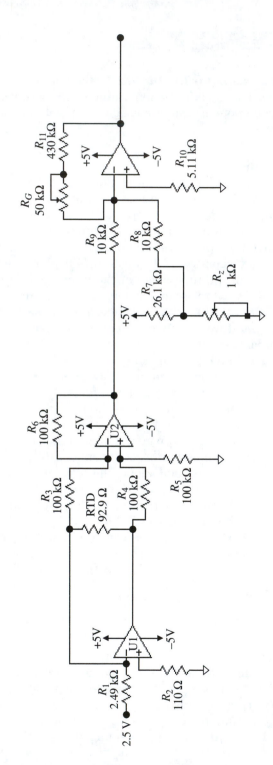

▲ **FIGURE 9–36**
Input section design

Transformer Selection

The preliminary design for the power supply was completed in section 8–5 of Chapter 8. The transformer calculations call for the input voltage to be 115 V AC rms and that the secondary be rated at 19.78 V AC rms at 275 ma with a center tap. A standard split-bobbin construction transformer is available with 115 V AC rms input, 20 V AC rms output at 300 ma with a center tap. This is very close to the actual requirements.

Active Devices Selection

The active devices to be selected for the digital thermometer include the 2.5 V voltage reference, four op amps, the +5 V regulator, the –5 V regulator, the A/D converter, and the seven-segment LED displays.

2.5 V Reference: The voltage reference will determine the amount of current that the current source puts through the RTD. The calibration adjustments can compensate for values other than the specified 1 ma. But this value, however, should not be exceeded by much or self-heating of the RTD could be a problem. The primary concern of the voltage reference is the stability of the reference as the temperature changes. A 2.5 V reference is available from many manufacturers. The one selected for this project has an overall tolerance of $\pm1\%$ and a temperature coefficient of ±10 ppm.

Op Amps: The amplifiers shown in the schematic diagram are all 741 op amps. A 324 op amp will actually be used. The 324 family of op amps is simply four 741s in one package. The A and E specifications for the 741 are utilized in selecting the actual 324 component. The 324 is selected with a plastic DIP through-hole package.

Voltage Regulators: The digital power supply requires 5 V at 250 ma. The analog supplies are ±5 V at 25 ma. The next largest linear 5 V regulator available is a 500 ma. The 500 ma size will be used for the digital power supply and the 100 ma version for the analog supply. A 5 V negative supply is available at 100 ma also. All the regulators chosen are available in the TO-92 package.

A/D Converter: The type of A/D converter has already been selected as the 7117 A/D converter with a seven-segment decoder driver output. The package selected is the plastic through-hole 40-pin DIP.

Seven-Segment LED Displays: High-efficiency, common cathode, red seven-segment LEDs, 0.5″ high are needed to display the digitized temperature. These are also available in a standard DIP. The parts list for the digital thermometer is shown in Figure 9–37.

Item #	Description	Company Part Number	Manufacturer	Manufacturer Part Number
1	2.49 k ohm, 1% Metal Film, 1/4w Resistor	10000000	Res. Inc	2.49kX
2	110 ohm, 1% Metal Film, 1/4 w Resistor	10000001	Res. Inc	110X
3	100 k ohm, 1% Metal Film, 1/4w Resistor	10000002	Res. Inc	100kX
4	26.1 k ohm, 1% Metal Film, 1/4w Resistor	10000003	Res. Inc	26.1kX
5	10.0 k ohm, 1% Metal Film, 1/4w Resistor	10000004	Res. Inc	10.0kX
6	430 k ohm, 1% Metal Film, 1/4w Resistor	10000005	Res. Inc	430kX
7	5.11 k ohm, 1% Metal Film, 1/4w Resistor	10000006	Res. Inc	5.11kX
8	1 M ohm, 1% Metal Film, 1/4w Resistor	10000007	Res. Inc	1MX
9	470 k ohm, 1% Metal Film, 1/4w Resistor	10000008	Res. Inc	470kX
10	1k ohm, 10 turn Trimpot	10000009	TP Inc.	101tp1
11	50 k ohm, 10 turn Trimpot	10000010	TP Inc.	502tp1
12	10 k ohm,10 turn Trimpot	10000011	TP Inc.	102tp1
13	.1 uF, 50 V, Ceramic Capacitor	10000012	Cap. Inc.	P1101
14	.01 uF, 50 V, Ceramic Capacitor	10000013	Cap. Inc.	P1102
15	100 pF, 50 V, Ceramic Capacitor	10000014	Cap. Inc.	P1106
16	.1 uF, 50 V, Mylar Capacitor	10000015	Cap. Inc.	M1101
17	.22 uF, 50 V, Polypropylene Capacitor	10000016	Cap. Inc.	P3224
18	.047 uF, 50 V, Polypropylene Capacitor	10000017	Cap. Inc.	P3473
19	3300 uF, 16 V, Electrolytic Capacitor	10000018	Cap. Inc.	P5144
20	330 uF, 16 V, Electrolytic Capacitor	10000019	Cap. Inc.	P5140
21	3 1/2 Digit LED A/D Converter	10000020	IC Inc.	TC7117CPL
22	5 V .1 A Voltage Regulator	10000021	IC Inc.	LM320LZ-5.0
23	5 V.1 A Negative Voltage Regulator	10000022	IC Inc.	LM340LAZ-5.0
24	5 V 1 A Voltage Regulator	10000023	IC Inc.	LM340T-5.0
25	Quad Operational Amplifier	10000024	IC Inc.	LM324P
26	Voltage Reference- 2.5 Volts	10000025	IC Inc.	LM4040CZ-2.5
27	115/20VAC,.3A Center Tapped Transformer	10000026	IC Inc.	MT2111
28	Seven Segment .56" Red LED Display	10000027	IC Inc.	67-1463
29	Printed Circuit Board	10000028	PCB Inc.	NA
30	Terminal Block -5 Position	10000029	Conn. Inc.	W5500

▲ **FIGURE 9–37**
Digital thermometer parts list

▶ **References**

Harper, C. A., ed. 1977. *Handbook of Components for Electronics.* New York: McGraw-Hill.

Warring, R. H. 1983. *Electronic Components Handbook for Circuit Designers.* Blue Summit, PA: Tab Books.

▶ **Exercises**

9–1 A 1% tolerance, 1 kΩ, metal film resistor is rated at 25°C. It is used in an application in which the ambient temperature will go from 0°C to 65°C. Assuming a temperature coefficient of 100 ppm/°C, what is the worst-case resistance range expected for this resistor value after combining the effects of the temperature coefficient and the resistor tolerance?

Component ID	Quantity	Price per 1	Price per 100	Price per 1000	Total Cost per 1	Total Cost per 100	Total Cost per 1000
R1,R13	2	0.5	0.04	0.018	1	0.08	0.036
R2	1	0.5	0.04	0.018	0.5	0.04	0.018
R3--R6,R15	5	0.5	0.04	0.018	2.5	0.2	0.09
R7	1	0.5	0.04	0.018	0.5	0.04	0.018
R8,R9	2	0.5	0.04	0.018	1	0.08	0.036
R11	1	0.5	0.04	0.018	0.5	0.04	0.018
R10	1	0.5	0.04	0.018	0.5	0.04	0.018
R12	1	0.5	0.04	0.018	0.5	0.04	0.018
R14	1	0.5	0.04	0.018	0.5	0.04	0.018
RZ	1	1.75	1.14	0.95	1.75	1.14	0.95
RG	1	1.75	1.14	0.95	1.75	1.14	0.95
RR	1	1.75	1.14	0.95	1.75	1.14	0.95
C1--C4	4	0.32	0.21	0.12	1.28	0.84	0.48
C5	1	0.32	0.21	0.12	0.32	0.21	0.12
C6	1	0.32	0.21	0.12	0.32	0.21	0.12
C7	1	0.48	0.32	0.2	0.48	0.32	0.2
C8	1	0.54	0.38	0.23	0.54	0.38	0.23
C9	1	0.54	0.38	0.23	0.54	0.38	0.23
C10	1	1	0.67	0.42	1	0.67	0.42
C11, C12	2	0.17	0.13	0.085	0.34	0.26	0.17
U1	1	4.05	3.5	3.1	4.05	3.5	3.1
U2	1	0.42	0.34	0.28	0.42	0.34	0.28
U3	1	1.33	1.1	0.9	1.33	1.1	0.9
U4	1	1.05	0.85	0.6	1.05	0.85	0.6
U5	1	0.8	0.6	0.54	0.8	0.6	0.54
U6	1	2.17	1.8	1.67	2.17	1.8	1.67
T1	1	8.37	6.25	5.25	8.37	6.25	5.25
D1--D4	4	2.31	1.85	1.65	9.24	7.4	6.6
NA	1	7.25	6.25	5.5	7.25	6.25	5.5
J1	1	1.85	1.65	1.5	7.25	6.25	5.5
					Total Cost/1	Total Cost/100	Total Cost/1000
					59.5	41.63	35.03

▲ **FIGURE 9–37**
Continued

9–2 For each circuit shown in Figure 9–38, calculate the possible ranges of the voltage V_{OUT}. $V_{IN} = 10$ V.

9–3 In the circuits shown in Figure 9–39, calculate the smallest resolution values for the voltage V_{OUT}. $V_{IN} = 10$ V.

9–4 When selecting the actual resistors to be used in a circuit, what are the two key performance parameters that should be used to determine whether a carbon film or a metal film resistor is selected? Use the resistor selection chart shown in Figure 9–1 on page 186.

9–5 Which key design factors promote the use of a wire-wound resistor instead of a metal film resistor in a particular circuit?

9–6 If a design requirement can be accomplished by using higher tolerance resistor values or adjustable trimmer potentiometers, which is likely to be more cost-effective?

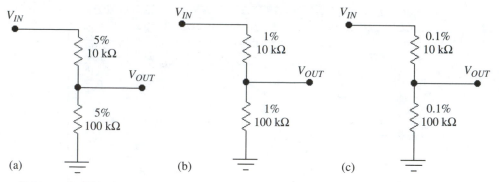

▲ **FIGURE 9–38**

One percent and 5% voltage divider circuits (see Exercise 9–2)

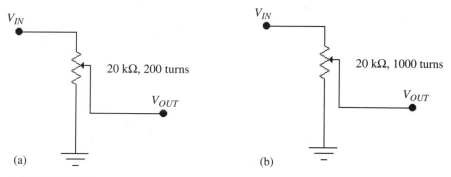

▲ **FIGURE 9–39**

Potentiometer voltage dividers with number of turns (see Exercise 9–3)

9–7 Calculate the equivalent series resistance (ESR) for a 0.1 μF capacitor that has an insulation resistance (IR) of 100,000 MΩ at a frequency of 60 Hz. Assume that series resistance and inductance equal zero.

9–8 Define in your own words the meaning of the term *dielectric absorption*. When is it an important consideration?

9–9 Define the terms *dissipation factor* and *quality factor*. How are they related?

9–10 What type of capacitors would you select to perform power-supply decoupling of various integrated circuits on a printed circuit board?

9–11 A capacitor is to be used in a precision low-pass filter circuit for an audio signal. What are the key parameters of concern when selecting the capacitor to be used in this circuit? Based on using the best available capacitor to meet these requirements, which type of capacitor would you select?

9–12 A capacitor is to be used in a 60 Hz, sample-and-hold circuit that is connected up to an A/D converter. What is the most important parameter to be considered in the selection of the type of this capacitor?

9–13 A bipolar capacitor with a value of 10 µF is needed for a circuit. List the different types of capacitors and methods that you can use to achieve this overall capacitance value.

9–14 Show the schematic symbol for the following switches:

   a. SPST maintained switch

   b. SPST momentary switch

   c. SPDT maintained switch

   d. DPDT maintained switch

9–15 What are the three different types of applications for electrical connectors?

9–16 When replacing any type of electrical contact, which two electrical parameters must be known?

9–17 When selecting active components, list the maximum value parameters that should be considered.

9–18 List and define each category of information that must be included on a parts list or bill of material.

10 ▶ Step Four: Execution (The Design Breadboard)

▶ Introduction

The design project is now ready to take on a physical shape. Up to this point, the results of our efforts have been mostly on paper. A myriad of complicated specifications, schedules, schematics, and simulations have been produced, but no components have yet been wired in place. Now the preliminary design takes on a physical form as the circuit schematic is breadboarded and tested. In this chapter we review component procurement, electrical safety, and the different methods of breadboarding circuits. We will also discuss intermediate design issues, such as ambient temperature, noise considerations, design for manufacturability and serviceability, and developing initial cost estimates. The specific chapter topics are as follows:

- ▶ Component procurement
- ▶ Safety
- ▶ Electronic tools
- ▶ The soldering process
- ▶ Breadboarding
- ▶ Breadboard and prototype methods
- ▶ Breadboard testing
- ▶ Design for manufacturability
- ▶ Design for serviceability
- ▶ Cost analysis

10-1 ▶ Component Procurement

The actual selection of the components completed in Chapter 9 resulted in the identification of manufacturer's part numbers for all the circuit components. Company part numbers are issued for each component using the company's part numbering system. All this information is included on the development parts list discussed in Chapter 9. Now the circuit components must be procured to complete the breadboarding and prototype stages of the project.

Before buying any components, determine those that are currently being used by your company or organization. These should be available from stock unless needed to support manufacturing. Check with the production planner to determine availability of any components from stock. Any components that are not available from inventory stock need to be ordered or requested as samples.

Before we proceed with procurement, it is important to understand the relationship and the perspective of the various sales organizations with which you are likely to deal. The business establishments that you approach about buying electronic components are either distributors or the manufacturers themselves, represented by either manufacturers' representatives or direct sales representatives. Representatives of each of these groups will slap you on the back and tell you funny stories. It is important to understand the differences between them. In addition, the sales network for future or current employers likely is one of the following three types.

Manufacturers' Representatives

Manufacturers' representatives are sales organizations who represent various manufacturers of electronic components. They are often called *manufacturers'* or *sales reps*—or just "reps." The sales rep usually represents a number of manufacturers that produce complementary, noncompeting products. Sales reps are efficient at representing companies because their sales calls present their whole line of products to customers that likely have a need for those products. For example, if a circuit board assembly company currently buys resistors to assemble circuit boards, then it is likely it will also need to buy capacitors. A typical electronic sales rep may handle eight to ten manufacturers of various lines, such as integrated circuits, resistors, capacitors, and inductors. The sales rep is not a direct employee of the manufacturer but receives a commission for all sales of the manufacturer's products into an assigned sales territory. The commission is paid whether or not the sale was induced or made directly by the sales rep. The sales rep does not buy and sell the manufacturer's product. The product is actually purchased from the manufacturer through the sales rep or from a distributor that sells the manufacturer's products. The primary job of the sales rep is to manage the assigned sales territory by signing up and training distributors and calling on large direct customers called *original equipment manufacturers* (OEMs) that will purchase directly from the manufacturer. The sales rep cannot promote sales outside his or her assigned territory.

Direct Sales Representatives

Direct sales representatives are identical to the manufacturers' reps except that they are direct employees of the manufacturer. The other primary difference is that direct sales representatives represent only the manufacturer's product line. Direct sales reps are cost effective when the manufacturer is a large company with a very broad product line. The operating mode and objectives of the direct sales rep are almost identical to those of the manufacturers' rep.

Distributors

Distributors buy components from many manufacturers at a significant discount for the purpose of reselling to customers. Distributors typically stock a manufacturer's product for immediate shipment to meet customer requirements. Distributors have salespeople who make sales calls just as the sales reps do, but, because distributors take possession of the product, they can sell it anywhere. So distributors, unlike sales reps, have no territorial limits. Distributors will usually handle competing product lines as well. Distributors will only make money if you buy the product from them. Figure 10–1 shows a comparison of the sales rep and the distributor.

Original Equipment Manufacturers (OEMs)

A company that is described as an OEM is a manufacturer of equipment for resale. The OEM category evolved as sales organizations tried to categorize their sales to customers. OEMs manufacture equipment for resale to their own customers, and they compete against other OEMs making similar products. The following statements are usually true about OEMs:

▶ **FIGURE 10–1**

Sales rep and distributor comparison

	Sales Rep	Distributor
Territory	Defined territory	Unlimited territory
Product lines	Noncompeting	Competitive
Primary focus	OEM sales and supporting distributors	Supporting end-user sales
Source of income	Commission on all territory sales	Buying and selling at discount and profit

1. They buy in volume so that they represent a high level of sales that usually continue every year.

2. They are very price sensitive.

3. They demand high levels of quality and support.

Considering the magnitude of OEM sales, combined with their price and service sensitivity, most manufacturers prefer to take OEM sales directly. They accomplish this through the sales rep. *OEM* is a relative term. When a manufacturer views a company as an OEM, it really means that the company represents a high volume of ongoing sales to the manufacturer. Because of this they offer the OEM company special pricing. Companies that manufacture original equipment in relatively low volumes are technically OEMs but may not be considered one by manufacturers. Lower dollar-volume sales are usually directed through distributors. Low-volume sales are usually called "user" sales. The term *user* is meant to imply that the component is used internally by the customer, not installed in equipment and resold. It really means that the customer is not an OEM and is usually a small volume customer.

Example 10–1, OEMs

To better understand the OEM concept, let us take the example of a sales rep as he (or she) makes sales calls on two different prospective customers. The first sales call involves ABC Cable TV, and this company is an OEM of cable TV tuner boxes. They are in need of a particular integrated circuit in quantities of 50,000 units per year. The list price of the integrated circuit is $1. The sales rep is excited by the opportunity and discusses the application with the engineering department. After the sales rep determines that his line of components meets the needs of this customer, he indicates that he will discuss this sales opportunity with the manufacturer in question. Then he will respond with a quotation that will represent the company's best level of OEM pricing.

On the next sales call, the sales rep visits another OEM customer, XY Medical, a small manufacturer of specialized electrocardiograph machines. After discussing the application, the sales rep identifies that the same integrated circuit used by ABC Cable TV ($1 list price) meets the requirements of XY Medical. When price, volume, and availability are discussed, the customer indicates that a volume of 500 per year is needed for the component in question. The sales rep indicates that this volume is not high enough to warrant OEM pricing and that the component should be available in stock from a distributor. The sales rep offers to get samples of the component for the customer and says that the distributor will issue a quotation on a variety of quantities of the component. In this case XY Medical did not qualify as an OEM to the integrated circuit sales rep.

Another sales rep, representing a chart recorder manufacturer, calls on XY Medical. XY Medical will require 500 chart recorders per year. The list price of the chart recorders is $100 each. This sales rep recognizes this level of business, potentially $50,000, as high enough to warrant OEM pricing and indicates that a quote will be issued soon reflecting the highest level discount for OEM pricing.

It is important to understand the difference between sales reps and distributors because, as an electronics professional, you will need to develop a working relationship with each. These are the types of requests that should be directed at sales reps:

1. Requests for component data sheets and data books.

2. Requests for sample quantities of components.

3. Requests for technical support. Manufacturers usually have a field application support engineer assigned to different sales territories, and the sales rep can request support and training from him.

4. Requests for OEM pricing and factory delivery. This usually ranges from 2000 to 100,000 units per year, depending on the value of the component.

5. Requests to determine the distributors selling the manufacturer's product.

View the sales rep as a territory manager whose goal is to have you use the manufacturer's product, either as an OEM or a user. You should realize that if you buy the product from a distributor outside of the sales rep's territory, the sales rep receives no commission on the sale.

These are the requests that should be directed to a distributor to:

1. Determine price and delivery of some quantity of components.

2. Purchase components in low to medium volumes.

3. Determine the product lines handled by the distributor.

4. Determine the sales rep for a certain product line.

Look at the distributor as a source for supplying your component requirements in low- to medium-volume levels. The distributor will sell you any brand of products they carry, but it will try to direct sales toward product lines that are large sellers. The distributor usually enjoys a higher discount from these product lines, making them more profitable to sell.

It should be noted that not all products have sales reps. For example, most resistor manufacturers do not have sales reps. Because of their relative simplicity and low cost, products such as resistors and other similar products are usually marketed solely through distribution without reps of any kind.

Requisitions and Order Quantities

To procure components for the breadboard and prototypes, they have to be placed on order with a distributor. To do this, most companies use a form called a *requisition*, which is completed by the person requesting the components. Completing the requisition requires listing the supplier along with the manufacturer's name and part number. The quantities will have to be listed for each component.

To determine the total quantity of components to order, consider the number required for breadboards, prototypes, and even pilot production quantities. If the cost of a component is relatively low, ordering enough for all the prototypes and pilot production units is suggested. If the lead time for a component is very long, order enough to include the pilot production, even if the component has a high cost. There is a chance that changes will occur that make certain components not usable, but the gamble is better than reaching the pilot production run and not having the components to complete it.

The requisition is completed and then given to someone in purchasing, who is usually called a *buyer* or *purchasing agent.* The buyer then contacts the supplier indicated and places a purchase order for the components. It is important at this time to request that the buyer obtain pricing information in different quantity levels (e.g., 25, 100, 1000, and 10,000) for each component as it is ordered. As long as the supplier must be contacted anyway, it is an opportune time to acquire accurate cost data on the components in different quantity levels.

10–2 ▶ Safety

Before beginning to use electronic tools and wiring up circuits for testing, we should discuss the issue of safety. The environment in most college laboratories is very safe. Administrators and instructors are typically very conscious about safety. Consequently, they tend to restrict laboratory activities to exclude exposure to high voltages. In industry there can be much exposure to high voltage. Some companies are very safety oriented, while others are not. As you begin your career, it is your responsibility to emphasize your own safety.* Promoting safety is simple, and like other disciplines it requires good habits. To develop safety in the electrical and electronics workplace, develop the following work habits:

1. Develop a respect for the serious danger that electricity poses. The human body has a resistance between 800 Ω to 1000 Ω. A current of as little as 16 ma is enough to prevent a person from letting go of a live conductor, which can cause serious injury or death.

2. Do not wire up circuits that are energized.

3. Do not touch the conductors of any circuit over 24 V while energized.

4. Develop neat, systematic, and organized test procedures when working with any electrical circuit and follow them all the time. The test area should be neat and organized.

5. Make all electrical connections very secure in all electronic and electrical circuits.

6. Verify that the power is cut off to a circuit by taking a voltage measurement. Do not rely on "knowing" that the power is cut off to a circuit before touching the circuit. *Verify it with a meter!*

---

*Over my career I have received a number of electrical shocks. While each case was preventable, I had followed enough of my own safety habits to make the experience survivable, and I learned something each time.—J.S.

7. Be aware of test equipment where one of the test leads is connected to building AC ground and chassis ground. This is the case with oscilloscopes. The negative test lead of the oscilloscope is connected to AC ground and the scope chassis. In this case care must be taken because any circuit point, where the negative lead is placed, becomes AC ground. This will not cause a problem if the point is isolated from AC ground or, if it has an equal potential, to AC ground. Otherwise a short circuit will be created. (See Appendix B for a more detailed discussion of taking oscilloscope measurements.)

8. Be aware of the danger of projectiles when cutting component leads and from exploding components. Wear safety glasses during circuit assembly work and when powering up new circuits. Review the voltage, current, and power ratings of all components used in a circuit.

9. Develop a special respect for 220 V and higher circuits. Not only is the voltage higher, but it is important to realize that neither leg of a 220 V circuit is at ground potential. A familiarity with 110 V AC circuits and the fact that the common side of the 110 V circuit is at ground potential is sometimes misleading. Neither side of 220 V AC circuits is connected to ground. It is never safe to touch either side of an energized 220 V circuit, as the potential present is actually 110 V between the 220 V lead and AC ground.

10. In difficult test situations, use a ground fault circuit interrupter (GFCI) as a protective device. GFCIs work by measuring the current present in each leg (the send and return legs) of a circuit. If there is a difference between the two currents of more than 4 ma to 6 ma, the GFCI shuts off power to the circuit. In this case the GFCI is assuming that the difference between the two currents indicates a flow of current through some other path to ground. That other path could be an individual. The GFCI is designed to shut off power to the circuit in roughly 0.1 seconds.

10–3 ▶ Electronics Tools

Whether involved in product development or product support, the electronics professional can become involved in a wide variety of activities. The tools required vary, but it is best to be equipped for almost anything. The following is a discussion of the tools most needed by electronics technicians and engineers. This section discusses the use of those tools used in the breadboarding and prototyping process, soldering, desoldering, solderless breadboards, and wire wrapping. (See Appendix B for a discussion on the use of test equipment.)

Soldering Iron, Holder, Sponge, and Tips

Soldering irons are available with different wattage ratings and an assortment of tips. The application will determine which soldering iron and tip combination to use. For most circuit board applications a small soldering-pencil-type tip with a 25- to 35-watt iron is usually the best. For applications requiring more power, such as soldering

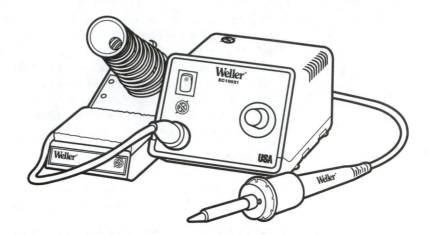

▶ **FIGURE 10–2**
Soldering iron assemblies
(Courtesy of Cooper Tools)

wires onto terminals, a chisel-type tip with a 40-watt iron or larger may be required. Soldering guns are usually used for very high wattage applications. Usually an assortment of interchangeable tips can be purchased for any particular soldering iron but the pencil and the chisel tips are the ones most often used. Use the pencil tip for all assembly work and the chisel-style tip for all desoldering. With the soldering iron, some sort of iron holder will be needed as well as a wet sponge and holder for cleaning the tips during soldering. These can be purchased for the least cost, individually, but are available as one assembly that includes the soldering iron, holder, and sponge with holder. Optional power control is available also. These complete units are more expensive but very efficient, versatile, and easy to use (see Figure 10–2).

Desoldering Tools

Two types of desoldering tools should be included: the hand desoldering pump and braided solder wick material. The hand desoldering pump is quicker and easier to use, but may not be completely effective (see Figure 10–3). For more difficult applications, where the solder must be removed almost completely, the braided solder wick material should be used as a secondary process (see Figure 10–4).

Wire Strippers

A wire stripper is an important tool. To complete a reliable electronics assembly, the wire stripper must strip back the wire without piercing the conductor. If the conductor is pierced, this will weaken the wire, and it will likely break after being bent

▲ **FIGURE 10–3**
Desoldering pump *(Courtesy of Cooper Tools)*

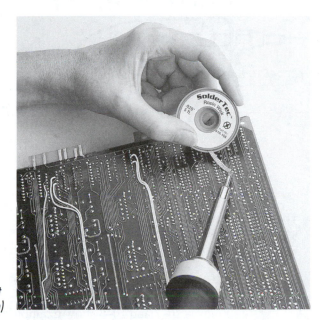

▶ **FIGURE 10–4**
Solder wick *(Courtesy of Contact East, Inc., www.contact-east.com)*

just a few times. Utilization of the proper size and style stripper is critical. There are a variety of wire strippers available. Some are adjustable to different wire sizes, while others have a series of standard wire sizes they can accommodate. In any case it is imperative to use a quality, sharp, and properly sized wire stripping device for all wire stripping operations. Figure 10–5 shows a wire stripper that can accomodate standard wire sizes.

Wire-Wrap Tools

Other than the wire-wrapping materials to be discussed later, a wire-wrap gun and an unwrapping tool, both shown in Figure 10–6, will be needed for any wire-wrapping project. Wire-wrap guns are available that are electrically or hand powered. The wire-wrap gun wraps the wire around the pin in a clockwise direction. The unwrapping tool should be turned in a counterclockwise direction to undo the wrap.

▶ **FIGURE 10–5**
Wire stripper *(Courtesy of Contact East, Inc., www.contact-east.com)*

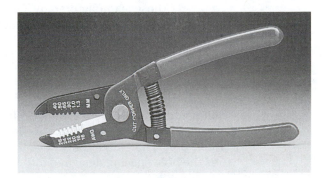

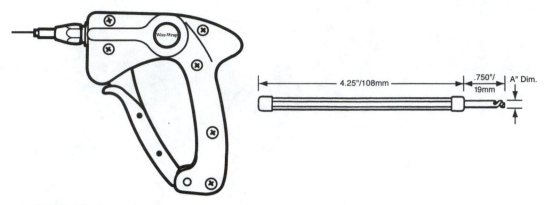

▲ **FIGURE 10–6**
Wire-wrapping tools *(Courtesy of Cooper Tools)*

Miscellaneous Small Hand Tools

Figures 10–7 and 10–8 show examples of the assorted hand tools that are needed and described below:

Screwdrivers: Full set of small to medium flat-blade and Phillips screwdrivers and set of miniature screwdrivers

Allen wrenches: Set of assorted sizes

Nutdrivers: Set of asssorted sizes

Needle nose pliers: Miniature with long nose, medium size, and a more rigid nose for bending

Side cutters: Miniature for most circuit board work and a larger size for cutting wire

Files: Set of small assorted sizes

Trimpot adjuster: Invaluable for making trimpot adjustments

Razor-blade knife: There are a variety of uses for the razorblade knives such as those manufactured by X-ACTO. The primary use is to cut runs on printed circuit boards to implement circuit modifications. There are a variety of cutting blades and holders to choose from (see Figure 10–8).

Resistor Lead Former

This is a very handy tool (see Figure 10–9) when building prototype circuit boards. It is used to bend resistor leads to the proper length for insertion into the printed circuit board holes. Not only does it help assemble the prototypes but it also verifies the hole spacing designed into the circuit board.

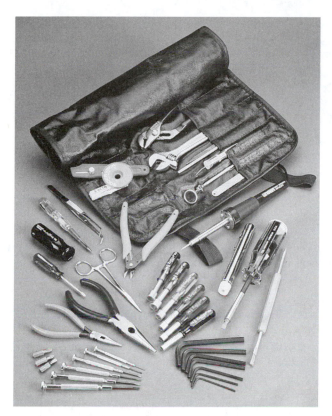

▲ FIGURE 10–8
X-ACTO knife *(PhotoDisc, Inc.)*

▲ FIGURE 10–7
Assorted hand tools *(Courtesy of Contact East, Inc., www.contact-east.com)*

▲ FIGURE 10–9
Resistor lead former *(Courtesy of Contact East, Inc., www.contact-east.com)*

Integrated Circuit Extraction Tools and Static Protection

There are many specialized integrated circuit extraction tools that are designed to preclude any damage to the integrated circuit during its removal or installation. Through-hole and SMT devices use significantly different tools. While handling many integrated circuits (MOS-type semi-conductors), it is important to consider damage caused by static electricity. To preclude any damage, use a conductive, grounded mat at all workstations and use a grounded wrist strap when handling MOS-type components. Figure 10–10 shows examples of extraction and insertion tools for DIP packages and conductive mats.

Circuit Board Holder

The circuit board holder (see Figure 10–11) is a valuable tool any time both hands are required for an operation. This occurs often when desoldering components and sometimes when assembling a board. The circuit board holder can be a helpful, though not required, electronics tool.

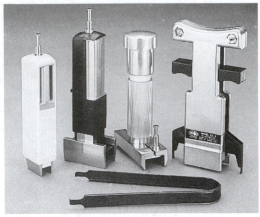

▲ **FIGURE 10–10**
Integrated circuit extraction tools and conductive mats *(Courtesy of Contact East, Inc., www.contact-east.com)*

▶ **FIGURE 10–11**
Circuit board holder *(Courtesy of Contact East, Inc., www.contact-east.com)*

Hand Drill and Set of Bits

The hand drill (see Figure 10–12) is very useful when doing work on printed circuit boards. It can be used to drill out the holes on a printed circuit board or to make modifications. For drilling holes in printed circuit boards, a very small set of drills will be required starting with #70 through #50. The hand drill may be most useful as a slicing tool to remove circuit board runs when modifications must be made. The accessory bits are important for these slicing or complete copper removal operations.

Solderless Breadboard

The solderless breadboard (discussed in detail later on page 250) is an effective tool for checking the operation of circuits in the very early stages. A power supply of

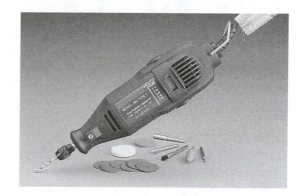

▶ **FIGURE 10–12**
Hand drill, bits, and accessories
*(Courtesy of Contact East, Inc.,
www.contact-east.com)*

some sort is required to operate the circuit to be tested. Prestripped wire, cut to a variety of lengths, can be purchased to make breadboarding very quick.

Test Leads

A set of test leads, which include oscilloscope probes, a BNC to banana jack, and banana jack leads plus a set of good "easy-hook" leads, are very handy (see Figure 10–13). The easy-hook leads can be used to clip around component pins to allow for taking continuous measurements of a circuit point.

Digital Multimeter

Today's multimeters (see Figure 10–14) are an incredibly versatile tool and include many new built-in functions. Features such as autoranging, logic probe, diode test, frequency counting, and maximum and minimum measured values are a good sample of some of the advances in multimeter technology. Multimeters are discussed in detail in Appendix B.

▶ **FIGURE 10–13**
Test leads *(Courtesy of Contact East, Inc.,
www.contact-east.com)*

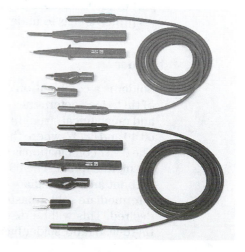

▶ **FIGURE 10–14**
Multimeter *(Courtesy of Fluke Corporation. Reproduced with permission.)*

10–4 ▶ The Soldering Process

Soldering is one of the most important steps of electronics assembly and is a key part of the breadboarding and prototyping process. It is important for all electronics professionals to understand the process of making a good solder connection.

Solder

Solder is a combination of tin and lead material that is actually an alloy. The ratio of tin to lead determines the characteristics of the solder, such as its melting point and mechanical strength. Solder is available in different tin-to-lead ratios, such as 60/40 and 63/37, where the numbers represent the relative percentage of tin to lead. The 63/37 combination is the most popular because at this ratio the alloy has its lowest melting point. The 63/37 solder alloy also changes directly from solid to liquid without an intermediate solder paste stage. All other tin-to-lead ratios have some intermediate solder paste stage. In SMT soldering applications, the paste stage is desired. This will be discussed later. Another consideration is that with a lower tin-to-lead ratio, solder has a better wetting action. It can be more easily spread over a metal surface area. For applications such as solder plating a printed circuit board, a lower tin-to-lead ratio is desirable. The cost of solder increases with the tin-to-lead ratio so the solder should be selected with the application and the relative cost in mind.

Solder is available in wire, bars, or paste forms. Solder wire is used for all hand wiring. Solder bars are utilized in wave and dip-soldering processes. SMT processes employ solder paste. You should select the solder wire size that is most appropriate for the components to be connected. For larger components use a larger-diameter solder wire and vice versa.

Flux is a material that is contained within most solder. Flux is needed because most metals oxidize when exposed to the oxygen in the air. This oxide layer will prevent a solder joint from being formed between two components being soldered. To make matters worse, oxidation increases as temperature rises. So, as a soldering iron is applied to two component leads being soldered, oxidation is encouraged by the increase in temperature. Flux is provided to dissolve the oxide layers present before—and those induced during—the soldering process. The remains of the chemical reaction between the flux and oxides collect on the surface of the solder joint, forming a varnish-like residue around the joint. This residue can be cleaned off later with a cleaning process that will depend on the type of flux material used.

Flux removal is desirable because the flux residue can be corrosive at higher temperatures. It can conduct small amounts of leakage current, and it is unsightly, too. The chemicals previously used by most circuit board assemblers included some form of trichlorotriflourethane (commonly called "Tri-Chlor"). These chemicals are very good cleaning agents, but are in that class of complex fluorocarbons that are detrimental to the environment. Consequently, manufacturers can no longer use them. To counter this, flux agents have been developed that are water soluble and can be cleaned with soap and water. Many manufacturers now use industrial dishwashers for small circuit board cleaning operations. In addition, a variety of "no-clean" fluxes have also been developed. These no-clean fluxes form a hard varnish residue after soldering that represents a high impedance and is non-corrosive at high temperatures. Although the residue is still unsightly, there is no functional reason to remove no-clean flux.

Hand Soldering

Hand soldering involves using a soldering iron to heat the two components to be soldered and then applying solder to the joint area. Before soldering, be sure that all points being soldered are clean. To complete a good solder joint, the following steps should be completed (see Figure 10–15):

1. Clean all areas to be soldered.

2. Heat both of the metal points to be soldered with a soldering iron.

3. Apply solder to the metal point of the joint area furthest from the soldering iron until the solder melts and flows.

4. Remove the solder from the joint.

1. Clean areas to be soldered.

2. Heat both points being soldered.

3. Apply solid solder to metal furthest from soldering iron.

4. Remove solder after enough has melted to form the joint.

5. Continue applying heat until solder flows over entire joint area. Then remove soldering iron.

▲ **FIGURE 10–15**
Soldering process

5. Leave the soldering iron in place until the joint flows. This should only take a second. Then remove the soldering iron from the joint. Care should be taken to make sure that neither component being soldered moves during this stage. Any movement at this point will likely cause what is called a *cold-solder joint*. A cold-solder joint results when an alloy does not develop between any one of the components and the joint area. Movement of a component when the solder joint is being formed will likely cause a cold-solder connection.

6. Clean off the soldering iron tip frequently with a damp sponge soaked with warm water. This is to remove flux and other residue that will build up on the soldering iron tip.

Other concerns when soldering are:

1. Do not cut off excess leads until the solder joint has sufficiently cooled after about 10 seconds.

2. When soldering integrated circuits and other temperature-sensitive components, make sure to use a smaller wattage soldering iron, and be careful not to hold the iron on the joint excessively long.

3. Be sure to use a grounded wrist strap to handle static sensitive components such as any MOS or CMOS components.

When properly done, the solder process creates an alloy that results from the combination of the tin, lead, and metal materials being soldered. The solder joint will provide very low resistance and good mechanical strength.

Desoldering

Many times during the breadboard process, it is necessary to remove components that have been soldered. Desoldering is also required to repair printed circuit boards when a component must be replaced. There are two methods of desoldering: flowing solder away from the joint or the use of vacuum suction. Vacuum suction can be applied with the hand desoldering pump shown in Figure 10–3 on page 240 or the pneumatic operated desoldering units shown in Figure 10–16.

To flow solder away from the joint, braided desoldering material (solder wick) is placed over the joint and the braided material is heated until the solder flows. When properly done, the solder will flow into the braiding and leave the joint area free of solder.

The other method involves the use of a soldering vacuum device, either a hand desoldering pump or a vacuum desoldering station. Whichever tool is utilized, the process involves heating up the solder joint with a soldering iron and sucking the solder away from the joint with either vacuum device. Care must be taken when desoldering circuit board components to prevent overheating the circuit board pad. Continued overheating will cause the pad to pull away from the circuit board, making reliable repair difficult or impossible.

10–5 ▶ Breadboarding

Breadboarding is the process of experimentally constructing a schematic circuit from the actual components selected for the design. The purpose of breadboarding is to test the preliminary design of a circuit or module. The breadboard

▶ **FIGURE 10–16**
Desoldering units *(Courtesy of Cooper Tools)*

is usually temporary in nature and is more easily modified than a printed circuit board. There are many different methods for completing a breadboard. The method selected depends on the goals of the prototype phase and the complexity and circuit technology of the circuit to be breadboarded. The breadboarding technique discussions that follow do not include all breadboarding methods but do include the methods most viable today.

Solderless Breadboard

The solderless breadboard is one that most electronics students have had experience with in the laboratory. This breadboard consists of groups of circuit connection holes called *points* that are all located on a 0.1″ grid. These circuit points are arranged in groups that are connected together underneath the plastic surface of the breadboard assembly. Horizontal rows of five points are connected together and are arranged into a vertical column. A center channel that is provided for mounting a standard dual in-line package integrated circuit separates two such columns. The integrated circuit straddles the center channel, making connections to connecting points on each side of the channel. With the integrated circuit mounted in place, the laboratory breadboard provides four connection points for each pin on the integrated circuit. Other components, such as resistors and capacitors, can also be mounted across the center channel. There are also separate vertical columns where the points are connected together vertically instead of horizontally. These are commonly used for the purpose of bussing power supply voltages and ground to various circuit points. There are a variety of solderless breadboards available, so to be sure about which points are connected to which, use your multimeter to verify the connection scheme of the breadboard you are using.

To construct a circuit with the laboratory breadboard, components are inserted into the connection points so as to implement all the connections shown on the schematic. This is accomplished by connecting two components to the same connection group (i.e., five points that are tied together) if they are electrically connected. To connect the remaining circuit points together, use 22- to 30-gauge wire that is stripped back about 3/16″ to 1/4″. Precut and prestripped wire is available in various lengths for use in these breadboards. The solderless breadboard is limited in the current that can flow in the connections and the wire typically used to make them. The conductive grid of most solderless breadboards is rated at 4 amps; be sure the appropriate wire size will handle the expected current flow in the circuit. The inherent capacitance of the connections and potential cross talk between conductors also limits the frequency of operation of the solderless breadboard system. The practical frequency limit of most solderless breadboards is 10 MHz, but short, neat lead connections and careful breadboard layout are required to achieve this frequency.

Universal PCB Breadboard

Universal PCB breadboards (see Figure 10–17) are printed circuit boards that have holes all located on a 0.1″ grid system. There are many varieties of these universal boards. In general, some of the holes in the grid system have copper pads to allow for soldering and some of these pads are connected together. The variations come from the location of the holes that have copper pads and which of the copper pads

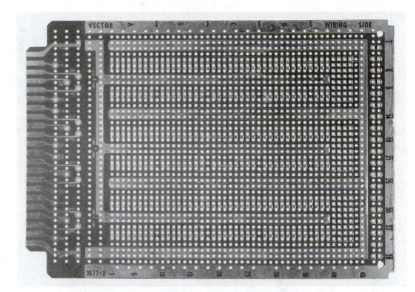

▶ **FIGURE 10–17**
Example of a universal breadboard

are connected together. Some versions are more suited to large numbers of discrete components, and others are geared toward integrated circuit applications. PCB-edge connections are even provided on some models. Whichever variety is used, there are areas where the configuration of the universal circuit board requires modification with a hand drill to cut away circuit points that are connected together.

Prototyping a circuit with the universal circuit board method is very tedious. After selecting the proper universal circuit board variety, the components are soldered in place. Next, the circuit connections shown in the schematic must be implemented either by soldering wires that connect all the circuit points or soldering together circuit pads with a continuous conductor to make a circuit board run. Finally, some areas of the copper connected pads may need to be disconnected. This can be accomplished by cutting away the connecting copper runs with a razor-blade knife or a hand drill fitted with a slicing tool. If carefully planned out and neatly implemented, the result can be very close to the eventual printed circuit board in function, physical size, and layout. This method can be combined with wire-wrap methods and SMT technology to be discussed next. Figure 10–18 shows an example of a breadboard circuit that was developed with a universal breadboard scheme. Both the top and bottom views are shown. Notice how the circuit can be made to closely resemble the eventual printed circuit board.

Surface-Mount Technology

Surface-mount technology provides unique challenges in breadboarding circuits. Again, there are many choices to be made. The most obvious is to breadboard the circuit with the through-hole equivalents rather than the actual SMT components. Avoiding the use of SMT components at the breadboard stage is the best choice, if possible. Most SMT components are available in a through-hole package. If not, the SMT component can either be mounted on an SMT socket or on an SMT "carrier"

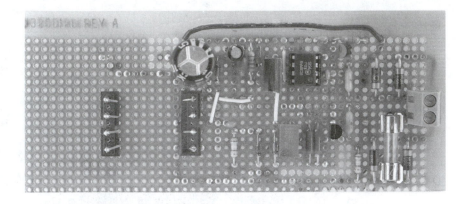

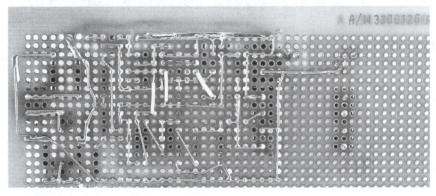

▲ **FIGURE 10–18**
Universal printed circuit board example

board. Figure 10–19 shows an example of an SMT socket on a prototype board with extra area allotted for other through-hole circuitry. Figure 10–20 shows an example of an SMT carrier board that can be soldered into a universal breadboard prototype. SMT carrier boards allow the mounting of the SMT components where sockets may not be available or practical. The SMT carrier board can then be mounted to a universal printed circuit board and wired in the circuit with solder connections that go between the two boards. Soldering the SMT component to the carrier board may take some advanced soldering skills, as the space between connections is small. Use an extra-fine tip for these applications and apply a thin coating of solder to the SMT pads.

It is best to build the breadboard with standard through-hole resistors and capacitors instead of SMT chip resistors and capacitors. The SMT chip components are extremely small, making them hard to handle and solder.

Wire-Wrapping

Wire-wrapping is a solderless technique in which circuit connections are made by small wire connections that are tightly wrapped around pins called *wire-wrap pins*. Wire-wrap wire is generally 28- or 30-gauge wire. Wire-wrap pins, which feature

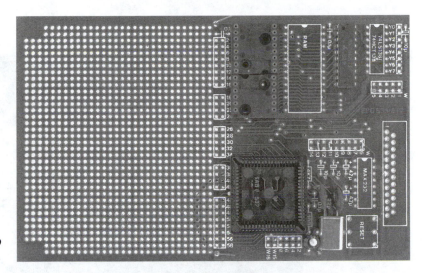

▶ **FIGURE 10–19**
SMT sockets board

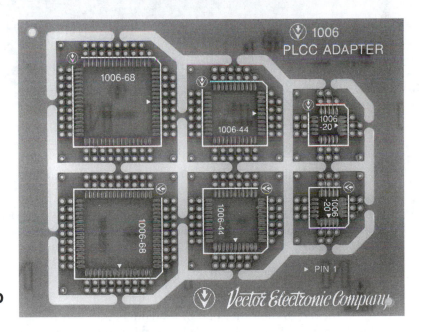

▶ **FIGURE 10–20**
SMT carrier board

right angle corners, are incorporated into integrated circuit sockets or part carriers or used as individual wire-wrap posts. Figure 10–21 shows an example of a wire-wrapped circuit. The wrapping is accomplished with a device called a *wire-wrap gun* or *wire-wrap tool*. To develop a breadboard utilizing the wire-wrap process, wire-wrap sockets are utilized for all integrated circuits, and part carriers are used for nonintegrated circuit-type devices, such as resistors, capacitors, diodes, and transistors. The wire-wrap sockets and part carriers are lightly glued to a perforated phenolic board. Circuit connections are made point-to-point, as follows:

▶ **FIGURE 10–21**
Wire-wrap application

1. Strip back the wire-wrap wire about 3/8″.

2. Insert the bare-wire end into the outer hole on the wire-wrap gun and bend the wire at a right angle with the axis of the wire-wrap gun.

3. Insert the center hole of the gun over the wire-wrap post to be connected, making sure that the end of the gun is flush with the back of the phenolic board.

4. Squeeze the lever on the wire-wrap gun and the bare wire will be wrapped clockwise around the wire-wrap pin.

Proceed to the other end of the connection and repeat the process. A reliable connection is actually made by the wire and the square edge of the wire-wrap pins as the wire is wrapped around the pins.

The Nonbreadboard

There are many times when the breadboarding process is actually implemented with a printed circuit board. This occurs most often with microprocessor-based digital boards, where the time and complexity of breadboarding is very high. In this case the engineer ends up with only one wire-wrapped board, which is of questionable value after it is complete and made operational. Usually each software engineer will need a circuit board for software development, so numerous breadboards are often required in the early project stages. This situation promotes going directly to a printed circuit board. In this case the circuit board schematic is checked out and sent directly to the circuit board layout person, usually an electrical designer from the drafting department. Circuit board layout will be discussed in detail in Chapter 11. The circuit board is laid out and prototype quantities are ordered with a priority lead time. When the circuit boards have been received, they are carefully built up and tested one section at a time. When functional problems are encountered, they are analyzed and corrected with a combination of "cuts" and "jumpers," component value changes, and by adding components. In this case the first generation of the printed circuit board becomes the breadboard. The breadboard stage is not necessarily skipped, but it is replaced with the first version of the printed circuit board. It is desirable to make the prototype board a printed circuit board when one or more of the following is true:

1. When the circuit is a complicated circuit board with embedded software

2. When the design is close to being a standard design, like well-known bus-oriented structures for microprocessors; in other words, the schematic has a good chance of being functional

3. When testing of the prototype is not possible until some basic software is developed, and that is not scheduled to be complete for a couple of weeks

4. When more than one prototype board is required as soon as possible

5. When the prototype board must be reliable

6. When engineering and technician time is at a premium and drafting layout time is more available

The high degree of error checking that results from schematic capture software and circuit board layout programs has promoted the use of the nonbreadboard technique. When using complementary schematic capture and board layout packages, the most significant benefit is that the artwork will exactly equal the schematic. If the schematic is correct, then the artwork will be also. With this kind of accuracy, it is a fact that the actual printed circuit board has a higher chance of being equal to the schematic than any breadboard. This does not mean that the prototype board will function as required, which is why breadboards are still favored in many cases. The breadboard is more readily changed or just scrapped when major design problems develop.

10–6 ▶ Breadboarding Methods

This section discusses some key methods for performing operations that are often required during the breadboarding stage.

Implementing a Circuit Schematic

Whichever method of breadboarding is selected, a schematic circuit is put into some physical form. To accomplish this accurately, an organized and methodical process must be followed. The following procedure produces very good results:

1. Organize all the breadboarding materials and tools on a bench-top area to be used for the duration of the process.

2. It is imperative to start with a neat and orderly schematic. The schematic should be complete and include wire connections for *all* components that are breadboarded. All of the components on the breadboard should have a component number (e.g., R_1, C_1, and the like). The schematic should be laid out in an easy-to-understand format with inputs on the left-hand side and outputs on the right side.

3. Make two copies of the schematic. One copy will be the breadboard master copy and should be initialed, dated, and filed with all breadboard documentation. As the breadboarding proceeds, use a highlighter marker to religiously highlight each connection as it is made. When this process is used, the most common error, missing connections, is eliminated.

4. Start out wiring power to all points on the board where it is required. Make sure that the power and ground runs are made from heavier wire. Then follow the schematic, left to right, wiring the circuit in logical groups. For example, wire up all the power supply components and check out the power supply. Then wire all data connections from one component to the next component. This tends to minimize errors, because an error, such as wiring to an incorrect pin, will be noticed when an attempt is made to make the correct connection to that pin.

5. Make the connections as short and as neat as possible. Keep in mind to provide easy access for later inspection and the connection of test leads.

6. Many times there are integrated circuits where multiple components are located on one chip. Be sure to note which component on the integrated circuit is used for which schematic function by noting the pins and using a component subdesignation on the breadboard master schematic. Take the example of a Dual 4 Input NAND gate that is given a component designation of U_1. There are two 4 Input NAND gates on U_1; one will be designated U_{1-A} and the other U_{1-B}. Also, be sure to properly terminate all unused integrated circuit components as required in their specifications. Unused TTL gates should have their inputs tied high (+5 V) for example.

7. Make a sketch of the assembly showing the relative location of the components with their designation.

By performing this procedure, you may be surprised to find that the circuit will work the first time—that is, if you remember to turn the power on. Of course, the schematic design must be a "working" design to begin with.

Assembling Printed Circuit Boards

At some time in the development process, the assembly of a printed circuit board will be performed. Whether you are a design engineer or a technician working on the project, it is important to get involved with circuit board assemblies. A great deal about the printed circuit board can be learned from its assembly, and all engineers should take the opportunity to assemble one of the prototype boards. A circuit board is not difficult to assemble, but some methods are easier than others are and more productive.

During the development process, the assembly of the prototype boards should be looked at as an opportunity to experience any assembly problems first hand, as well as a chance to check out the current documentation. To begin, obtain copies of the current documentation package. This consists of the schematic, assembly drawing, bill of material, and drill code. These should be labeled as "prototype masters" and be initialed and dated by the assembler. These will be marked up in red by the assembler and will be maintained for use when the next round of board modifications is performed. The purpose of each document is listed below:

Schematic—The schematic shows the electrical connection between every component on the printed circuit board. The primary purpose of the schematic is to document the connections of the circuit and to provide for an understanding of the circuit operation.

Assembly drawing—This drawing is a pictorial drawing of the assembly, which locates each component on the printed circuit board by the component designation shown on the schematic drawing as well as the bill of material.

Bill of material—This is a list of all the parts that are contained on the printed circuit board. The company part number with the component designation is provided. The component designation ties the schematic, assembly drawing, and bill of material all together. The supplier part number may be included as well.

Drill code—This is a mechanical drawing of the printed circuit board. It shows the physical size of the board as well as the size and location of all holes in the board. It is not actually required for board assembly, but it should be checked out and verified as part of the prototype assembly procedure.

If the board being assembled has never been breadboarded, it should be built up as if it were a breadboard, one section at a time. The power supply should be assembled first and checked out. Then, each other functional circuit should be built up and tested, one section at a time.

The assembly for standard prototype boards should proceed as follows:

1. Gather all components and organize by component type: resistors with resistors and so on. Preform the resistor leads (i.e., bend the leads at a right angle to the resistor body) using a preforming tool for the spacing between the mounting holes.

2. Start assembling the board by installing the lowest profile components, the resistors first. Install all of one value as one step in the process. This is how it will be done in manufacturing. Bend the leads slightly to hold the resistor in place until soldering. Check off each location of the component on each drawing, verifying the accuracy of the information.

3. Next, install all integrated circuits. Bend the corner pins slightly to hold the integrated circuits in place before soldering.

4. Install the capacitors and all other components next.

5. During assembly, make notes on the drawings of any situation that should be improved. This can be anything from relocating a component to promote easier assembly to simply changing hole spacing or sizes.

6. Keep track of the time it took to assemble the board and note this also.

Performing the assembly process in this manner provides not only a working prototype, but also more accurate assembly documentation and the efficient assembly of the board by the manufacturing department.

DIP Integrated Circuit Removal

It is a difficult process to remove a dual in-line package (DIP) integrated circuit from a printed circuit board. The process is worth describing as a first step to experiencing it. The obvious problem is the removal of all solder from each pin or keeping all of the solder on all pins molten at the same time. Because this is seldom accomplished, the integrated circuit must usually be pried out of its location while heating up a majority of the pins. The integrated circuit usually does not survive the process intact, let alone the potential for thermal damage to the internal components. The primary goal is usually the removal of the integrated circuit without damaging the runs on the printed circuit board. There are special solder tips that heat all the pins of a dual in-line package simultaneously. To remove integrated circuits, the following process is suggested:

Method 1—Emphasizes reuse of the integrated component:

1. Use a hand desoldering pump to remove as much solder from the bottom side of each pin on the integrated circuit. Next use the braided solder wicking material to clean out as much of the remaining solder as possible.

2. Insert a small screwdriver underneath one end of the integrated circuit on the top of the board if possible. While heating up the pins on that edge of the integrated circuit, attempt to pry up the integrated circuit. If successful, try this on the other end of the integrated circuit. If, at any time, the pads become damaged, discontinue this process and proceed to Method 2.

3. Clean up all the solder and any flux residue from the pad areas.

Method 2—Emphasizes reuse of the circuit board:

This process is more time-consuming but there is less chance of damaging the circuit board pads.

1. Simply cut the integrated circuit out of the circuit board using side cutters. The cut should be made along the integrated circuit body on the component side of the board. Try to leave as much of the lead as possible so there is something to grab onto when trying to pull out the lead later.

2. With the integrated circuit body removed, heat up each pin individually while using needle-nose pliers to pry the lead out of the hole. Be careful not to pry out the lead before the joint is sufficiently heated, as this will damage the pad.

3. Use the desoldering pump and braided solder wick to clean out the holes.

There are occasions when the purpose for removing the integrated circuit is to use it on another circuit board when a few integrated circuits are needed to complete the assembly of some prototype boards. The integrated circuits were scavenged from other circuit boards. When removing integrated circuits for this purpose, Method 1 must be utilized, as the emphasis is placed on preventing damage to the integrated circuit. Otherwise, Method 2 should be used.

10–7 ▶ Breadboard Testing

The goal of breadboard testing should be to check out as many of the functional areas and as many aspects of the design's variability as possible. The specifications can be used as a guide at this point, but there are areas that are simply impractical to test at this stage. A good example of this would be the performance of shock and vibration tests on a solderless breadboard circuit. The solderless breadboard would not be expected to provide reliable connections in a shock and vibration environment. On the other hand, the circuit should be exposed to the ambient temperature variations specified in order to check out the effect of temperature on the key circuit output parameters. This can be done in an environmental test chamber.

Operational Tests

The basic operational tests should be done first. These are the tests that will determine whether the circuit functions properly in a lab environment (i.e., at room temperature, with nominal AC input voltage and so on). Each function that is listed in the specifications should be performed and verified for proper operation. All other aspects of the specifications should be verified as well. This includes all power supply voltages and responses to all inputs and outputs.

Ambient Temperature Considerations

At this point it is important to verify the temperature sensitivity of the outputs of the device under development. It is also important to consider the expected ambient temperature rise that will occur in the enclosure with all of the circuits operational.

This should be simulated, if possible, and internal temperature measurements taken at key hot points. This is to verify that all components are kept within their specified ranges.

Ambient temperature variations are considered in the preliminary design, but now is the time to determine how well the circuit will actually handle temperature variations. To test ambient temperature sensitivity of a circuit or module, consider the following:

1. Determine the key outputs of the circuit and develop a way to monitor these parameters while the circuit is placed in an environmental test chamber.

2. Determine the components and their parameters that will be most susceptible to temperature variability. Select components with the worst-case range of those parameters and perform repeated testing with different components in place in the circuit.

3. After the testing, tabulate and review the data to see if the results are within the specification's requirements. If not within the specifications, design modifications must be considered. The modification could be something as simple as using a component with a better temperature drift rating, or it could involve significant design changes.

Electrical Noise Considerations

The amount of electrical noise testing that can be performed at this time is dependent on the type of circuit and the breadboarding scheme utilized. For example, an analog circuit built up on a solderless breadboard would be subject to many poor connections. With connecting wires strewn in many directions, analog circuits will pick up and radiate many unwanted noise signals. Detailed electrical noise testing of this kind of breadboard would not be productive. At this stage, universal PCB breadboards and actual printed circuit boards (the nonbreadboard) offer a much better opportunity to perform noise tests. They better represent the final working circuit board, with all the inherent noise immunity that one provides.

The most important noise tests to be considered at the breadboard stage are input noise tests and input power noise immunity. Input signal noise testing can be performed to determine the relative ease with which noise signals can be passed into the circuit with an input signal. Power supply noise immunity can be checked by placing a noise signal on the input power to the breadboard and determining if the noise signal passes into the output of the voltage supply of the breadboard circuit.

10–8 ▶ Design for Manufacturability

It is never too early to start thinking about the ease of manufacture of a product, but this stage of the project is an opportune time to think more seriously about this issue. Previously, the lack of any physical form is a limiting factor in visualizing many manufacturing design considerations. As the various modules begin to take shape, make a list of the process steps that will be needed to assemble the product. Here are some key questions and issues that should be considered:

1. Can each module be completely tested on its own?

2. Can the final product be completely tested before being installed in its enclosure?

3. Only those manufacturing adjustments that are absolutely necessary should be included in the product.

4. Try to automate any testing and calibration that must be performed on the product.

The circuit designer must become familiar with the needs and methods of manufacturing and experience the assembly process firsthand. It is also important to involve people in the manufacturing departments, such as manufacturing engineers, floor supervisors, and the assemblers, in the design process. It is the combination of all these sources of knowledge working together with the designer that achieves the best results. As discussed previously, teamwork and concurrent engineering principles promote the involvement of the manufacturing department early in the project and a feeling of ownership by the members of that department. They make the process work, and they will stand behind the product because they had a part in its development.

10–9 ▶ Design for Serviceability

The service requirements should be stated in the specifications for the project. This is an area marketing people tend to gloss over so the service engineering group and the design engineers will need to add some focus to this topic. Serviceability can range from complete field repairability (repairable to the component level) to no serviceability at all (a throwaway unit). Very seldom are today's products repairable to the component level. Most assemblies are so small and packed together so intricately that disassembly usually damages the product. If the circuit board uses SMT components, it is usually impractical to replace those components. A typical intermediate position is to provide field reparability down to the board level, where circuit boards are changed out and the failed board is deemed not repairable and scrapped. Whichever the case, it is important to consider the serviceability of the product more seriously at this time. If the unit must be repairable to some degree, it must use packaging hardware that will allow disassembly and reassembly in the field. Many times the issues of ease of manufacturability will go against the ease of serviceability.

Example 10–2

A good example of the conflict between ease of manufacture and serviceability involved the method of mounting printed circuit boards to a plastic front panel on an industrial temperature controller. The initial design utilized threaded inserts that were pressed into plastic tabs that were part of a plastic front panel. The circuit boards were attached to the front bezel very securely with screws. As a cost-saving measure, the manufacturing department desired to eliminate both the time-consuming insertions of the threaded inserts as well as the screwing operations. The design of the plastic front panel was changed to include plastic clips that would attach the circuit boards to the front panel. These design changes were integrated into the tooling for the plastic front panel and the change was implemented. The result of this cost-saving measure was a significant reliability and serviceability problem. The plastic clips did not hold the boards in place securely. This made the interconnections between the boards intermittent in high-vibration applications. The plastic clips also made it very difficult to remove the circuit boards once they were attached to the plastic front panel.

The result of this poorly implemented project was an increase in field failures. Also, customers and field service people were very unhappy with the new process for removing circuit boards from the front panel. At a great cost to the company, the design was changed back to the original design that utilized mounting screws.

10–10 ▶ Cost Analysis

This is the point in the project when costs are first reviewed. A development bill of material should be completed using a spreadsheet program for each assembly within the project. In addition to the normal bill of material information (component designation, part number, quantity per assembly, and the like), cost information for the component should be added. This document can perform as both a parts list and a cost spreadsheet for any circuit board or any assembly. Figure 10–22 shows an example spreadsheet used for cost analysis.

The total material cost allotted in the cost budget should be included on the development bill of material and compared to the actual cost tabulations. If there are major discrepancies between the two, then either the total cost goal was incorrectly divided among the assemblies or a cost problem exists for the development project. Any cost problem must be resolved before continuing the project. One of the four following scenarios should result:

1. The allotment of the total cost should be changed so that it more correctly reflects the actual cost breakdown.

2. The design should be modified to reduce costs in line with the projected cost goal.

3. The previous cost goals stated in the specifications are determined to be erroneous, and new cost goals are established that correctly reflect the market situation.

4. The results of the cost analysis show that the cost goals cannot be obtained; consequently, the project is canceled.

	Digital Thermometer Parts List	Cost Analysis									
	Final Assembly										
Item #	Description	Company Part Number	Manufacturer	Manufacturer Part Number	Component ID	Quantity	Price per 1	Price per 100	Price per 1000	Total Cost per 1	Total Cost per 100	Total Cost per 1000
1	Main Printed Circuit Board Assembly	10000090	NA	NA		1	59.50	41.63	35.03	59.50	41.63	35.03
2	Enclosure	10000201	NA	NA		1	11.50	10.85	9.85	11.50	10.85	9.85
3	Line Cord Assembly	10000202	NA	NA		1	1.00	0.90	0.85	1.00	0.90	0.85
4	LED Filter	10000203	NA	NA		1	1.50	1.25	1.10	1.50	1.25	1.10
5	Front Panel Label	10000204	NA	NA		1	1.85	1.76	1.55	1.85	1.76	1.55
6	Total Assembly and Test	10000205	NA	NA		1	40.00	40.00	40.00	40.00	40.00	40.00
										Total	Total	Total
										Cost/1	Cost/100	Cost/1000
										115.35	96.39	88.38

▲ **FIGURE 10-22**
Cost analysis spreadsheet

▶ **Summary**

The breadboard phase is an extremely important phase. It should result in a working breadboard that performs as many of the required operations as can be practically simulated. The greater the degree of testing performed at the breadboard stage, the better are the chances for overall success of the project. In many projects only minimal testing can be completed at this stage. In these cases there is clearly a higher risk of serious problems occurring later in the project. This phase should support the continuation of the project and the likelihood of its successful completion. The potential to meet the specifications should be supported by the results of the testing and the cost analysis. If there are any doubts or any weak areas of performance, now is the time to address them. Breadboards are much easier to modify than a completed printed circuit board or a design that is ready for manufacturing.

During the breadboard phase, some considerations have been given for the manufacturability and serviceability issues. It is important to note them and incorporate them into the design of the prototype. The cost analysis should indicate that the cost goals can be met. If there are serious cost issues, they should be addressed before continuing on with the project. The concurrent engineering concepts that were discussed earlier are summed up and repeated here as follows:

Consider all the aspects of the design from the start of the project to the finish.

This means consider not only the product's performance, but its quality, manufacture, service, customer use, and profitability from the first day of the project as well. The project manager and the project team should make sure that these issues are considered. With these issues adequately addressed, the design project is now ready to proceed to the prototype stage.

Digital Thermometer Example Project

Breadboard Testing the Digital Thermometer
The digital thermometer was breadboarded using the universal printed circuit board method described earlier in this chapter. This breadboard type was selected to promote the completion of more testing at the breadboard level. The power supply circuit was breadboarded first and checked out with the specified loads. The ripple and output voltages each were within specifications and supplied the required load current. The input section, A/D circuit, and displays were added to the breadboard in order and each was checked out before proceeding further.

Digital Thermometer Operational Tests
The operational performance of the digital thermometer was tested by providing a simulated RTD input to the circuit and monitoring the displayed temperature value. The RTD input was simulated to the circuit by a precision resistance decade

box. At an ambient room temperature of approximately 25°C and nominal AC line conditions (115 V AC), the RTD resistance's for 0°F to 200°F were applied to the digital thermometer circuit on 20°F increments. The results of these tests are shown in Figure 10–23.

The graph shows a slight bending of the overall signal. This is due to the fact that the RTD is not perfectly linear. The graph shown represents the actual RTD curve accurately. All data taken were within the required specifications.

Test Point	Simulated Temperature (°F)	Actual Temperature Reading (°F)
1	0	0.10
2	20	20.20
3	40	40.25
4	60	60.30
5	80	80.35
6	100	100.40
7	120	120.38
8	140	140.35
9	160	160.27
10	180	180.23
11	200	200.15

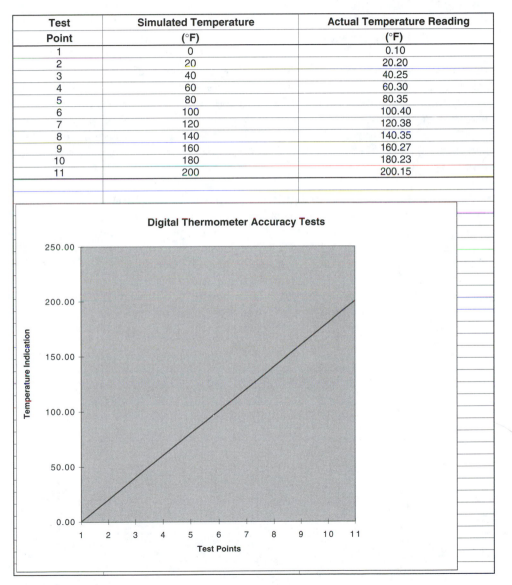

▲ **FIGURE 10–23**
Digital thermometer operational test data

Digital Thermometer Ambient Temperature Tests

This testing was completed by placing the breadboard circuit in an environmental test chamber. The breadboard was powered with a variable AC power supply located outside the test chamber along with the RTD simulator. The RTD simulator was set to the RTD resistance that equated to 100°F and the power supply was adjusted to supply the nominal 115 V AC value. Testing commenced at a temperature of 25°C, and the circuit was allowed to settle for one hour. The indicated value on the display of the digital thermometer was recorded, and the test chamber setting was increased to 35°C. After settling for one hour, the new displayed value was recorded. This process was repeated at 45°C and 55°C, and then the test chamber was cycled back down to 45°C, 35°C, 25°C, 15°C, 5°C, and 0°C. Then the test chamber was brought back up to room temperature stopping at 5°C, 15°C, and 25°C for measurements. Using this process, two data points were recorded for each ambient temperature, approaching the ambient temperature from each direction. The test was then repeated with the line voltage on the variable AC supply increased to the high line condition of 126.5 V AC. This was done to produce the maximum self-heating within the digital thermometer. The data from this testing is shown in Figure 10–24.

EMI Immunity Testing the Digital Thermometer Breadboard

The digital thermometer was subjected to some basic EMI immunity tests that involved first placing voltage spikes on the power supply to see if any failures or erroneous operation occurred. The next step involved placing the sensor wires in close proximity to a fluctuating electrical field and observing any affect on the operation of the digital thermometer. When the induced voltage spike was increased to levels above 500 V, some short-term changes in the displayed value were observed. Also, when the input signal wires were placed very close to an induced electromagnetic field, short-term changes in the display were again noted. Both of these situations were viewed as acceptable.

Ambient Temperature (°C)	Expected Temperature Indication—(°F)	Actual Temperature Indication—(°F)	Error (± 1°F)	Within Tolerance
25	200	200.2	0.2	Yes
35	200	200.5	0.5	Yes
45	200	200.9	0.9	Yes
55	200	201.1	1.1	No
45	200	201	1	Yes
35	200	200.8	0.8	Yes
25	200	200.3	0.3	Yes
15	200	199.9	−0.1	Yes
5	200	199.4	−0.6	Yes
0	200	199	−1	Yes
5	200	199.2	−0.8	Yes
15	200	199.6	−0.4	Yes
25	200	199.8	−0.2	Yes

▲ **FIGURE 10–24**

Digital thermometer ambient temperature test results

▶ References

Reis, R. A. 1999. *Electronic Project Design and Fabrication.* Upper Saddle River, NJ: Prentice Hall.

▶ Exercises

10–1   What is the primary difference between how the manufacturers' representative (i.e., the sales rep) and a distributor receive income for sales? What is the difference between their geographical territories?

10–2   When ordering parts for the completion of the prototype, what are the criteria for determining the quantity to be ordered?

10–3   What are the primary differences between an OEM and an end user?

10–4   From a safety perspective, what is the unique difference between 115 V AC and 230 V AC?

10–5   What is a GFCI and how does it work?

10–6   What fact is most important to note about the negative ground clip on an oscilloscope probe?

10–7   Discuss the two distinctly different methods of removing solder from a connection.

10–8   When solder is specified as being 60/40, what does this mean? What is the realizable difference between 60/40 and 63/37 solder?

10–9   What is flux and what is its purpose? If a flux is called "water-soluble flux," does it need to be removed and what is the process of removal?

10–10   What is a cold-solder joint and how is it formed?

10–11   List the four different methods of completing a breadboard and list one primary advantage and disadvantage of each.

10–12   Discuss the importance of using the proper-sized wire stripper in preparing electrical connections.

10–13   While wiring a breadboard schematic, describe the best way to ensure that the wired breadboard will be identical to the actual schematic.

10–14   If you are assembling a printed circuit board, which two drawings and documents are required?

10–15   If you are troubleshooting a printed circuit board, what drawing would you need more than any other drawing? What would be the second and third most needed drawings?

10–16 Two methods of removing DIP integrated circuits are discussed in this chapter. What was the basis for using one method over another?

10–17 List the three general types of tests that should be performed on the breadboard circuit.

10–18 After the breadboard has been completed, a cost analysis should be performed to determine the status of the design relative to the cost goals. List the alternatives that can be considered if the breadboard design is already over the identified cost goal.

11 ▶ Step Four: Execution (Prototype Development)

▶ Introduction

The breadboard is complete and has been tested as much as possible. The next stage of the project develops the design into what is called a *prototype*. A prototype is the stage after breadboarding, when all the circuits and modules are connected together to form a reliable working system. The prototype should look, act, feel, and assemble like the eventual finished product. The prototype may consist of very reliable breadboards, but it almost always consists of printed circuit boards.

The prototype phase of the project should not begin unless all major issues or problems have been addressed adequately. Most unsuccessful projects fail because they enter the prototype stage prematurely with problems that are very hard to fix later. During the breadboard stage, the design is fluid and changeable. After the prototype stage begins, however, decisions about many design issues are made that solidify a number of aspects of the design.

In this chapter we will discuss the development, testing, and documentation of the prototype:

- ▶ Documentation accuracy
- ▶ Printed circuit board development
- ▶ Prototype development
- ▶ Prototype documentation
- ▶ Prototype testing

11–1 ▶ Documentation Accuracy

At this point in the project, there are two documents that define the electrical design of the project: the circuit schematic and the parts list. The accuracy of these documents is extremely important as we move into the project's next phase, because all of the drawings and documents developed next are based on them. Now is a good time to check over and update the schematic diagram and the parts list.

Design Master Drawings

It is good practice for whomever has design responsibility for an assembly, module, or system to maintain one set of all the drawings for the design, designated as "Design Master Drawings." The *design master drawings*—or simply "design masters"—are hard copies of the latest revision of all documents that define the design. They should be labeled in red as design masters with the date and initials of the responsible design engineer. The purpose of design masters is to accumulate all of the modifications to be made to the drawings in one well-assembled document as errors are found and problems resolved. Without drawings designated as design masters, you will soon find yourself with many copies, notes, and scraps of paper listing important changes that may or may not be passed on when final changes are made to the design drawings. Utilizing design master drawings will improve the accuracy of design documents and greatly improve the efficiency of the project engineer. At the beginning of the prototype phase, the schematic and the parts list are checked and modified to correct any errors or implement changes made as part of the breadboard phase.

11–2 ▶ Printed Circuit Board Types

The invention of the printed circuit board was a significant factor in the development and growth of the electronics industry. Its development has been just as important as the transistor and the integrated circuit. Over the years the methods utilized to lay out and manufacture printed circuit boards have undergone significant change as large-scale integration and computer technology have been applied to the process. To develop a background for circuit board technology, the discussion begins with manual layout and taping methods and then proceeds to the current computer software process.

The likelihood of laying out a printed circuit board depends on the career path followed. Students seeking a technology degree are more likely to complete a board layout than students seeking an engineering degree. In any case there is a need to understand the process, as the printed circuit board is a key component in any electronic design. The design engineer will review the printed circuit board layout from an electrical perspective: grounding, component location, bypass capacitors, and the like. Before discussing the layout methods, let us review the makeup of the printed circuit board and the different types currently available.

Laminates

Every printed circuit board starts out with what is called the *laminate:* the copper-clad material that is etched, plated, and drilled to complete the bare circuit board assembly. The laminate consists of a base material that has a copper foil applied to one or both sides. The typical base materials consist of paper, glass cloth, or glass mat combined with phenolic or epoxy resins. A base material and copper foil are combined to form a particular grade of laminate. Laminate grades have been established by NEMA (National Electrical Manufacturer's Association) and military specifications. NEMA type G-10 (MIL Spec type GE) is a very popular general-use grade of laminate made from glass cloth bonded with an epoxy resin. Other laminate grades include the NEMA prefix FR that stands for flame retardant. The FR grades are favored for use on printed circuit boards used in products that must meet approval agency flame retardant requirements. NEMA type FR 4 is also very popular and is similar to the G-10 material with the addition of a flame-retardant epoxy. The mechanical strength, ease of machining, adhesion of the copper foil material, and the ability to withstand changing environmental conditions determine the ultimate quality of a laminate material. Laminates are available with single-sided or dual-sided copper foils, and very thin laminates can be layered together to form multilayer circuit boards.

Printed Circuit Board Manufacturing Process

Traditionally, the printed circuit boards have been fabricated from what is called a *subtraction process*, the removal of copper from a laminate material. Processes that add copper runs have been developed and are becoming increasingly popular. The subtraction process is still predominate and is the process described here. Figure 11–1 shows a flowchart of the process. The basic subtraction process fabrication of a printed circuit board involves the following:

1. The laminate is drilled and the holes are plated with a copper flash process.

2. The laminate material is thoroughly cleaned and dried.

3. The laminate is covered with a thin adhesive-backed sheet of material called a *photoresist* that is applied to the copper foil on the laminate. This photoresist material can be altered with the application of light to resist certain solvents called *developers.*

4. The photoresist material is exposed to fluorescent lights through the printed circuit board negative of the artwork to be etched. Where the negative allows the light to pass through, the photoresist is altered such that a developer solvent removes it from the laminate. Where the negative blocks the light, the photoresist material will not be sensitive to the developing solvent and remains on the laminate.

5. The laminate, with the developed photoresist material attached, is developed by placement in a developing solvent that strips away the sensitized photoresist material.

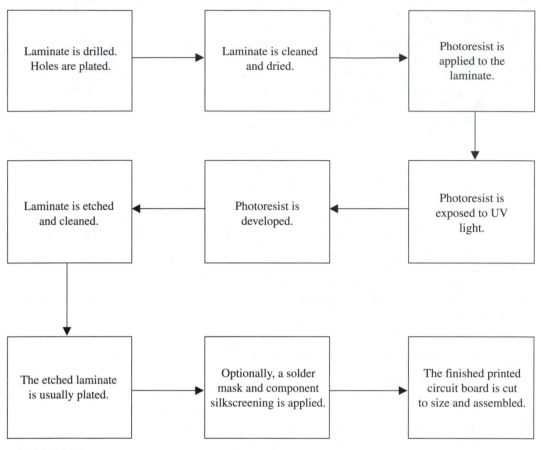

▲ **FIGURE 11-1**
Printed circuit board fabrication process flowchart

6. The result is the initial laminate with the remaining photoresist material covering the laminate where the copper runs are desired. The laminate is placed in a copper etching acid solution that strips away the exposed copper areas.

Single-Sided Printed Circuit Boards

Single-sided printed circuit boards are the simplest variety, because the copper runs exist on only one side of the laminate. For through-hole technology components, the side of the board with the copper runs is called the copper side. The other side is designated the component side. For surface-mount technology circuit boards, the components and the copper runs are present on the same side. At the present time, single-sided boards are used only for very simple circuits not required to be a minimal size. Because all the connections must be made on a single side, the layout is more complicated, and more space is needed for the copper runs. Also, the holes that are drilled into the printed circuit for mounting through-hole components do not need to be plated through as on double-sided boards. This is because the connection of the component to the copper run can be assured by the solder connection. A single-sided printed circuit board will be cheaper but physically larger than the double-sided alternative.

Double-Sided Printed Circuit Boards

The double-sided printed circuit board has copper runs on both sides of the board laminate. The circuit connections can be made much more easily, because there are two surfaces on which to make them. Double-sided boards also provide the ability to transfer a connecting run from one side of the circuit board to the other. This is accomplished through a hole in the printed circuit board called a *via* or *feed-through*. The via is a hole in the printed circuit board, the only purpose of which is to transfer the connecting run from side to side. Consequently, components are not mounted in the via holes. The via hole must be copper plated to ensure connection between both sides of the board. Plating of the holes also improves the solder connection made when the components are installed into the board. Copper runs can be connected to either side of a component hole so circuit connections can be passed from one side of the board to the other through them as well. The double-sided printed circuit board results in a smaller, denser circuit board.

Multilayer Circuit Boards

Multilayer circuit boards have additional thin laminates that provide circuit foils that can make additional circuit connections. Multilayer boards are utilized when complex circuit connections are required in a minimal space. Each layer is aligned with and sandwiched between the outer layers of the circuit board. Plated-through via holes and component holes are used to transfer connections from one layer to another. If no connection is made at a particular layer, then the via or component hole is isolated from making a connection at that layer. The most typical application of multilayer boards today is the provision of two inner layers to make all the power supply connections. In this case one of the layers becomes power supply ground, and the other makes all the positive power supply voltage connections. This is an optimum situation for noise immunity, as the power supply ground layer is one large ground plane that serves as a ground shield for the entire board. Additionally, with the positive power supply voltage on one side and the ground layer on the other, the inner laminate material acts as a dielectric. With a dielectric between them, the ground plane and the positive supply circuit runs act like a distributed capacitor, providing very noise-free power to the entire circuit.

The four-layer variety of the multilayer printed circuit board is the most common circuit board style currently being utilized. This type of circuit board has two inner layers, one that provides power supply ground and another that supplies the nominal 5 V for most digital systems. The outer component and copper sides of the circuit board make all the other interconnections. These boards cost more than double-sided boards but provide exceptional noise immunity and better circuit densities. The increased utilization of plastic enclosures in the electronics industry has promoted the need for a shielding ground plane in place of the metal enclosures that had once accomplished this. The increased use of plastic electronic enclosures requires that the circuit boards themselves contain some shielding.

11–3 ▶ Printed Circuit Board— General Design Considerations

In sections 11–4 and 11–5 ahead, the specific method of circuit board layout for manual or computer methods is described. In this section the general considerations for printed circuit board design are discussed as they are applied to both methods.

Circuit Board Design Considerations

The following is a list of design considerations for circuit board design:

1. Connecting Runs

a. In general, make all circuit runs (see Figure 11–2 for a summary) as short and as thick as reasonably possible while providing as much space between them as possible.

b. Provide clearance on all sides of a printed circuit board. An area of about 3.8 mm to 10 mm wide (0.15″ to 0.40″) is recommended. Components and

1. Make all circuit runs as short and as thick as reasonable.

2. Do not locate components within 0.15″ to 0.40″ from the edge of the circuit board.

3. Make circuit pads larger than the connecting runs.

4. Insert zero-potential runs between signal runs where crosstalk can occur.

5. Keep signal and return runs adjacent as they would be in a cable.

▲ **FIGURE 11–2**
Printed circuit board circuit runs—layout summary

circuit runs should not be located in these areas. These clearance areas are needed to avoid interference with board handling fixtures, guidance rails, and alignment tools.

c. All circuit pads should be larger than the connecting run to prevent the flow of solder away from the solder connection.

d. Keep in mind the possibility of crosstalk between long adjacent runs. Insert a 0 V potential run between them to minimize the potential for crosstalk.

e. Try to keep signal and return runs together as they would be in a cable. The equal currents flowing in opposite directions will minimize the inductive effects.

2. Circuit Board Perspective

When developing circuit board artworks, negatives, and silk screens and while fabricating prototype boards, be aware of the proper perspective for the situation. (Is the view from the bottom or top side of the board?)

3. Grounding and Shielding

The information presented in Chapter 8 on grounding and shielding should be applied to the circuit board layout.

a. *Ground plane*—A ground shield should be utilized wherever possible. This means that the circuit board ground conductor should contain as much copper area as possible. The ideal situation is the one described earlier for multilayer boards, where one entire layer of the board is allocated to ground plane. In single- or double-sided circuit boards this concept is implemented by making large portions of the circuit board available as a solid copper ground plane. Since most through-hole circuit boards are wave soldered, the ground plane is configured in a crosshatch scheme as shown in Figure 11–3. This is done because large areas of solid copper absorb heat from the stream of wave solder, resulting in an uneven and lower solder temperature in that area of the board. This degrades the quality of the solder connection. The circuit board with a large exposed ground plane also has a tendency to warp because of the uneven absorption of heat caused by the large mass of exposed copper. The crosshatch ground plane resolves both of these issues.

b. *Ground distribution*—Ground should be separated and grouped into the following types: low-level analog signal grounds, low-level digital signal grounds, higher level switched circuit grounds, and a chassis ground (enclosure or card-cage ground). These ground types should be connected at only one point. Within each type of ground, ground connections should be distributed to subgroupings of the appropriate circuitry in parallel, as shown in Figure 11–4 and discussed in Example 11–1. This is done to preclude the

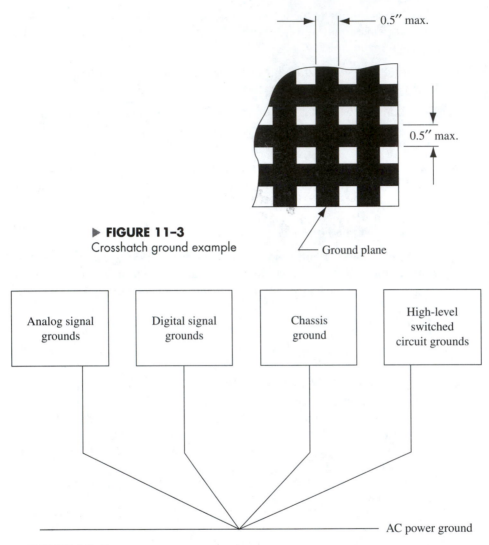

▶ **FIGURE 11–3**
Crosshatch ground example

▲ **FIGURE 11–4**
Ground distribution system

effect of one long ground loop, where the current return flow from one area of a circuit can affect the operation of another circuit.

c. *Guard rings*—These are used with operational amplifier circuits to minimize leakage current that can occur at the input terminals to the op amp. The potential error, induced by this leakage current, increases dramatically when the signal source impedance is large. Guard rings are copper traces placed along each printed circuit board surface where the input terminals make contact. On a double-sided printed circuit board with a through-hole package op amp, the guard ring should be placed on both sides of the board.

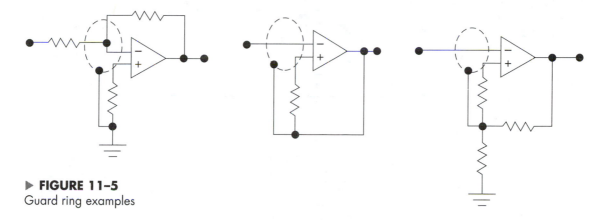

▶ **FIGURE 11–5**
Guard ring examples

The guard ring should circle around the sensitive op amp inputs and be connected to the same potential as the positive and negative inputs. See Figure 11–5 for examples of guard rings and their connections for various op amp circuits.

Example 11–1

A circuit schematic shows 15 digital integrated circuits to be laid out on a double-sided printed circuit board. The power supply ground connections are being planned for these integrated circuits. The problem is to determine a practical way of making these ground connections that will minimize the creation of a ground loop. According to the circuit board design considerations just discussed, ground connections of a similar type of circuitry should be grouped together and a separate ground connection should be made to subgroupings of that type of circuitry. In this example, all of the circuitry is low-level digital circuitry so these circuit grounds should all be kept together. One extreme approach is to have 15 individual parallel runs connecting to the common digital ground. This would require a large amount of circuit board area and make other connections very difficult to make. The other extreme is to provide one continuous ground connection to all the integrated circuits. A ground loop results when components, attached to the end of the loop, are at a higher ground potential than those at the beginning. Also, ground current flowing from the components at the end of the loop affects the ground level of those components at the beginning. The most practical solution is to break the 15 integrated circuits into three groups of five and provide parallel ground connections to each group of five, as shown in Figure 11–6.

4. Decoupling Capacitors

As power is distributed throughout a printed circuit board, the circuit runs exhibit some amount of inductive reactance. Inductive reactance will oppose a change in the current flow through the runs. At lower frequencies of operation, the inductive reactance is insignificant because

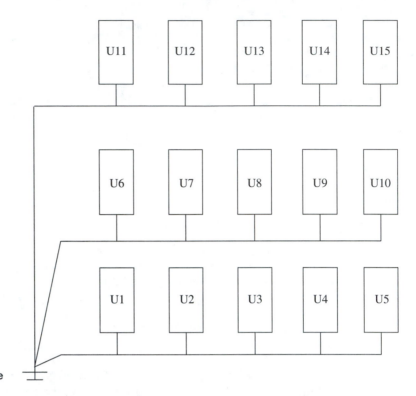

▶ **FIGURE 11–6**
Ground distribution example

the switched components have enough time to complete their switch transition. They can accommodate a delay in the availability of current caused by the inductive reactance. However, in high-frequency applications, such as typical digital circuits, the inductive reactance of the runs is more critical. It provides a delay of the additional current needed for the device to make the switch transition in time for the circuit to function properly. In these cases, a decoupling capacitor is used to counteract or decouple the power supply run from the effect of the inductive reactance. A 0.1 µF ceramic disc capacitor is typically used for this purpose. The 0.1 µF capacitor stores enough charge in reserve to supply the requirements of the switched component and enable it to switch in the required time. As a rule of thumb, one decoupling capacitor is used for every two integrated circuits. It is important to locate the capacitor as close to the component as possible with thick, short runs. Locating any decoupling capacitor on thin runs away from the component completely defeats its purpose (see Figure 11–7).

5. Component Placement and Orientation Guidelines

The orientation and placement of components are important parts of any circuit board layout and varies depending on the type of package technology (through-hole technology or SMT) and the soldering process.

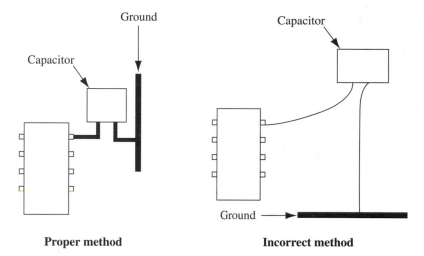

▶ FIGURE 11–7
Decoupling capacitor bad
example

Proper method   **Incorrect method**

a. *Through-hole guidelines*—Through-hole components are usually either hand soldered or wave soldered. Wave soldering is the method most often used in a manufacturing environment, and it is the process in which component orientation can become important. Wave soldering is an automatic method of soldering in which liquid solder is pumped through a spout continuously to form a well-defined wave. The solder temperature can be tightly controlled over the surface of the wave. The circuit board to be soldered is passed over the wave and all the solder points on that side of the board are soldered. The advantages of wave soldering are as follows:

   1. Short solder times.
   2. Reduced temperature distortion of the circuit board. This is because only a portion of the board is exposed to the wave at any time.
   3. There is a continuous flow of fresh solder returned to the wave. Any flux or other residue is filtered out of the process within the solder flow loop.

   If the circuit board is wave soldered, the orientation of the board as it flows through the wave should be determined. In determining which direction to pass the circuit board through the wave-solder machine, the primary concern is the maximum width circuit board that the wave-solder machine can process. If both dimensions of the circuit board are less than the maximum width for the solder machine, then the board can be passed through the wave in either orientation. If the machine can accommodate only one side of the board, then the longer dimension of the board must be parallel with the direction of flow, as shown in Figure 11–8.

   If neither dimension of the board fits through the wave, then either a larger wave-solder machine must be used, or the mechanical design must be redone to reduce the board size. This is not the type of information that one wants to learn about during the initial production run.

Wave solder flow direction ⟶ ▶ ▶ ▶ ▶ ▶ ▶ ▶ ▶

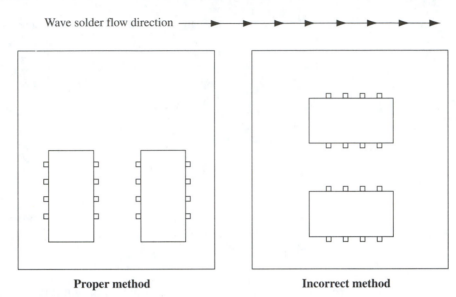

Proper method   **Incorrect method**

▲ **FIGURE 11–8**
Wave solder flow direction

When placing DIP through-hole packages, the orientation of the main body of the integrated circuits should be perpendicular to the intended flow of the wave solder. The connecting runs on the component side of the board should be in the direction of the wave flow. This is to preclude solder spilling over from one DIP connection to the next. The connecting runs on the component side of the board should be run perpendicular to the runs on the solder side of the board.

b. *SMT guidelines*—SMT circuit boards require a lot more planning in their design. The land patterns or circuit pad sizes are critical for the completion of a reliable solder joint. Accordingly, the circuit pads used, either computer generated or stick-on circuit "puppets," should be of the proper sizes as recommended by the Institute for Interconnecting and Packaging Electronic Circuits (IPC) in their standard IPC-SM-872. The clearances between the components, which provide for all manufacturing aspects of the circuit board, are shown in Figure 11–9.

SMT circuit boards can be soldered automatically in two ways: wave or paste-reflow soldering. Because SMT technology has a completely different set of processing issues, the component placement criteria are different. The square SMT packages present a special problem for wave soldering because leads are located on all sides of the package. It is impossible to place these components so that all pads are in the direction of the wave flow. For this reason a *solder mask* is recommended with most SMT circuits. A solder mask is a coat of epoxy resin covering the entire printed circuit board except where the pads will require soldering. Solder

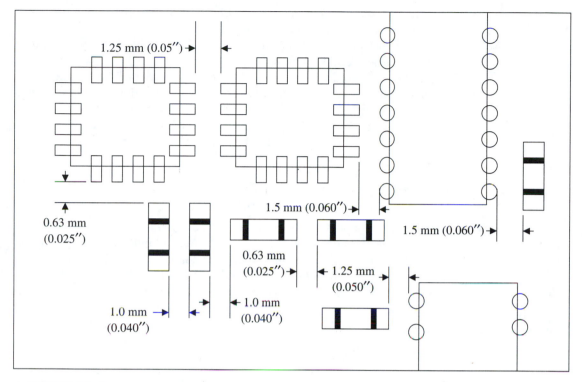

▲ **FIGURE 11–9**
SMT land patterns and clearances

masks are used to eliminate solder bridging between adjacent conductors during wave soldering. Solder masks are currently used with most printed circuit boards, through-hole or SMT type, as circuit run densities have increased dramatically.

Otherwise, the placement of SMT components should be as follows:

1. All passive components should be mounted parallel to each other.
2. All integrated circuit packages should be mounted parallel to each other.
3. The longer axis of any integrated circuit package and that of passive components should be perpendicular to each other.
4. The longer axis of the passive components should be perpendicular to the direction of travel of the board through a wave-solder machine.

If paste-reflow soldering is to be utilized, the placement of SMT components is not that critical.

6. *Tooling Holes*

To provide for mechanical alignment on any parts placement or testing apparatus, a minimum of two (preferably three) unplated holes should be located in the corners of the circuit board. The actual hole diameters depend on the actual equipment being utilized, but they are generally between 2.5 mm and 3.8 mm (0.10″ to 0.15″).

For SMT circuitry, optic targets are needed in addition to the tooling holes to orient and register the component pads to the center of the device. This is accomplished with *fiducials*, which are optical alignment targets that are silk-screened onto the board. Three fiducials should be placed on a known grid in the corners of the circuit board to form a three-point datum system. Figure 11–10 shows an example of the application of tooling holes and fiducials. This figure shows two different types and sizes of fiducials that can be used.

7. Consider Circuit Board Testing Requirements and the Need for Test Points

Review preliminary circuit board test plans to determine the circuit points that will require access during manufacturing and field-testing. Each point must be accessible for testing and provide a means for attaching meter and oscilloscope test probes. Test points are available as a standard component that can be soldered into the board. Test points are more critical on

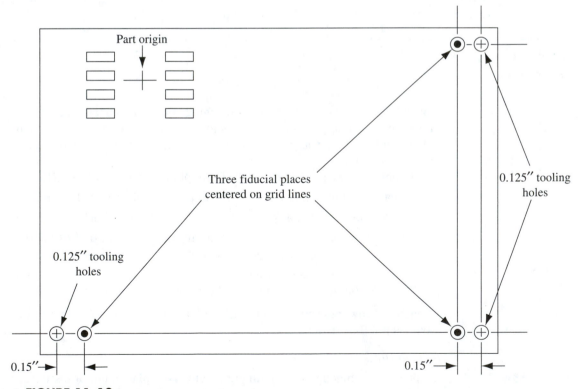

▲ **FIGURE 11–10**
SMT fiducials and tooling-hole examples

SMT boards, because it is almost impossible to clip probes onto SMT components. With through-hole circuits, test leads can often be attached to component leads without the use of a purchased, assembled test point. If test points are not planned for, it is difficult for manufacturing and field service to test boards and will result in many artwork changes later in the project.

8. Large or High-Power Components

Large or high-power components also require specific attention to design details. It is important to ensure that large components are affixed to the circuit board with the appropriate mechanical strength to support their weight. Most large components designed for circuit board mounting have some means of mechanical mounting incorporated into their design other than the solder connections. Be sure to utilize the manufacturer's recommended circuit board mounting scheme. Do not rely on just the solder connections to mechanically hold a large component onto a printed circuit board.

Components that utilize higher levels of power should utilize heat sinks when necessary. Be sure to determine this before laying out the circuit patterns by providing for the mechanical mounting of any heat sinks. When a component does not require a heat sink, but generates more than 1 W of power (i.e., a 2 W power resistor), be sure to consider its location relative to other temperature-sensitive components. Also, mount the device up off of the surface of the printed circuit board so that heat can radiate evenly in all directions.

11–4 ▶ Printed Circuit Board Layout—Manual

The manual circuit board layout method is one that has been used since the invention of the printed circuit board. The manual layout method is seldom used in industry today, since the process has been, for the most part, replaced with computer software layout programs. However, there is a benefit to understanding this process before discussing the computer methods. Manual circuit board layout involves the use of layout templates to draw the components into position on a layout drawing that is usually a semitransparent medium of some kind. A mylar material is recommended, with one side having a matte finish for the layout drawing, because it erases cleanly and produces a strong, clear pencil image. The layout and the eventual artwork should be done over a precision grid background so that all the holes in the board can be located on the grid. A typical grid background is one with 1 mm spacing. The layout and the eventual taping are usually done at a scale of two times the actual size of the components, although standard 1X, 2X, and 4X templates and circuit puppets are available. A 4X scale, for example, would be used on small, dense circuits. *Circuit puppets* are adhesive-backed pads that conform to the various circuit components, as shown in Figure 11–11. The layout is usually done as a positive. The dark areas define where the copper runs will be. Negative circuit puppets are available if one wishes to develop an artwork directly as a negative.

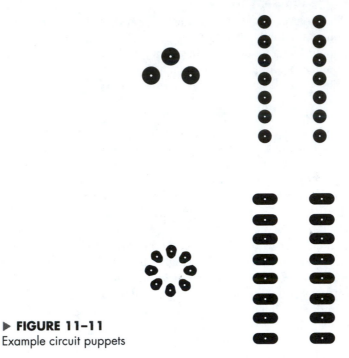

► **FIGURE 11–11**
Example circuit puppets

The Layout Drawing

Developing the layout drawing is the most difficult and critical part of the layout process. Its development will ultimately determine how well the circuit board functions and how easily it will be manufactured, tested, and serviced. Completing the layout drawing involves selecting the location of all components on the board and defining the pattern of copper foil connections that will complete all the required connections. To complete the layout drawing, use the following process:

1. *Circuit board mechanical design*—Determine the desired type, size, and mounting configuration of the printed circuit board. This includes the length, width, thickness, and type (number of sides and layers). The choices for mounting usually include using mounting pads or standoffs, card cages, or other specialized hardware that affix to the board edge. These decisions must be made in conjunction with the overall mechanical design for the project as well as the selection of interconnection methods that are discussed below. Be sure the equipment used to process the board in manufacturing can accommodate the circuit board size selected.

2. *Board interconnection*—Determine the method and optimum location of all connections to the printed circuit board. The actual connecting method must be selected and all relevant information about the connection must be determined.

3. *Component locations*—At this point the mechanical outline of the board and the location and space requirements for its mounting and interconnection have been defined. Next, select the ideal location of the components on the board while considering all of the following:

   a. Keep connections as short as possible. Keep components that have connections between them close together.

   b. Keep the functional blocks of the circuit together.

   c. Maintain an orderly flow of any signal from input to output.

   d. Make the thickness of the runs appropriate for the signal that they carry. Give special consideration to power supply and ground runs. These should always be as direct and as thick as possible. High current runs should be very thick.

   e. At the same time, consider the voltages present on adjacent runs and try to keep the runs as far apart as possible. Approval agencies often have spacing requirements for runs carrying voltages in excess of 30 V.

   f. Leave room for components that will dissipate a lot of power and consider the need for a heat sink.

   g. Attempt to keep component configurations as consistent as possible (i.e., have all integrated circuits going in the same direction, all polarized devices in the same orientation, resistors adjacent and in line, and so on).

   h. Provide access to testing for key circuit areas. Consider the eventual testing method that will be employed and the need for test points.

   i. Consider access to any adjustments or the need to remove any components from the board or the need to remove the board from the assembly.

   j. The board should present a professional and high-quality appearance.

4. *Layout copper runs*—The layout drawing should now include the desired location of all the components in addition to the complete mechanical profile of the board with mounting and interconnection hardware. The process of actually determining the connecting runs is the most difficult part of this process. The process usually takes a number of cycles, so it is recommended to place another layer of matte-finished mylar over the layout drawing that has been completed up to this point. This is done so as not to waste the efforts completed thus far, which occurs when the layout drawing must be redrawn after the first attempt to lay out the connections is abandoned for a better way. One way is to lay out the connections on the top layer of mylar. As the process evolves, simply replace the top layer and start over. The process is an iterative one that involves learning the best way to make the connections for a given circuit. The more experienced designer requires fewer iterations. At this stage, using a light table will make it easier to see through the different levels of mylar.

Draw in the connections with a simple line that will represent the actual circuit run. For double-sided boards, use a red pencil for one side of the board and a blue pencil for the other. For the most difficult connections, utilize a via hole to transfer a connecting run from one side or layer of the board to another. Via holes should not be overused. Each via requires that another hole be drilled in the board adding a small cost to the board. Also, via hole connections are slightly less reliable than a solid copper run. It is best to make all power supply connections first and then proceed making the other connections.

5. *Layout design check*—After a number of attempts at completing the connecting copper runs, a successful layout drawing is complete. Before starting the taping process, it is important that whoever has design responsibility for the circuit board check all aspects of the layout design. This includes the mechanical size and shape, manufacturing and testing issues, and the location length and size of all circuit connections. This should be done before the taping process is started.

6. *The pad master*—After the layout design is checked, the layout is ready for taping. This involves the use of the circuit puppets, pads, and artwork tape to implement the pencil layout as a taped positive. The taping should be completed on what is called *clear taping film*. There are usually at least two layers (except for single-sided boards, where there is one) representing the layout so there must be a way to register or locate the sheets of taping film on top of each other. This is usually done with what is called a *pin bar*, where the pins in the pin bar line up with the taping film that is prepunched to the pin size and spacing on the pin bar. To align the different layers when not on the light table, crosshair puppets are added to each layer of the artwork to allow proper and accurate alignment.

   The taping is started for circuit boards with more than one side by generating what is called a *pad master*. The pad master is simply one layer of taping film that has pads marking the location of each hole that will be on the board. Generate the pad master by placing one layer of taping film over the layout and placing pads and puppets as appropriate over all of the via holes and mounting pads. Figure 11–12 shows an example of a pad master.

7. *Artwork taping*—The artwork is completed by placing another layer of taping film over the pad master and layout drawing. This layer of taping film represents the copper runs for one side of the circuit board. The copper runs are completed on the taping film by using special artwork tape available in many precision widths. Be sure that the tape overlaps the connecting areas; do not stretch the tape as it is applied. A layer of connecting runs is completed for each side of the board. A completed double-sided circuit board artwork will include a pad master and one layer of connecting runs each for the top and the bottom. A four-layer, multilayer board artwork consists of one pad master and four layers of connecting

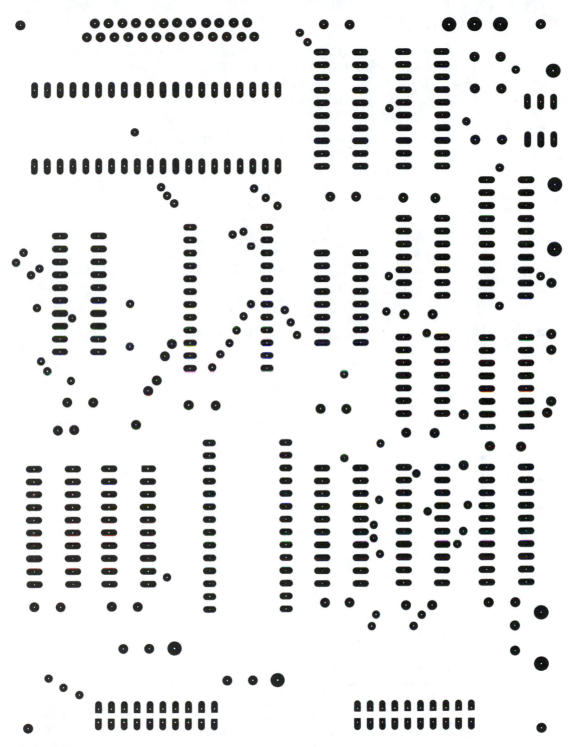

▲ **FIGURE 11–12**
Example pad master

run taping films. Figure 11–13 shows one layer of circuitry for the same board pad master shown in Figure 11–12. Figure 11–14 shows Figure 11–13 aligned with the Figure 11–12. Figure 11–14 shows all copper areas on the top side of the circuit board. It represents the complete artwork for that side of the board.

8. *Artwork checking*—The artwork is now complete. If the board design is complex, there is a good chance that there are errors in the artwork. All it takes is one missing bit of tape or one hole that was forgotten. The detail included in a complex board is incredible. The possibility of an error increases proportionally with complexity. It is best to have two people check the artwork: one who reviews the schematic and another who checks the artwork. The board designer, for example, should review the schematic, while a competent person who is not the designer checks the artwork. This is because the board designer has spent many hours staring at this same artwork, and he or she will have difficulty getting the objective distance needed to see any errors in it. The person reviewing the schematic calls out each connection while the artwork checker verifies the connection. The schematic reviewer highlights each run as it is checked and continues until all of the runs have been verified.

9. *Artwork photography*—The completed and verified artwork is now ready to be photographed for reduction and developing a negative. The artwork is sent to a company that specializes in precision photography and reduction. The pad master is combined with the various layers, reduced, and photographed. This results in a 1:1 negative of each side or layer of the board. The negative and a document called a *fabrication* drawing or a "drill code" are sent to a circuit board manufacturer for fabrication. A *drill code* defines the mechanical aspects (i.e., size, laminate, plating, and hole sizes) of the circuit board.

11–5 ▶ Printed Circuit Board Layout—Computer

There are two computer methods that can generate a circuit board artwork: custom software designed to lay out circuit boards or CAD software to draw the layout of the board. The latter method is identical to the manual process except that the computer is used as a drawing tool. Circuit board layout software packages are most often used today, so this discussion focuses on them. There are many software layout packages available and currently in use. The discussion that follows is general enough to describe the process but may not be entirely accurate when applied to a specific software package.

As in the manual layout process, the schematic is the source of defining the circuit connections. The computer layout software requires what is called a *schematic capture file* to define the schematic. This is simply the schematic keyed into a schematic capture program, which is in a format compatible with the layout software package. All software layout programs have mating schematic capture

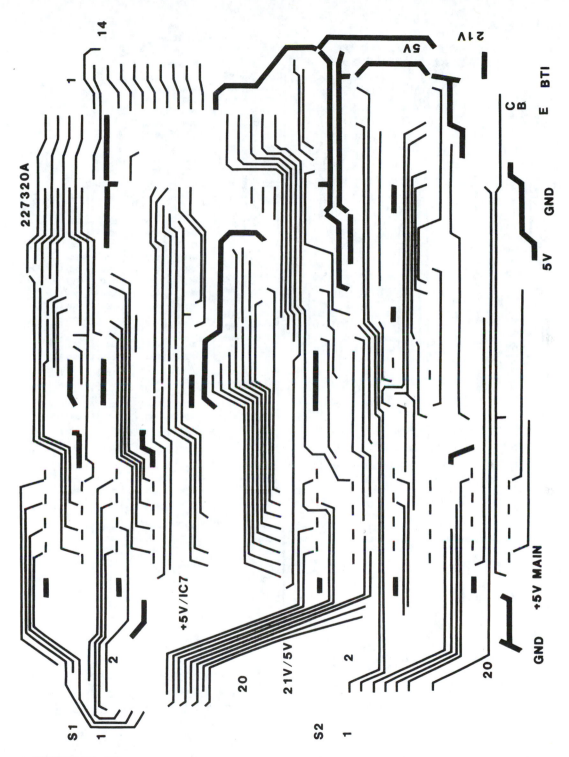

▲ **FIGURE 11–13**
Single-layer taping example

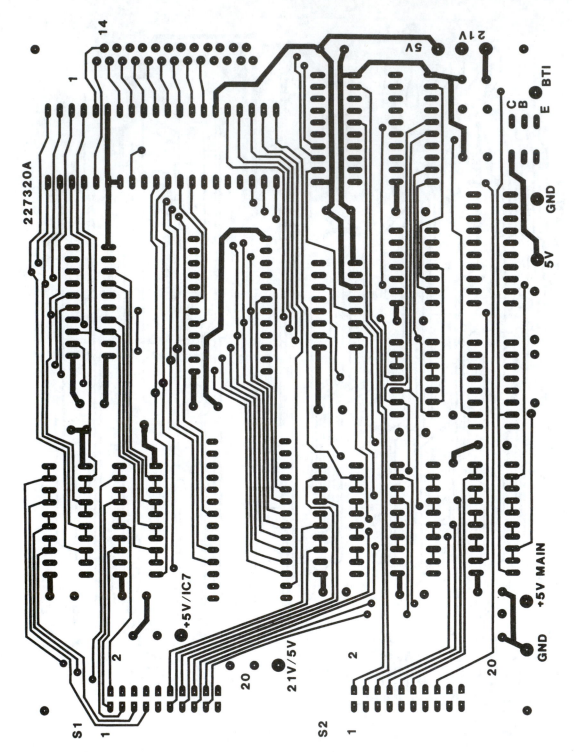

▲ FIGURE 11–14

One-side copper artwork (pad master and copper runs)

software, and many accept other schematic capture files as well. When the schematic is created, it may be necessary to add or create a *device library* for components that may not be present in the standard device libraries in the software. The device library contains all of the pertinent information about the device, the type of component, number of connections and the label for each connection, and the physical package definition. Make sure the device library is complete for all of the components on the schematic, because the package information is needed to begin the computer layout process.

The computer layout process begins with the same two steps as the manual layout process: the layout of the mechanical outline and the interconnections. The result is an outline drawing that is the mechanical design of the board and its interconnections. Next, the components are positioned on the board layout by dragging them into position. The result is identical to the pad master drawing generated in the third step of the manual process.

The circuit designer has to make a choice at this point whether or not to use the layout program's autorouter feature. An *autorouter* is a software routine that determines the path of the connections required by the schematic. It is analogous to the trial-by-error process described earlier in the manual layout method. The computer will attempt many circuit paths for making the connections and will choose the ones that its software intelligence determines are the best. The autorouter is a key part of a software layout package. The quality and quantity of the intelligence included in it determine its performance. The price of the software layout package is usually indicative of the amount of intelligence included and the quality of the autorouter, or, in other words, the higher the price, the better the autorouter.

The decision for using the autorouter is based largely on the experience one has with it. If you have no experience with a particular autorouter, then you can develop some through experimentation. Even the best autorouters are not perfect and cannot possibly possess all of the design criteria for a particular design nor the human insight to make design decisions. There are at least three areas where autorouters produce undesirable results:

1. Power supply and ground runs are not direct or large enough.

2. Connection runs are placed adjacent to areas where there is potential for picking up interference.

3. The utilization of too many via holes.

Many board designers have approached the autorouter decision by connecting up the power supply and ground connections manually. Then they engage the autorouter to make the rest of the connections and modify any undesirable results manually. This is probably the best way to use the autorouter function. There are many times when an autorouter will not be able to make all the connections, and these connections will have to be completed manually.

After completion of the computer layout process, whether it is accomplished manually or with an autorouter, the computer will save files that contain the artwork for each layer of the printed circuit board. The benefit of this process is that the artwork is guaranteed to be accurate and reflect the connections defined in the schematic capture file. If the schematic capture file is correct, then the artwork is as well. There is no need to check and verify that all the connections have been made as was necessary with the manual layout. However, it is still important for the design engineer to review the artwork to make sure that the design will meet the overall requirements for the circuit board.

Another real benefit of software layout programs is that they generate the other drawings needed to fabricate and assemble the circuit board. Photography is unnecessary. The artwork and drill code files are simply sent to a circuit board fabricator on a floppy disc or any computer network—including the Internet. While software layout programs are very powerful and offer many benefits, they are somewhat difficult to learn and use. They are as sophisticated as most CAD drawing software programs. Every attempt is made to make the software as easy to use as possible, but there are simply too many complex features and functions that require knowledge and training to use them properly. These are the types of programs that require consistent use to develop the expertise to use them effectively.

An ideal way to use the electronic design software tools available today, simulators and layout programs, is to have the design engineers perform schematic design on a simulator that is compatible with the layout software being used. When the breadboard and simulation is complete, the file containing the updated schematic is handed over to the board designer. Using the schematic in conjunction with the layout program, the board designer completes the layout and all relevant drawings. The result is a very fast and accurate circuit board development process.

11–6 ▶ Printed Circuit Board Documentation

With the artwork complete, it is time to put together the complete documentation package that will define both the unpopulated and the completely assembled printed circuit board. To define the unpopulated or bare printed circuit board, a document called a *fabrication drawing* or *drill code* must be generated.

The Fabrication Drawing

A fabrication drawing is required for each printed circuit board as it defines the mechanical requirements of the printed circuit board. The fabrication drawing, combined with all of the board artwork, completely defines the unpopulated circuit board. The fabrication drawing specifies the mechanical shape and dimensions of the printed circuit board as well as the size and location of any holes. The fabrication drawing must also point out details such as the board laminate material, tolerances, plating, and other optional requirements, such as solder masks and silk-screening. The following is a detailed list of the issues that should be considered to note on the fabrication drawing:

1. The board laminate material

2. Requirements for a solder mask and silk screen

3. Reference to the artwork number and revision level

4. Reproduction of artwork tolerances on the circuit board

5. Plating specifications

6. All mechanical tolerances

The following is an example of the notes included on a typical fabrication drawing:

1. Material: FR4 glass epoxy, 1/16″ thick with 1 oz copper each side per Mil Spec #MIL-P-13949. All holes to be plated through with a minimum of 0.001″ thick copper. After plating the holes, the surface copper should have a total copper thickness of 2 oz.

2. Apply solder mask and silk screen per artwork drawings #12345678, Revision A.

3. For circuit artworks use drawing #12345678, Revision A.

4. Defects:

   a. Circuit run defects such as holes, nicks, and scratches shall not reduce the conductor width by more than ± 0.002″.

   b. Maximum allowable line reduction shall not exceed 0.005″.

5. Plating: The board should be plated with tin/lead 63/37 ± 5% plating to a total thickness of 0.0004″ to 0.0006″. All solder plated areas to be subject to hot oil solder reflow process.

6. Hole diameter tolerances: +0.005″, –0.002″.

Figure 11–15 is an example fabrication drawing.

Solder Mask

A solder mask is a coat of epoxy resin that covers the entire printed circuit board except where solder connections are to be made. The application of a solder mask is optional. Its purpose is to prevent solder from bridging over adjacent circuit runs and pads during automated wave-solder operations. Increasing circuit densities and the use of wave soldering has made the use of a solder mask very common. On through-hole-only technology boards, the artwork for the solder mask can be simply the pad master drawing discussed earlier. When surface-mount technology is used, the pads for all of the SMT components must be included on the solder mask artwork as well. In any case an artwork should be completed and labeled as "Solder Mask" for the subject fabricated printed circuit board. Most computer circuit board layout packages will generate a solder mask on request after the layout has been completed. The circuit board fabricator will apply the solder mask if called for in the fabrication drawing.

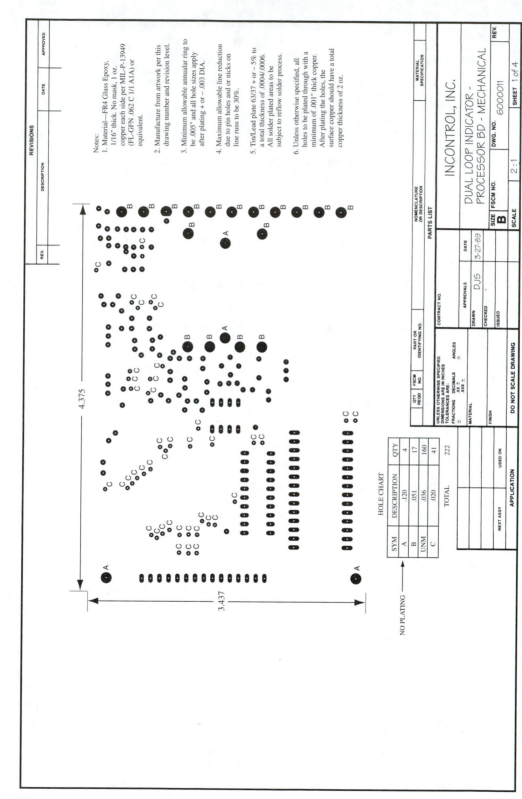

▲ FIGURE 11–15
Example fabrication drawing

Silk-Screening

Silk-screening is another optional process that involves the marking of the board with component and other reference numbers. Silk-screening is an aid to manufacturing and service personnel in the field and provides a professional appearance. If silk-screening is required, a silk-screen artwork must be completed separately that includes the outline and orientation of all component packages, their reference designations, and other pertinent information such as test points and the like. Most computer layout programs will generate a silk-screen drawing automatically after the board layout is complete. The fabrication drawing must specify that the board is to be silk-screened and refer to the drawing number for the silk-screen artwork.

Solder Paste Screen

The solder paste screen is used as a screen for the application of solder paste to surface-mount technology circuit boards where either vapor phase or infrared reflow processes are utilized to make the solder connections. Therefore a solder paste mask is necessary only for surface-mount circuit boards that will be manufactured with one of these reflow solder processes. The solder paste screen is similar to the solder mask, except that it includes only the surface-mount pads or connections on a given side of the circuit board. If through-holes exist on the circuit board, they are either vias or used for mounting through-hole components. In either case solder paste should not be applied to these holes. For that reason all through holes should not be included on the solder paste screen. Solder paste screens will be used by whomever assembles the printed circuit board. These artworks can be generated on request by most computer layout programs. The board fabricator will not have any need for the solder paste screen. The actual silk screen itself will have to be ordered from a silk-screen fabricator per the solder paste screen artwork.

Assembly Drawing

The assembly drawing is a pictorial drawing that will show the physical representation of the circuit board. The assembly drawing will show the location and reference designation of all components that are to be assembled to the board. It will include special notations about the assembly process and refer to the following other related documents:

1. Parts list and bill of material

2. Circuit board specifications

3. Circuit board artworks

4. Circuit board schematic

5. Circuit board assembly and test procedures

Figure 11–21 on page 310 shows an example assembly drawing.

Parts List and Bill of Material

The circuit board parts list is essentially the same as any other parts list or bill of material. It should include the manufacturer and its part number, the company part number, reference designations, and the quantity used per assembly. The manufacturer's part number is the number assigned to the component by its supplier. The purchasing company assigns the company part number to the component. The component will be stocked and tracked by the company part number. Reference designations are numbers that tie a component from the schematic to the parts list and the assembly drawing. For example a resistor is given a reference designation of R1 on a schematic. That same designation (R1) will be used to show the location of the resistor on the assembly drawing and to specify the resistor on the parts list. Figure 11–22 on page 311 shows an example Parts List Document.

11–7 ▶ Prototype Development and Documentation

The completion of the circuit board documentation package signifies the completion of the prototype circuit board design. At this time a purchase order is usually placed for the bare circuit board. Then the board design can be completely assembled and tested. In parallel with the circuit board prototype design, the package and mechanical design of the product—and possibly software development—are proceeding. In order to test the product prototype, both of these activities must be complete, and the resulting mechanical parts and software must be available.

The Mechanical Design

It is just as important for the mechanical design to be thought out, tested, and completed at the mock-up stage before entering the prototype phase. This is the point where the electronic and mechanical designs fit together. The mechanical design must be completed before the printed circuit board can be laid out, because the mechanical design will define the size, mounting, and interconnection of the printed circuit boards. It is very important that the design team coordinate their efforts during the preliminary design phase. At the end of the preliminary design phase, the mechanical design should include a set of preliminary drawings for each mechanical part and a mock-up of the mechanical assembly.

The completion of the mechanical portion of the prototype design involves the definition of every mechanical component included in the product. If a mechanical component is a custom fabricated part (i.e., designed especially for the product), then a drawing must exist that completely defines its design and fabrication. If the part is a standard part and is available from one or more manufacturers, then the primary supplier's part number should be specified as the design part number for the component.

All of the mechanical components and circuit boards must be broken down into assemblies that will be defined with assembly drawings, parts lists, and assembly and test procedures. The method for determining how to break the prototype into assemblies is not always straightforward. The correct subdivision will depend on how the product will be assembled in manufacturing. It is important to involve manufacturing in these decisions as they will have to contend with the results. Involving manufacturing and industrial engineers and other manufacturing personnel in the development of the product's assembly structure (often called the *product structure*) will save making engineering changes later.

To complete the prototype, all the custom parts must be fabricated as called for on the engineering drawing and made available for assembly. All standard mechanical parts must be specified and available as well. The initial product assembly structure should be defined and the parts list for each assembly must be complete. It is desirable, but probably not practical, for all of the product assembly drawings to be complete at this time.

Software

Software can be one of the more difficult tasks to complete on schedule. This is probably due to the mind set inherent with its name. The fact that software is easily changed will entice the project manager to put off its definition and development until later in the project, after the hardware has been defined. What is often missed, however, is the link between the software and hardware design decisions made in the early stages of a project. The most common example is a product requiring some number of keys to perform the various operations required. It is important to completely define these operations and the key sequence that will be used to manipulate them. For example, think of a project in which it has been decided that five keys are needed to perform the required operations without specifying the exact steps of each operation. Later on it is learned that some detail was missed, and either more keys are needed, or a coded depression of two keys simultaneously must be implemented to perform an operation. Either solution is undesirable and represents a serious project problem at this late point. Adding one key will cause major hardware design changes to circuit boards as well as mechanical assemblies and artwork. Using the depression of two keys simultaneously is the easy way out for the designers. But the end user of the product will pay the price: The product will not be as easy to use as it otherwise would have been. This goes back to the importance of defining the problem and the solution plan completely before starting the solution. Because software is thought to be so easily changed, less attention is paid to defining the detail of how the software will function.

In order to test a product that incorporates some level of software, be it a computer system or a microprocessor-based product with embedded software, enough software must be completed in order to test all the key aspects of the prototype hardware. An example follows.

Example 11–2

A team has been assigned to develop the base requirements for prototype software that will allow for sufficient testing of a temperature indicator. This development team is working on a two-channel, microprocessor-based temperature indicator.

The indicator will be programmable to accept one of six different types of inputs on either of its two inputs. The input signal types have different analog signal ranges. The microprocessor will control all of the operations within the indicator and, after power is applied, both input temperatures will be sampled, converted to digital value, and displayed. The operator can program the indicator to perform some mathematical operations, such as averaging the two inputs and displaying that average.

In order to perform tests on the prototype hardware, some level of the software must be complete and functional. The specific level of functionality must be described to the programmers and identified on their project schedules to coincide with the completion of the prototype. The best way to determine the functionality needed is to consider worst-case scenarios. In this product, the worst-case scenarios are played out with the signal range of the inputs and the mathematical manipulation that will be required of the two input samples. It was stated that the indicator must accept two of possibly six different types of inputs. Which types have the smallest and largest signal range? Of the math operations that could be selected, which will be the most difficult for the processor to complete? In this case it is decided that the prototype software for the indicator must be able to sample each temperature input, with one input selected to sample the smallest input signal range type, while the other input will sample the largest. After sampling, the two inputs will be averaged and displayed as this was determined to be the most intensive math calculation for the microprocessor to perform. The brief write-up for the prototype software for the temperature indicator would be as follows:

> Temperature indicator prototype software: The indicator shall sample and display the average value of two inputs. The input signal of channel 1 will be a type T thermocouple (the smallest signal input type) and the input to channel 2 will be a type J thermocouple (the largest signal input type). The resulting input values shall be averaged and the result displayed on the seven-segment LED display.

This write-up is included in the initial specifications as a requirement for the prototype tests.

Prototype Assembly

When all the components for a particular assembly of the product being developed are available, prototype assembly can begin. It is important to use the assembly of the prototype as a test of the documentation that has been developed. It is also a test of the initial process that defines how the product goes together. Remember to use design master drawings to keep track of all changes.

When the prototype bare circuit boards are available, they should be assembled using the existing documentation. The actual procedure for assembling a prototype circuit board was discussed in Chapter 10. The primary goals are to verify that the circuitry is correct and that it will function in the final assembly as called for in the specifications. The ease of manufacture and serviceability will also be verified.

The same goals are applied to all of the other assemblies and the final product. Particular attention should be paid to the following:

1. Performance that varies with the tolerances of components

2. Performance that changes with environmental conditions

3. Significant design issues that affect the product's manufacturability or serviceability

4. Significant issues that affect the customer or end user's ease of use of the final product

There is usually a give-and-take scenario between many of the issues just described. For example, the accuracy performance of a product can usually be improved if additional steps and components are added for the calibration process. These additional steps and components will negatively affect the manufacturability by further complicating the manufacturing process and increasing costs. These are the most important decisions considered on most projects. An error in judgement here can defeat an otherwise brilliant product idea. In these matters, try deciding in favor of higher quality and ease of use unless the associated costs are overwhelming.

11-8 ▶ Prototype Testing

After the prototype is assembled, it must be evaluated and tested by the engineering department using the current revision level of the specifications as a guide. The prototype's operational performance should be compared with the key parameters in the specifications. Environmental requirements should also be tested. Then it must be asked: Is the prototype performing all of the functions with the accuracy required in the specifications? This is sometimes a difficult question to answer. There are often situations concerning the application of a product that cannot be completely simulated in the laboratory. These situations make it difficult to completely evaluate the prototype. Every effort should be made to check out each aspect of the operational requirements of the prototype. Problems and failures will occur sooner or later. It is much better to find them sooner. The prototype testing should be very thorough, and formal test reports should be completed and reviewed.

Performance Testing—Reference Conditions

Review the specifications to determine all the reference conditions. For most products this will include the power supply voltage level, its frequency, and the ambient temperature to which the product will be exposed. A specific input signal type and range may be included as well. Develop the reference conditions in the laboratory and proceed to verify each of the specified accuracy tests for each parameter over the required range of input signal.

After the accuracy tests are complete, review the operational requirements of the specifications. Any operational areas that are functional at this point should be tested. As mentioned earlier, when software is also being developed for a product, there may be operational aspects of the product that are not yet complete at the prototype stage. That is why it is important to identify the product operations that are expected to stress the design of the product the most. These are the operational features that should be included by the programmers in the prototype software for evaluation. All operational features that are complete should be tested. Any problems or discrepancies should be detailed and included in the test report.

Performance Testing—Environmental Conditions

These tests are designed to determine the impact of variations in environmental conditions on the performance of the product. The performance of the product is looked at from a broader perspective and includes areas such as surviving its shipment to the customer (shock and vibration testing). In these situations, performance over the range of specifications may be simply that the product continued to function properly after exposure to the environmental condition.

The environmental conditions that are most often of concern are the input voltage, ambient temperature, humidity, electrical noise immunity and emission, and shock and vibration.

1. Input Voltage Testing

These tests will simply vary any of the power input voltages over the range called for in the specifications. This includes both voltage and frequency. All of the operational specifications will be measured and compared to specification requirements.

Example 11–3

This is a good example of the result of inadequate product testing. An engineering manager at a company began receiving complaints about products being received at a customer's semiconductor plant in Thailand. The customer had tested the company's product using the company's accuracy specifications and found the temperature indication readings inaccurate. The Thai customer had received the product as part of equipment supplied by the company's largest OEM customer. Both the end customer and the OEM were not pleased with the performance of the product, and they wanted answers.

The customer returned samples of products to the company that they found inaccurate. Performance testing completed at the home office found them to be well within accuracy specifications. A service call was made to the customer's location after all other attempts to rectify the problem failed. The field engineer brought with him a product sample of known accuracy as well as the recommended factory-test equipment. The sample product, when installed at the customer's location, showed the same inaccuracies as initially reported on the other units. After verifying the customer's test equipment and methods, it was determined that the inaccuracies the customer was seeing were real. The same products were being tested with the same procedures and equipment with different results at the two locations. The field engineer consulted with engineers at the main plant to analyze the situation as everyone searched for the key difference between the two installations. It did not take long before the input power became suspect as one of the few possible differences. It was determined that the supplied power was 115 V AC, 50 Hz instead of the usual 60 Hz. This turned out to be the key difference between the two installations. The engineers at the home plant then performed accuracy tests on the product with 115 V AC, 50 Hz power applied and found the accuracy results were identical to what had been experienced in Thailand. The product specifications called for the product's operation over power input frequencies of 50 Hz to 60 Hz. Further study revealed that both the prototype and final product tests had not included variations in frequency, only variations in voltage amplitude.

It took three months of testing and software changes and then another trip to Thailand to retrofit all products previously received. The cost of correcting this problem was significant, but the most negative result was the lack of confidence in the company's product, which was felt by both the end user and the OEM customer.

2. Ambient Temperature and Humidity

Ambient temperature and humidity testing require some sort of environmental test chamber to supply the range of ambient conditions called for in the specifications. During these tests, the product to be tested is connected to the nominal input power and input signal. The product is placed in the chamber and exposed to the range in temperature and humidity required. All of the key parameters are monitored and recorded during the test. This can be done visually by taking readings through the chamber access window or by taking measurements of parameter values by extending wires into the chamber that connect to the points being measured.

The change of temperature in the test chamber should be done in a routine way. This is the process that is usually recommended for testing over a range of 0°C to 55°C, 0% to 90% humidity:

Step 1—Maintain the unit under test in the test chamber at 25°C with 0% humidity for one hour. During this initial test run, the humidity will be kept at 0%. Record all parameter measurements.

Step 2—Increase the temperature to 35°C and allow to settle for one hour. Record all parameter measurements.

Step 3—Repeat Step 2 for 45°C and 55°C.

Step 4—Decrease the temperature in steps and allow one hour before taking measurements. Take readings at 45°C, 35°C, 25°C, 15°C, 5°C, and 0°C.

Step 5—Increase the temperature again and take measurements for the points 5°C, 15°C, and 25°C.

Step 6—Repeat Steps 2 through 5 with the humidity set for 50%.

Step 7—Repeat Steps 2 through 5 for 90% humidity.

This procedure provides consistent settling times and approaches all but the end temperatures from both directions. The resulting data for each parameter being monitored includes three readings at the starting temperature, one reading each at the two end temperatures, and two readings for all other points. To evaluate the design for worst-case situations, these tests can be run for different combinations of environmental conditions. For example, high ambient temperature can be combined with high input line voltage. This situation will create the highest temperature inside the enclosure and the maximum stress on components. Low ambient temperature can be combined with low input line voltage to simulate the lowest temperature inside the enclosure. This can be varied depending on the project and the requirements listed in the design specifications.

3. Electrical Noise Immunity

The level of electrical noise testing will depend on the requirements of the specifications and may require some elaborate test equipment and a screen room. A *screen room* is a room that is completely shielded with wire mesh to keep out all electromagnetic fields. It is also called a *zero electromagnetic field test site*. Electrical noise immunity testing will involve testing for electrostatic discharges, radiated electromagnetic fields, induced magnetic fields, and conducted power line noise.

a. *Electrostatic discharge testing*—This test utilizes a special electrostatic discharge (ESD) simulator device with an adjustable high-voltage supply with variable source capacitance and resistance settings. The ESD simulator is adjusted over the range of voltage, resistance, and capacitance called for in the specifications. The ESD simulator voltage is then applied to various points on the outside surface of the equipment. All surfaces should be tested with special emphasis on obvious trouble spots such as keypads and input and output areas. Before testing, the definition of a test failure should be understood. Take the example of a product that is susceptible to a static spike applied to a keypad. After the spike is applied, the unit momentarily locks up (i.e., the unit does not respond to any key depressions), but the software quickly recovers from the error, and there are no other errors or damage to the product. Would this be considered a failure? The answer would depend greatly on the type of equipment being tested. If the device was a critical piece of medical equipment, then any susceptibility at all during the testing would be unacceptable. On the other hand, for a commercial computer game product, this type of recoverable susceptibility may be acceptable.

b. *Immunity to radiated electromagnetic fields*—To complete these tests, the required electromagnetic fields, as listed in the specifications, must be generated. The product must be exposed to these fields and any susceptibility measured. Typical specifications call for electromagnetic fields of 1 V per meter for frequency ranges of 150 kHz through 25 MHz and 1 V per meter for frequencies of 25 MHz to 1 GHz. Transmitters are available that will generate these kinds of fields in a test environment. Ideally, these tests should be performed in a controlled environment, such as a screen room, to prevent any unknown electromagnetic fields from affecting the testing.

c. *Immunity to induced magnetic fields*—Typical specifications require products to be unaffected by induced fields generated by 20 A of 60 Hz current flowing in a wire around the product. To complete these tests, a wire, carrying 20 A of 60 Hz current, should be looped around the product three times while the unit being tested is examined for any sign of susceptibility.

d. *Power line tests*—These specifications typically state that the unit shall not be susceptible to voltage spikes as high as ± 500 V for a duration of 50 ns when applied over 360°. To complete these tests, a voltage spike generator must be used that allows the selection of the desired voltage spike, the duration, and the phase angle at which it is located. A reasonable test would be to generate a 500 V, 50 nanosecond spike for phase angles that start at 0°, for every 6° through 360°.

4. *Electrical Noise Emission*

Electrical noise emission testing involves measuring the amount of electrical noise generated by a product that is emitted back into the power supply and the amount that is radiated into the environment. The test to determine the noise fed back to the power supply is called *conducted EMI testing.*

a. *Conducted EMI testing*—Conducted EMI testing measures that amount of EMI that appears on the power line connected to the equipment under test. The test setup requires an EMI receiver, something called a *line impedance stabilization network*, and preferably a screen room or conductive ground plane. The unit under test is powered and operated under normal conditions while the EMI measurements on the receiver are monitored and recorded. Figure 11–16 shows a complete layout of the conducted EMI test setup.

b. *Radiated EMI testing*—These tests require a dipole antenna, an EMI receiver, a ground plane, and preferably a screen room (Figure 11–17). The unit under test is set up and operated in a normal fashion while the dipole antenna is moved around to various locations while measurements are taken. The measurements are then compared to the requirements called for in the specifications.

5. *Vibration*

Vibration testing should be performed to ensure that the product can withstand any vibrations that might be incurred during normal use or shipment of the product. The specifications should call out the expected worst-case vibration in the form of a vibrating force and the frequency at which that force is applied. Vibration test equipment is usually large, expensive, and available only at companies that do a significant amount of vibration testing.

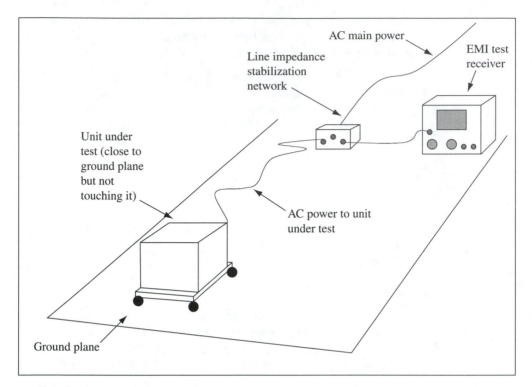

▲ FIGURE 11–16
EMI conducted noise test setup

An outside testing laboratory can complete the testing if the equipment is un-available. These tests involve attaching the unit to be tested to the vibration tester and adjusting the tester to induce the force levels and frequencies of vibrations defined in the specification. Usually the unit under test is not op-erational during the vibration tests; however, designs utilizing military-type specifications will require that. Most often, failure of vibration tests will be seen after the tests are complete, when the operational tests are redone to verify that the unit successfully passed.

6. Shock

Mechanical shock test equipment is similar in nature to the vibration testers: large and expensive. Most often the worst-case shock to which a product is exposed is during shipment. For that reason the specifications often used for these tests are the appropriate Interstate Commerce Commission (ICC) shipping specifications for the product size and weight that is being shipped. The companies that supply packaging materials are intimately familiar with these specifications. They usually have the equipment to perform these tests and do so as part of their package design and verification process. The specifications for mechanical shock are expressed in Gs.

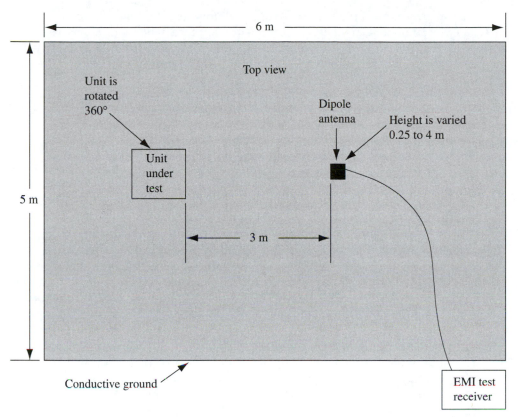

▲ FIGURE 11-17
Radiated EMI test setup and sample limits

If access to a shock testing device is available, then the unit to be tested should be attached to the shock tester while the specified forces are applied in each physical plane. Like vibration testing, it is important to verify the status of the unit to be tested before the shock tests are initiated. This is because the determination of the device's ability to withstand the mechanical shock applied will be made by its status after testing.

11-9 ▶ Cost Rollup

At each stage of the project, additional and more accurate cost information becomes available. At this point all of the parts for assembling the prototype are available, and quotations for manufacturing quantities should be available as well. A parts list or bill of material defines all of the assemblies. The most current cost information should be applied to the development bill of materials that were described earlier. The total estimated cost for the design, in the annual quantities forecast, should be totaled and compared to the cost goal in the specifications. If the estimates exceed the total, then some action must be taken to reduce costs. Otherwise, if management agrees, the specifications can be modified and the cost goal increased. The end result of increasing the cost goal is that the company will make less profit than planned, or the intended price of the product will be increased to maintain profitability at the higher cost.

▶ Summary

The successful assembly and testing of the prototype are a significant project milestone that signify the completion of the prototype step and the design phase. Step Four, the execution of the design, is now complete. The deliverables from Step Four, the Design Phase, are shown in Figure 11–18.

At this point all of the custom-designed components have been fabricated and assembled into the final product. Enough of the software is complete to verify all of the key operations of the prototype. Environmental testing is complete, and any problem areas have been resolved with design or specification modifications. The manufacturing costs have been totaled with the latest data and are either below the goals set initially or the goals have been modified.

During the prototype phase, the emphasis is on completing a design that will perform the functions called for in the specifications. The customer's ease of use, manufacturability, and serviceability are issues that are always highly considered in the design process. During the early stages of the project, it is very difficult to envision all of the manufacturability and customer use issues until the design takes on a real physical shape. After the prototype is complete, a natural shift occurs, as the emphasis of the project will focus on quality, manufacturability, serviceability, and customer use issues. In Step Five, Design Verification, the prototype is evaluated in each of these areas in addition to its functional performance.

▶ **FIGURE 11–18**
Step Four deliverables

1. Design drawings for all custom components

2. Parts lists for all assemblies

3. Software as required to complete product assurance testing

4. At least one complete prototype

5. A prototype test report

6. Manufacturing cost review

Digital Thermometer Example Project

The electrical and mechanical design of the digital thermometer has been implemented and the prototype has been developed, assembled, and tested. The circuit board artworks were completed using the manual layout methods described in this chapter. Only the electrical component drawings will be shown as examples here. The artwork for one side of the board is shown in Figure 11–19. An example fabrication drawing for the digital thermometer is shown in Figure 11–20. An example assembly drawing for the digital thermometer is shown in Figure 11–21. Figure 11–22 shows an example of the parts list for the digital thermometer.

Digital Thermometer Prototype Testing

For the digital thermometer the reference conditions are:

> 115 V AC, 60 Hz *at 25°C.*

> Reference accuracy testing of the digital thermometer involved simulating an RTD temperature sensor with a precision decade resistance box while recording the indicated temperature. The input RTD temperatures were simulated over the range of 0°F to 200°F in 5°F increments.

> For the digital thermometer, the only operational requirements are that the indicator shall flash 8s on the display when the indicated value is outside the range of 0°F to 200°F. This was verified in the testing.

Ambient temperature tests were performed on the digital thermometer. The testing was performed by the procedure described in this chapter. Figure 11–23 shows the test results.

In this case the digital thermometer specifications refer to the FCC class B document. The digital thermometer was tested successfully for radiated EMI per the test setup shown in Figure 11–17 on page 305.

The digital thermometer specification states that it should withstand a vibration force with amplitudes of 0.2 g applied at a frequency range of 0.3 Hz to 100 Hz. The digital thermometer was mounted in each of the three possible planes, at frequencies of 0.3 Hz, 50 Hz, and 100 Hz for a period of 12 hours each. The unit was not powered during the tests, but it was observed periodically for resonant frequencies. After testing, the unit was disassembled and inspected for weakened components.

The resulting printed circuit board for the digital thermometer is shown in Figure 11–24. This board was laid out, assembled, and tested using the procedures described in this chapter.

▶ References

Coombs, C. F. 1988. *Printed Circuits Handbook.* 3rd ed. New York: McGraw-Hill.

Mardiguian, M. 1987. *Interference Control in Computers and Microprocessor-Based Equipment.* Gainesville, VA: Don White Consultants.

Ott, H. W. 1976. *Noise Reduction Techniques in Electronic Systems.* New York: Wiley.

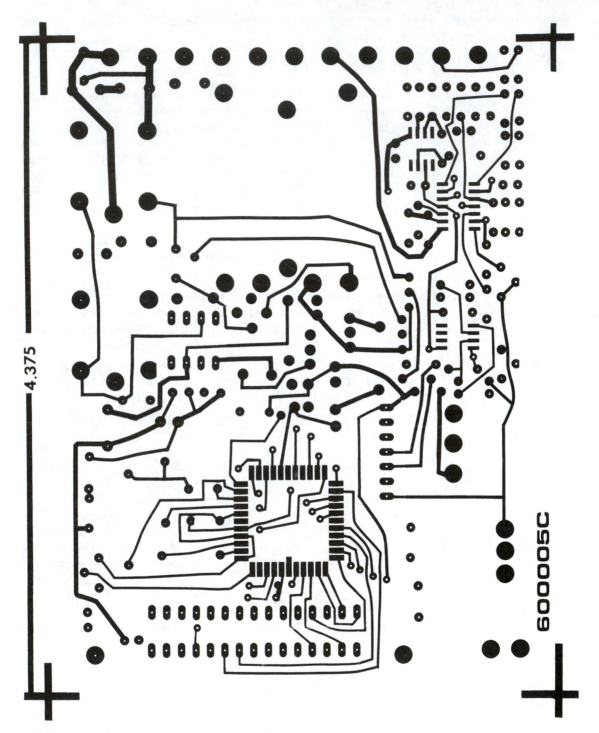

4.375

6000005C

▲ FIGURE 11–19
Digital thermometer artwork drawing

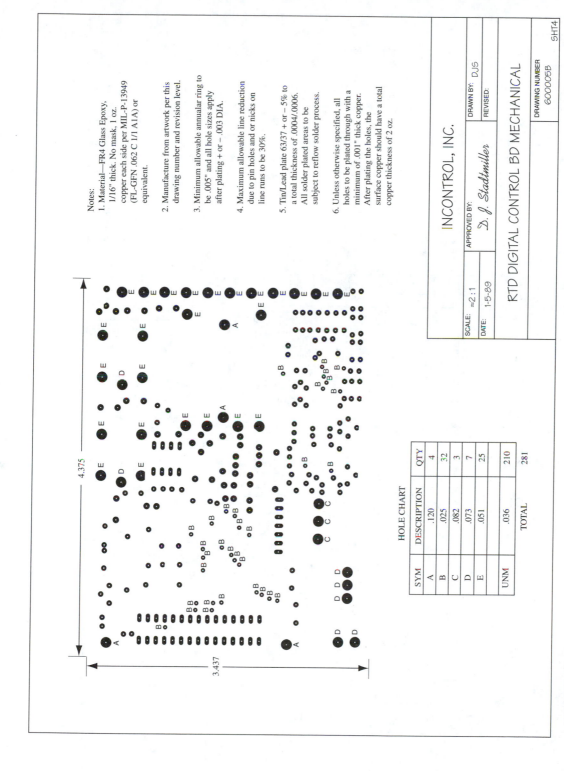

Notes:

1. Material—FR4 Glass Epoxy, 1/16" thick. No mask, 1 oz. copper each side per MIL-P-13949 (FL-GFN .062 C 1/1 A1A) or equivalent.

2. Manufacture from artwork per this drawing number and revision level.

3. Minimum allowable annualar ring to be .005" and all hole sizes apply after plating + or − .003 DIA.

4. Maximum allowable line reduction due to pin holes and or nicks on line runs to be 30%.

5. Tin/Lead plate 63/37 + or − 5% to a total thickness of .0004/.0006. All solder plated areas to be subject to reflow solder process.

6. Unless otherwise specified, all holes to be plated through with a minimum of .001" thick copper. After plating the holes, the surface copper should have a total copper thickness of 2 oz.

HOLE CHART

SYM	DESCRIPTION	QTY
A	.120	4
B	.025	32
C	.082	3
D	.073	7
E	.051	25
UNM	.036	210
	TOTAL	281

INCONTROL, INC.

SCALE: ≈2:1	APPROVED BY:	DRAWN BY: DJS
DATE: 1-5-89	*D. J. Stadtmiller*	REVISED:

RTD DIGITAL CONTROL BD MECHANICAL

DRAWING NUMBER
6O0005B

SHT4

▲ **FIGURE 11-20**
Digital thermometer fabrication drawing

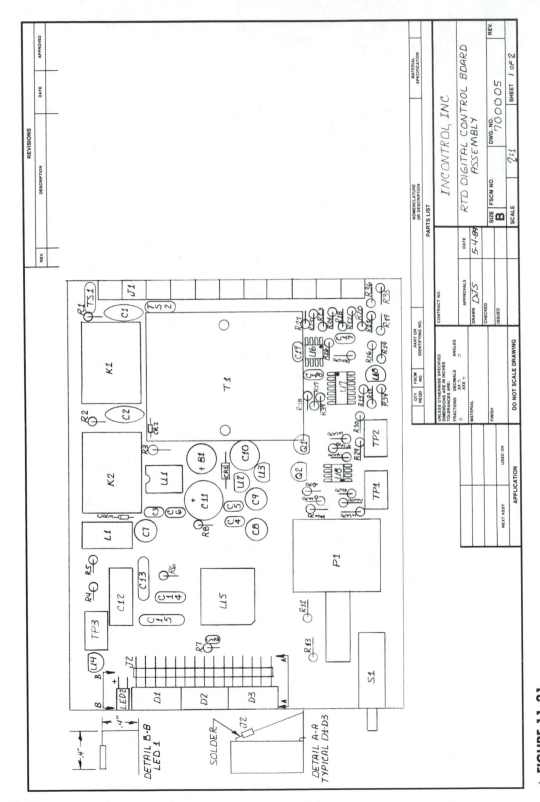

▲ **FIGURE 11–21**
Digital thermometer assembly drawing

Main Printed Circuit Board Assembly

Item #	Description	Company Part Number	Manufacturer	Manufacturer Part Number	Component ID	Quantity
1	2.49 k ohm, 1% Metal Film, 1/4w Resistor	10000000	Res. Inc	2.49kX	R1,R13	2
2	110 ohm, 1% Metal Film, 1/4 w Resistor	10000001	Res. Inc	110X	R2	1
3	100 k ohm, 1% Metal Film, 1/4w Resistor	10000002	Res. Inc	100kX	R3--R6,R15	5
4	26.1 k ohm, 1% Metal Film, 1/4w Resistor	10000003	Res. Inc	26.1kX	R7	1
5	10.0 k ohm, 1% Metal Film, 1/4w Resistor	10000004	Res. Inc	10.0kX	R8,R9	2
6	430 k ohm, 1% Metal Film, 1/4w Resistor	10000005	Res. Inc	430kX	R11	1
7	5.11 k ohm, 1% Metal Film, 1/4w Resistor	10000006	Res. Inc	5.11kX	R10	1
8	1 M ohm, 1% Metal Film, 1/4w Resistor	10000007	Res. Inc	1MX	R12	1
9	470 k ohm, 1% Metal Film, 1/4w Resistor	10000008	Res. Inc	470kX	R14	1
10	1k ohm, 10 turn Trimpot	10000009	TP Inc.	101tp1	RZ	1
11	50 k ohm, 10 turn Trimpot	10000010	TP Inc.	502tp1	RG	1
12	10 k ohm,10 turn Trimpot	10000011	TP Inc.	102tp1	RR	1
13	.1 uF, 50 V, Ceramic Capacitor	10000012	Cap. Inc.	P1101	C1--C4	4
14	.01 uF, 50 V, Ceramic Capacitor	10000013	Cap. Inc.	P1102	C5	1
15	100 pF, 50 V, Ceramic Capacitor	10000014	Cap. Inc.	P1106	C6	1
16	.1 uF, 50 V, Mylar Capacitor	10000015	Cap. Inc.	M1101	C7	1
17	.22 uF, 50 V, Polypropylene Capacitor	10000016	Cap. Inc.	P3224	C8	1
18	.047 uF, 50 V, Polypropylene Capacitor	10000017	Cap. Inc.	P3473	C9	1
19	3300 uF, 16 V, Electrolytic Capacitor	10000018	Cap. Inc.	P5144	C10	1
20	330 uF, 16 V, Electrolytic Capacitor	10000019	Cap. Inc.	P5140	C11, C12	2
21	3 1/2 Digit LED A/D Converter	10000020	IC Inc.	TC7117CPL	U1	1
22	5 V .1 A Voltage Regulator	10000021	IC Inc.	LM320LZ-5.0	U2	1
23	5 V.1 A Negative Voltage Regulator	10000022	IC Inc.	LM340LAZ-5.0	U3	1
24	5 V 1 A Voltage Regulator	10000023	IC Inc.	LM340T-5.0	U4	1
25	Quad Operational Amplifier	10000024	IC Inc.	LM324P	U5	1
26	Voltage Reference- 2.5 Volts	10000025	IC Inc.	LM4040CZ-2.5	U6	1
27	115/20VAC,.3A Center Tapped Transformer	10000026	IC Inc.	MT2111	T1	1
28	Seven Segment .56" Red LED Display	10000027	IC Inc.	67-1463	D1--D4	4
29	Printed Circuit Board	10000028	PCB Inc.	NA	NA	1
30	Terminal Block -5 Position	10000029	Conn. Inc.	W5500	J1	1

▲ **FIGURE 11-22**

Digital thermometer parts list

311

Ambient Temperature °C	Expected Temperature Indication—°F	Actual Temperature Indication—°F	Error (±1°F)	(+ or –1 °F) Within Tolerance
25	200	200.2	0.2	Yes
35	200	200.5	0.5	Yes
45	200	200.9	0.9	Yes
55	200	201.1	1.1	No
45	200	201.0	1.0	Yes
35	200	200.8	0.8	Yes
25	200	200.3	0.3	Yes
15	200	199.9	-0.1	Yes
5	200	199.4	-0.6	Yes
0	200	199.0	-1.0	Yes
5	200	199.2	-0.8	Yes
15	200	199.6	-0.4	Yes
25	200	199.8	-0.2	Yes

▲ **FIGURE 11–23**
Ambient temperature test data

▶ **Exercises**

11–1 What is the purpose of using design master drawings?

11–2 What is the purpose of a via hole? What process is necessary to provide conductivity through a via hole from one side of a circuit board to the other?

11–3 What is a laminate, and what do the different grades of laminate indicate?

11–4 What is the difference between single-sided, double-sided, and multilayer circuit boards?

11–5 On which type of circuit board can components be placed on both sides of the circuit board?

11–6 Compare double-sided and multilayer printed circuit boards as far as cost, size, and noise immunity are concerned.

11–7 How are power supplies and ground usually distributed on multilayer circuit boards?

11–8 What is meant by the term *ground plane*, and why should it be designed in what is called a *crosshatched pattern*?

11–9 Why are the grounds from analog and digital circuits usually kept separate as much as possible?

11–10 What is a guard ring, and why is it used?

11–11 What is a decoupling capacitor, and why is it used?

11–12 When using wave soldering, why is it important to know the direction of the wave when determining component locations on a printed circuit board?

▲ **FIGURE 11–24**
Digital thermometer printed circuit board

11–13 List and describe the documents and drawings required to completely document a printed circuit board.

11–14 Describe the purpose for a solder mask and what it consists of. How would you develop the artwork for a solder mask?

11–15 What is the purpose of silk-screening a printed circuit board? How would the artwork for one be developed?

11–16 List all of the drawings needed to specify a double-sided printed circuit board to a circuit board fabricator who supplies the bare printed circuit board. Assume that the board will have a solder mask and will be silk-screened.

11–17 Describe the purpose of a computer software board layout autorouter. What are the areas of performance that are a concern when using an autorouter?

11–18 List and describe the four types of EMI immunity testing. What are the legal requirements for immunity testing?

11–19 Explain the difference between the two basic types of SMT soldering processes: wave soldering and reflow soldering.

11–20 What is the purpose of test points, and why are they even more critical on SMT circuit boards?

12

Step Five: Verify the Solution (Design Verification)

▶ ## Introduction

At the end of the prototype phase, the engineering department completes prototype tests to verify the operation and performance of the design project. The design verification step verifies the design over the full limits of expected performance. This evaluation determines if the project fulfills the economic reasons for its development. Manufactured products are, after all, developed to meet planned profit projections. The determination of a product's ability to meet the profits projected can be made with the answers to the following questions:

1. Does the design meet all of the operational and environmental performance requirements?

2. Can the manufacturing cost goals of the design be achieved?

3. How do the total project costs compare to the original projections?

4. Will the design function reliably and be easy to use in its intended applications?

5. Have the field service objectives been met?

6. Can the sales targets be met?

The viability of an in-house design project for internal use is determined by the improvement in performance, capability, or efficiency intended and meeting the economic payback expected. In this case the following questions must be answered:

1. Does the design meet all the operational and environmental performance requirements?

2. Were the total project costs as projected?

3. Are the improvements in operation, performance, capability, or efficiency as planned?

4. What is the current payback period, and how does it compare to the payback period projected at the beginning of the project?

Economic payback is a term used to express the amount of time it will take to recoup the investment made in a project. It is determined by taking the total investment and applying the increased profitability resulting from a project toward that investment until the total investment has been replaced.

The design verification phase completes the verification process. This chapter covers the following topics that make up the design verification process:

▸ Product assurance testing

▸ Manufacturing process definition

▸ Manufacturing test and calibration

▸ Reliability

▸ Product release information

▸ Serviceability

▸ Manufacturing pilot run

▸ Design review

12–1 ▸ Product Assurance Testing

Product assurance testing includes the performance and environmental testing already completed by the engineering department during the prototype phase. Generally it is felt that the final verification tests should be completed independent from the engineering department. It is the same line of thinking that has the quality department inspect the work done by the manufacturing department. It is felt that the manufacturing department cannot be critical enough to inspect its work, and if errors are found, they will be resolved but not reported. The product assurance tests should include the final software to be released if the software has been developed for the project. The product assurance tests require significant technical capabilities, so it takes a well-staffed quality assurance department to complete them. Because only the largest companies can support a staff of this size, product assurance testing is usually accomplished in one of the following ways:

1. The quality department completes all product assurance tests.

2. The quality department oversees the performance and environmental tests performed by the engineering department during the prototype phase.

3. The quality department repeats only certain selected performance and environmental tests as the product assurance tests.

However the product assurance performance and environmental tests are accomplished, the procedures are identical to the prototype tests described in Chapter 11. The product assurance testing also includes the verification of the final software.

Final Software Evaluation

The final design verification must include the verification of the final release software for the project. A similar verification process, completed by the engineering department, should precede the quality department evaluation. Software testing is difficult to accomplish because most errors are found with combinations of data that seldom occur. These combinations must be identified and tested. Many times there are no obvious, logical reasons for the error condition. Because of this, software testing takes on a brute force methodology, where many unlikely combinations and conditions must be simulated and verified. The final software is evaluated along with the current revision level of the operator manual and software specifications. A test plan should be developed that covers all the key areas of operation. In complex systems, it is unlikely that all the possible data combinations can be tested.

It is important to use a system for defining the software version designation. A primary discriminator is one that indicates whether the software has been released. Released software has been completely evaluated and distributed for use. Unreleased software is still under evaluation and should never be distributed for use, except under special conditions such as controlled field tests. Software versions are usually configured as shown in Figure 12–1. There are many different systems but usually the most significant character, *V* in Figure 12–1, indicates whether the software has been released. In this example *V* means the software is released, and *P* means that it is preliminary, unreleased software. The next most significant character is called the *system release number*. The system release number indicates significant changes in functionality. The character after the decimal point is called the *version release number* and indicates relatively minor changes

▶ **FIGURE 12–1**
Software version number

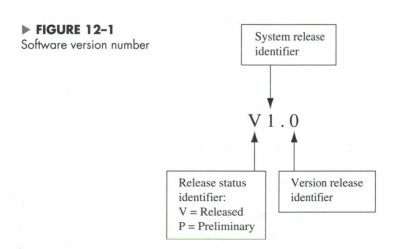

to the software. The version release number is used most often to correct errors in the software as they are found. It is important to tie software changes to any hardware changes that may be implemented. Software changes should not be distributed without undergoing a formal evaluation. The resolution of errors should be coordinated to minimize the number of releases and subsequent evaluations.

12–2 ▶ Manufacturing Process Definition

Now that the prototype is available, it is time to review and further define the manufacturing process. The manufacturing department leads this activity, but it needs much assistance from the design engineers. The definition of the manufacturing process involves the determination of the step-by-step assembly process. It also defines which assembly levels are tested and their test procedures.

There have been many changes in manufacturing ideologies over the last 20 years. The basis for this is a trend away from *batch processing* to a concept called *one-piece flow*. In batch processing, a large quantity of product is manufactured by assembling a quantity of each succeeding assembly level in different manufacturing areas, until the whole batch is completed. One-piece flow processing promotes the manufacture of a single-end product in a manufacturing cell until it is complete. This is also called *cellurized manufacturing*. Both processes are shown in the flowcharts in Figure 12–2.

Take the assembly of a personal computer as an example. A company desires to manufacture 100 PCs in a week. Batch processing (see Figure 12–2) of

▶ **FIGURE 12–2**
One-piece flow vs. batch processing

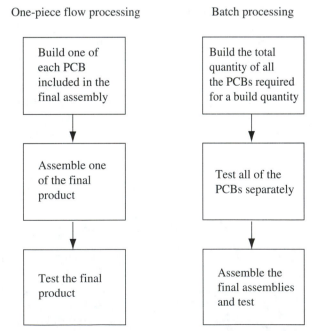

these 100 units requires building 100 PC circuit boards on Monday. Testing and doing further assembly of those 100 circuit boards would occur on Tuesday. Final assembly of the 100 units would be performed on Wednesday. Then the final assemblies would be tested, calibrated, and burned-in on Thursday. The final assemblies would be packaged and ready for shipment on Friday.

A pure one-piece flow, cellurized manufacturing process (see Figure 12–2) dictates that one set of PC circuit boards be assembled, tested, and completed into a final assembly. Then another and another would be assembled until the 100 units are completed. All of the work would be completed in one manufacturing cell.

The benefits cited for one-piece flow processing are as follows:

1. Because each unit is completed shortly after all of the operations have been performed, process errors that occur are quickly identified and can be rectified before being repeated in other units.

2. Shipment lead times are reduced because units are available for shipment sooner than if they were batch processed.

3. One-piece flow represents a streamlined process in which the handling and movement of components and subassemblies around the plant are minimized.

4. All of the workers in the cell become cross-trained and are able to perform other cell operations. Because of this, they understand the complete assembly process and are more able to recognize problems in their infancy.

These are all significant benefits over the results experienced with batch processing. The problem with the one-piece flow concept arises with highly automated operations, where expensive equipment is required. Take the soldering operation for the PC example discussed previously. If batch processed, all of the boards would be assembled and then transferred to a wave-solder area for soldering. A pure one-piece flow process would require a wave-solder machine for each manufacturing cell. This is an expensive and space-consuming proposition that usually results in a compromise between batch processing and one-piece flow. This is just an example of the types of decisions that must be made. The manufacturing department or group and the design engineers must work together concurrently, to determine the best methods for testing, calibration, and burn-in of all subassemblies and the final product.

12–3 ▶ Manufacturing Test and Calibration

When the manufacturing process is laid out, the stages in which assemblies will be tested are defined. These stages will depend on the product and the type of process being used: batch processing or one-piece flow. Manufacturing test methods have also seen significant change in the last 20 years. These changes are similar to those discussed for manufacturing assembly.

To better understand this change in manufacturing test philosophy, let us explore the differences between batch processing and one-piece flow as they pertain to circuit board assembly and testing. The following represents a typical batch process circuit board assembly and test operation for a high volume manufacturer.

Board Assembly and Test—Batch Processing

1. Incoming circuit board components are inspected before assembly.

2. Circuit boards are assembled in lots of 100 with autoinsertion machines. Some components are manually inserted before completing the assembly.

3. The assembled boards are sent to a wave-solder station, where they pass through the wave-solder process. The boards are cleaned and allowed to dry after soldering.

4. The quality department usually inspects each board or sample-inspects the boards for assembly errors and quality defects.

5. The inspected boards are sent to a test station, where an automated tester performs a series of preprogrammed tests on the particular circuit board.

6. If the board passes the tests, it proceeds to the next level of assembly. If not, a printout that includes information about the test failure is attached to the printed circuit board. The circuit board is sent to a test station, where a test technician will troubleshoot and repair the board. The repaired board is sent back to the automated test station for verification.

7. The final product is assembled using the circuit board under discussion in addition to other circuit boards that have experienced the same process. The final product is assembled and tested with a test fixture designed especially for the product. If the unit passes the final assembly tests, it is burned-in and readied for shipment. If not, the circuit boards that possess the malfunctions must be identified and corrected.

This process has been in use for many years. The problems associated and experienced with this process were as follows:

1. Assembly or process errors are often repeated throughout an entire lot of boards before being detected during inspection or testing.

2. Because the process is segmented, there are time delays between the operations. There is generally poor communication between the people performing the different operations. Consequently, problem resolution and process improvement are not promoted.

3. Sometimes circuit boards pass their specific automated tests, but they will not function together with other circuit boards when assembled in the final product. These are called *dynamic failures* and result from the inability of the circuit board tests to verify the circuit board's complete operation.

4. In general, because so much time passes between the assembly of a circuit board and its assembly and test in the final unit, there is poor feedback to resolve quality issues as they develop.

In order to address and resolve these issues, cellurized manufacturing for circuit board assembly and test processes utilizing one-piece flow concepts were developed.

Board Assembly and Test—Manufacturing Cells

The following process describes cell manufacturing:

1. Incoming components are batch inspected, organized in assembly kits, and sent to the manufacturing cell.

2. The circuit boards are assembled and soldered within the cell. Small wave-solder machines have been developed for this purpose. If possible, all of the circuit boards included within a particular product are palletized (see Figure 12–3). This means they are all attached to each other, assembled and tested as a unit, and then broken down into individual circuit boards for assembly. The individual boards are held together by a series of small laminate areas that are located between the individual boards. These areas are easily cut away or broken off to separate the boards.

3. Testing is completed with custom test fixtures within the cell. Only minimal testing is performed on the circuit boards until the final level of operation. This is easily accomplished using the palletized board approach. The testing philosophy applied here is to provide testing for a particular assembly only, when the cost of not testing the assembly is greater than the cost of testing it. In other words, an assembly should be tested only if the cost of finding a problem later is greater than the cost of testing the assembly.

The downside of cell manufacturing is realized when automated processes and equipment are applied. Because this type of equipment is expensive, it is hard to justify locating it in every cell. In these cases two or more cells may share equipment by locating them in close proximity to it. The implementation of cell manufacturing has promoted the use of small, dedicated custom-designed test fixtures for testing assemblies and final products.

The test and calibration plan is developed as part of the overall manufacturing process and depends on the degree to which batch processing and cellular manufacturing techniques are applied. The test plan defines all of the assemblies that are tested along with the purpose for the tests.

Test Fixture Development

Once the test plan has been developed, test fixture development can begin. Consider the test plan to be the specification for the design of the test fixture. When designing test fixtures, the design challenge is making quick and secure connections to points in the circuit where test measurements must be made. Measurement

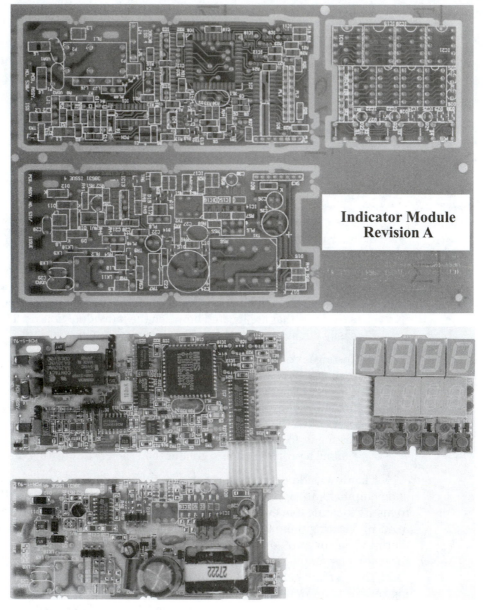

**Indicator Module
Revision A**

▲ **FIGURE 12–3**
Example palletized circuit boards

points that are available at the circuit board interconnections are easily accomplished by using a mating connector in the test fixture. For measurement points that are not accessible, a "bed of nails" approach must be used. A *bed of nails* is simply a test fixture that includes spring-loaded *pogo pins* (see Figure 12–4) that are located to make contact with specific pads on the bottom of the printed circuit

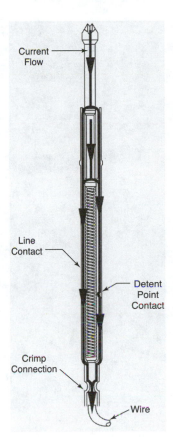

▶ **FIGURE 12–4**
Spring-loaded test pin *(Courtesy Interconnect Devices, Inc.)*

board. There are two types of bed-of-nails fixtures. In one type the board is pressed down onto the bed of nails and mechanically held against the spring-loaded pogo pins. The other type is called a *vacuum fixture*, where air pressure pushes the pogo pins against the board that is being tested. Companies that specialize in their design and construction usually develop vacuum fixtures. The design expertise and equipment to fabricate them are usually not available within most manufacturing companies.

Having determined a method for accessing test points, the fixture design can proceed. The test fixture is usually developed with an available programmable tester. Digital, analog, and combination circuit testers are available from a number of manufacturers (see Figure 12–5). Utilizing off-the-shelf automated test systems requires the development of high-level language programs that performs the tests.

However, as mentioned previously, cellurized manufacturing has promoted the use of small custom-designed test fixtures for use in manufacturing cells. These custom fixtures can become significant design projects on their own and should be approached as a separate design subproject. Many of these are microprocessor-based designs, for which custom software is developed and circuit boards must be designed and laid out. The documentation for any test or calibration fixtures should be documented as any other design project. This includes the specifications for

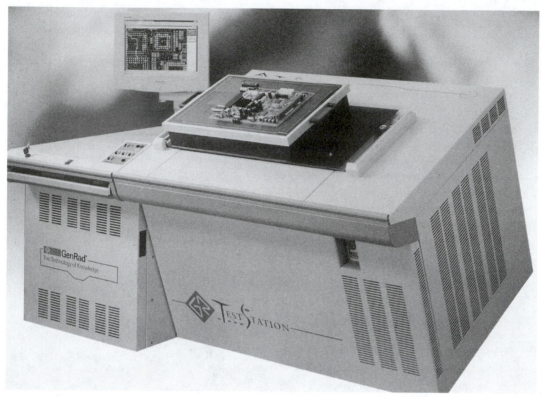

▲ FIGURE 12–5
Genrad tester *(Courtesy of Genrad, Inc.)*

the fixture design and all schematics and assembly drawings. In successful projects, there is often a need to construct additional fixtures for manufacturing, engineering, or field service. There may be the need to replace a fixture in future years. It is amazing how quickly one can forget the intimate details of a design after having toiled over it for months. It is important to make the effort to document the design while it is fresh in your mind.

Test and Calibration Procedures

Along with any test or calibration fixture, a procedure must be developed to document the process to be followed while using the fixture. These procedures should be concise and accurate. Any changes made to these procedures should be formally controlled as any other engineering document. (Methods for controlling engineering documentation changes are discussed in Chapter 13.) A test procedure is completed for each group of tests performed for an assembly and should reference the test fixture utilized. The test procedure should include the precise, step-by-step description of the tests to be performed in the sequence that they are to be completed. Completion of a test procedure often includes the compilation of test data and results for future reference. A blank sample of the test data sheet should be included in the test procedure. As with any manufacturing process, it is important to streamline all testing and calibration into a smooth process with minimal repetition of operations, while accuracy and quality goals are maintained.

12–4 ▶ Burn-In

A manufacturing process that has not yet been discussed is a process called *burn-in*. The purpose of the burn-in process is to minimize *infantile failures*. Infantile failures are premature failures that occur in weak components or connections. Infantile failures happen shortly after a product is put in use. Many companies define infantile failures as those that occur between 1 and 30 days after a product is first used in an application. The goal of the burn-in process is to induce any weak components to fail before shipment to the customer, thereby causing a pending infantile failure to occur where it can be repaired without any negative impact on the customer.

There are many theories about burn-in and as many different practical approaches to its implementation. In order to induce pending infantile failures to occur, the product should be powered and operated in conditions that equal the actual use by the customer. In order to accelerate the wear on weak components and accomplish burn-in quickly, the product is maintained at an elevated temperature. A more efficient burn-in is achieved if the temperature is cycled between room temperature, 25°C, and a temperature around 55°C. This causes the weak component to expand and contract as the temperature is cycled. If the lower temperature is set at 0°C instead of room temperature, the effectiveness of burn-in is further improved. The 0°C setting requires that the burn-in equipment have cooling capability. Finally, the on/off switching of power to the product is an additional stress that can often induce infantile failures.

Infantile failures are the result of weak or inferior components or connections. Many times components or connections are weakened by the manufacturing process. This is the case when a CMOS component is improperly handled and exposed to static electricity, thereby weakening it. Sometimes semiconductor components are weakened by exposure to severe temperature in a wave-solder process. Because components can be purchased preburned-in, many companies discontinue the burn-in process, neglecting the fact that the manufacturing process often weakens components.

Many companies believe in burn-in and perform it blindly, while others simply skip the process. The manufacturing department will always have a dim view of burn-in. It consumes much time and space and requires a significant capital investment. If a company has a solid manufacturing process that induces very few weak components, and purchases top-quality components from quality suppliers and checks the process constantly, then burn-in may not be necessary. Otherwise, burn-in is the only final check to weed out weak components. A sure way to measure the need or effectiveness of burn-in is to monitor the number of infantile failures that occur in customer applications. This will be discussed further in Chapter 13.

12–5 ▶ Reliability

Whenever a design is completed, there are always questions about its reliability. The reliability of a design is determined on average by how long it functions properly in the intended application without failure. In industry, the reliability of a design is measured by a term called the *mean time between failure* (MTBF). Statistics theory defines the arithmetic mean as an average. The MTBF is the average time between failures of a product design. There are two ways to project the reliability of a design: accelerated life testing and statistical mathematical reliability projections.

Accelerated Life Testing

Accelerated life tests expose a design to conditions in which failure-prone areas are stressed to accelerate their wear. Take the example of a control relay that is normally cycled on and off five times a day. Accelerated life testing would cycle the relay on and off five times an hour. The net effect of one day of testing would simulate 24 days in the intended application. To properly set up an accelerated life test, the design should be reviewed to determine all of the expected failure areas. The test developed should include some component that addresses each of the expected failure-prone areas. Continued exposure to high ambient temperatures is the only way to accelerate wear on many internal electronic components.

Statistical Reliability Projections

The statistical projection of reliability of an electronic design is accomplished by the use of reliability data that is available for many components. This reliability data has been developed with military specifications developed for the purpose of projecting reliability. As discussed previously, military equipment is subject to the most rigorous performance, environmental, and reliability specifications. The data supplied in military specifications lists a mean time between failure for components that is determined by the power, voltage, current, and ambient temperature at which the component is operated. The MTBF for each component utilized in a design is totaled. The grand total is divided by the number of components to determine projected MTBF for the complete design.

Software is now available that will take all of the components included within a design and will project reliability while considering the operational data supplied. Manual methods for completing these calculations can be cumbersome for a complicated design.

12–6 ▶ Product Literature

Almost all design projects require the development of some information describing their operation and use. For a product these documents are called *product literature*. Product literature is used to sell the product and supplies information on its operation and use. Product literature is usually developed by the advertising and publicity department, which reports to the marketing department in most companies. Product literature can include all of the following:

1. Formal product specifications

2. Product sales brochure

3. Operator and installation manuals

4. Service manuals

5. Advertising

6. Price sheet and ordering information

The development engineers will supply the technical information necessary to support the development of the product literature.

12–7 ▶ Customer Use and Serviceability

With price and quality being the same for many competitive products, it is important to design a product that is easy to use and repair. Field tests of the product are completed to evaluate the ease of customer use and serviceability. The best way to perform field tests is to have a service engineer supply the product, with preliminary product literature, to the customer test site. The service engineer should oversee and monitor the results of the installation process as the customer installs the product using the preliminary operator and installation manuals. Performing the field tests in this way tests both the product and the product manuals. Customers are usually agreeable to this process in exchange for keeping the product at no charge. It is best to locate the field tests close to the home plant so that periodic monitoring of the site can be accomplished. Field tests should be ongoing but the first few months of testing are the most critical. In general these are the questions that the field tests should strive to answer:

1. What problems are associated with the unpacking and installation of the product and what can be done to improve the product or the manuals to rectify them?

2. Does the product meet the operational requirements of the product at the test site?

3. How does the product rate in all performance areas when compared to other competitive products?

4. Were there any failures of the product that occurred after the product was installed and evaluated?

To give a broad and accurate picture of the product applied in the field, a field test site should be selected for every major variation of product application.

The formal time period for the field test should be between two and six months. Data should be taken during the field test, and a formal report should be written.

Example 12–1, Brewery Field Tests

It is always good to develop a working relationship with a local company that utilizes the products that your company designs and manufactures. It is much easier to install and monitor field tests when they are close to home. One company's customer, a local brewery, utilized the company's equipment and was most willing to serve as a field test site in exchange for free equipment. This particular plant was an exceptional field test site because the applications involved a heavy industrial and electrically noisy atmosphere with different ambient temperatures and humidity. These extremes served to expose the product to the varying environmental conditions that were simulated in the laboratory and could now be realized directly in an actual product application.

12–8 ▶ Manufacturing Pilot Run

The manufacturing pilot run is a dress rehearsal for the final manufacture of the product. The manufacturing pilot run should include a small lot of the product that is to be manufactured using the test and assembly fixtures and all of the manufacturing documentation that has been developed. The manufacturing documentation should include all of the following:

1. Specifications

2. Fabrication drawings for all custom parts

3. Assembly and subassembly drawings

4. Bill of materials for all assemblies

5. Schematic diagrams

6. Test, calibration, and final inspection procedures

7. Manufacturing process flow sheets

One set of all the documentation should be copied and marked as "Pilot Run Design Masters." These are used for accumulating the expected changes initiated during the pilot run. Before the pilot run, the people who assemble and test the product receive training on the new product. During the pilot manufacturing run, these same people will assemble and test the product using the fixtures, procedures, and documentation as a test of the entire development process. When the pilot run is complete, the manufactured products are completely tested and evaluated by the quality and engineering departments to verify their conformance with all specifications.

12–9 ▶ Design Review

After the product assurance, reliability and field tests, and the pilot manufacturing run have been successfully completed, a *final design review* should be conducted for the project. The purpose of the final design review is to gather the results of these tests and discuss the modifications to be made to the design before its formal release. In most organizations, a formal release of a design means the following:

1. *The design is released for use.* If the design is a one-time project to be used internally, the release signifies that the design can now be utilized. If the design is a product, that product can now be sold.

2. After release, *any future design changes* will be made under a formal *engineering change notice* (ECN) system. During the project, until the formal release, design changes are made on an informal basis. This is because the high number of changes that occur during a project are cumbersome to complete on a formal basis.

The final design review should include all of the project personnel and any of their support staff as appropriate. The agenda for the review should include all of the following:

1. *Product assurance tests*—Does the design meet all of the specifications? If not, either design changes must be implemented to correct this or the specifications must be modified to reflect the capability of the current design.

2. *Field tests*—Did the product meet the requirements of the application in an actual field-test site? Was the product easy to install and use? What design changes are recommended for further improvement?

3. *Manufacturing pilot run*—Did the units built during the pilot run conform to all test specifications? Are the projected goals for manufacturing build and test times achievable? What design changes should be made to improve the quality and or improve efficiency?

4. *Reliability tests*—Are the projected MTBFs 50% greater than the intended warranty time? What design changes should be made to improve the reliability of the design?

5. *Product literature*—Is the product literature accurate and does it correctly represent the product? What modifications must be made to the product literature before the release of the design?

6. *Product documentation*—Is the product documentation complete and accurate? What changes must be made to the documentation?

7. *Test fixtures and procedures*—Do the test fixtures promote the complete and efficient verification of the design in production? What changes can be made to improve performance?

1. Results of product assurance testing

2. Field test results

3. Results of manufacturing pilot run

4. Reliability test results

5. Product literature status

6. Product documentation status

7. Test fixture and procedure status

8. Financial review

▶ **FIGURE 12–6**
Design review agenda

Figure 12–6 shows a summary of the design review agenda. Such an extensive review, from so many different perspectives, results in a number of changes that improve the product design, performance, and efficiency. Many errors will be found in documents, procedures, and literature as well. Hopefully, there are no major design changes that must be made—only ideas about how to improve the design in the future. The project team must sort through all of the issues and determine the changes to be addressed before the release and those that can go on a wish list for the future. The results of the design review should be published in a formal report of the project that will be used as the release document.

▶ Summary

Step Five, the design verification, has been completed. If significant design changes result, then all or some of the design verification steps may need to be repeated on the modified design. Otherwise, the development project is complete and ready for use, or to be manufactured. By completing the design verification, we have confirmed that the design meets the criteria set forth by the specifications developed at the beginning of the project. If the specifications were put together properly, completing this step should mean the eventual success of the project. In the case of a new product, the specifications defined the features, functions, development costs, and manufacturing cost goals for the product. If these are well positioned in the marketplace, the product should enjoy strong sales and be profitable. In the case of a system for use within the company, the specifications defined the features, function, and development costs.

When properly defined and achieved, the system's improved performance provides increased efficiency that will recoup the investment in the amount of time projected.

What remains is to make the design modifications required, then formally release the design. At the design release meeting each document is formally signed off and released. However, the project is not yet complete. The project team should be kept together for one more step as the project or product is used in the intended application. After the design release, monthly review meetings of the project team are scheduled to review the progress of the project as it is being manufactured or used. As thorough as the design verification was, there are usually problems or situations that develop after release and must be addressed by the project team. After a period of about six months, the project team should reflect on the project performance and the design problems encountered and draw conclusions to be used in future design projects.

Digital Thermometer Example Project

To complete Step Five, design verification, for the digital thermometer project, the manufacturing process, test fixtures and procedures, accelerated life testing, and the design review must be completed. The example that follows discusses the development of test fixture and test and calibration procedures for the digital thermometer. The Six Steps are used to outline the development process.

Example 12–2, Digital Thermometer Test Fixture and Procedure Development

The optimum manufacturing process for the digital thermometer is to build up the two printed circuit boards as a panelized set of boards. The two boards can be tested together as a functional unit before assembly into their enclosure. In this example a test fixture will be designed to test and calibrate the digital thermometer circuit board final assembly. The Six Steps are highlighted as the design is completed:

Step One: Research, Gather Information
Gather and review all the necessary data and specifications about the digital thermometer, including all design drawings, manufacturing process flow diagrams, and specifications.

Step Two: Problem Definition
Design Requirements: Design a test fixture and procedure that will perform the following:

1. Verify that the digital thermometer final assembly meets all operational specifications. The test time for a working board should not exceed three minutes, including setup and teardown times.

2. The fixture and procedure should provide for the final calibration of the digital thermometer circuit board assemblies after the operational tests are complete.

3. If the board does fail a test, the fixture and procedure should provide for fault isolation testing to be performed to further identify the source of the problem.

Step Three: Plan the Solution

1. *Select Test Apparatus:* Determine the various types of testing devices that will be utilized from the types of tests to be performed (logic, analog, and the like).

2. *Complete Preliminary Design:* Determine a preliminary design that includes the method of circuit connection to all points to be measured as well as the measuring device.

3. *Complete Draft Test Procedure:* Develop a draft test procedure that will perform all of the operational testing required. Verify the test procedure by performing the tests on a number of circuit board assemblies using test lead connections in lieu of the test fixture.

4. *Complete Fault Isolation Tests:* Develop additional tests for fault isolation if the board fails the operational or calibration tests.

5. *Complete Manufacturing and Quality Review:* Review the completed draft procedure with manufacturing and quality control personnel assigned to the project.

6. *Complete Final Design:* Finalize the design for the test fixture, and then fabricate and complete testing.

7. *Complete Documentation:* Complete drawings that define the test fixture design and finalize the test procedure.

Step Four: Execute the Plan

1. *Select Test Apparatus:* The tests required involve taking analog voltage measurements in all cases. A custom test fixture is utilized that contains a number of digital voltmeters that can be selectively connected to the points required. This fixture is not be automated or under computer control initially.

2. *Complete Preliminary Design:* The test fixture needs to connect to the following points:

    a. +5 V digital

    b. +5 V analog

    c. −5 V analog

    d. +2.5 V reference voltage

e. +1.0 V A/D reference voltage

f. Bridge rectifier output

g. RTD voltage drop

h. Differential amplifier output

i. Linear combination amplifier output

j. Ground

Spring-loaded pogo pins are used to provide access to all of the required test points, except the RTD voltage that is accessible from the input RTD connection. The AC power input connector and the RTD input connector will both be utilized to make these connections to the test fixture.

The two boards are located on the test fixture by standoffs located where the boards will eventually be mounted in the enclosure. The boards are pressed down over the pogo pins and held in place by a small mechanical latch that is rotated into position. The AC input and RTD connections are made and the testing can begin. The test fixture has a main on/off switch and provide a selector switch to connect various RTD resistance values to the input. One digital thermometer is utilized and momentary push buttons are used to connect the proper test point to the voltmeter. The digital displays in the digital thermometer are a reference for the digital conversion of the temperature and are compared to the linear combination amplifier output.

3. *Complete Draft Test Procedure:* Use the following steps to perform the draft test procedure:

a. Install boards to be tested in test fixture and make AC power and RTD connections.

b. Select the RTD input to be 100°F and turn on the test fixture power.

c. The following voltages should be measured by depressing the appropriately labeled button on the fixture and verified as follows to be within the acceptance ranges:

+5 V digital: 4.95 V to 5.05 V
+5 V analog: 4.95 V to 5.05 V
–5 V analog: –4.95 V to –5.05 V
+2.5 V reference: 2.475 V to 2.525 V

Any measurements outside of the acceptance ranges cause a test failure at that point. The nonconforming measurement should be noted for the board troubleshooter.

d. Adjust the A/D reference voltage to be 1.00 V with trimpot R_{REF}. When complete, the digital display on the digital thermometer should equal the digital voltmeter display minus the decimal point when the push button that selects the linear combination amplifier output is depressed.

e. Perform 0°F calibration by selecting the 0°F reference resistance for the RTD on the test fixture input temperature selector. Adjust trimpot R_Z for the correct 0°F reading on the display.

f. Perform 200°F calibration by selecting the 200°F reference resistance for the RTD on the test fixture input temperature selector. Adjust trimpot R_G for the correct 200°F reading on the display.

g. Repeat Steps 5 and 6 until both values are within the ± 0.5°F.

h. Perform accuracy tests on the circuit board under test at input temperatures of 50°F, 100°F, and 150°F by using the input temperature selector. Each reading should be within ± 0.5°F.

i. Review test results. If all the data is within range, the board under test has passed the verification tests.

4. *Complete Fault Isolation Tests:* These additional fault isolation tests are completed upon failure of the board under test to meet the acceptance criteria. The performance of the fault isolation tests are to localize any failures into one of the following functional areas:

▶ Bridge rectifier

▶ +5 V digital power supply

▶ +5 V analog power supply

▶ –5 V analog power supply

▶ 2.5 V reference circuit

▶ 1 V reference circuit

▶ Constant current source circuit

▶ Differential amplifier circuit

▶ Linear combination amplifier circuit

▶ A/D display driver circuit

▶ Digital displays

These fault isolation tests include the following:

a. *Power supply failures*—If any of the power supply voltages are outside of the acceptance range, the bridge rectifier output is measured on the test fixture by depressing the bridge rectifier selector. The output of the bridge rectifier should be within the range 11 V DC to 15 V DC. If the bridge output is outside of this range, then the problem area is either the transformer or the bridge rectifier. Otherwise the power supply regulator or filter capacitor included within the faulty power supply output are at fault.

b. *2.5 V reference failure*—If the 2.5 V reference voltage is outside the specifications while the power supply voltages are within the acceptance range, inspect the circuit board for the proper value of R_1 and the proper assembly of the actual voltage reference integrated circuit U_4. If the correct components are assembled properly, then the problem is most likely within component U_4.

c. *Constant current source failure*—This results when the output of the constant current source is outside the specifications, when all other previous tests have passed. This means that the 2.5 V reference and all power supply voltages are correct. Check for proper components and installation of resistors R_2, R_3, and integrated circuit U_5. If the correct components are properly installed, the problem is likely within the integrated circuit U_5.

d. *Differential amplifier failure*—If the output of the constant current source is correct, while the differential output is not, check for the proper components and assembly of R_4, R_5, R_6, R_7, and U_5. If the correct components are properly installed, the problem is likely within the integrated circuit U_5.

e. *Linear combination amplifier failure*—If the output of the differential amplifier output is correct while the linear combination amplifier is not, check for the proper components and assembly of R_{12}, R_Z, R_8, R_9, R_{10}, R_{11}, R_G, and integrated circuit U_5. If the correct components are properly installed, the problem is likely within the integrated circuit U_5.

f. *1.0 V reference failure*—The +1.0 V reference voltage must be accurate for the A/D converter to function properly. If the +5 V analog supply is within specifications, then resistors R_{15}, R_{REF}, and integrated circuit U_7 should be checked for the proper component value and proper assembly. If all components are correctly assembled, then U_7 is the likely faulty component.

g. *A/D display driver failure*—If all power supplies and reference voltages are correct, as well as the output of the linear combination amplifier, the A/D converter should convert and display the value properly. If an incorrect value is displayed at this point, the problem is likely within the A/D functional area. Verify the correct component values and assembly of resistors R_{13}, R_{14}, R_{16}; capacitors C_8, C_{10}, C_{11}, C_{12}; and integrated circuit U_6. If all the components are properly assembled in place, then U_6 is the most likely faulty component.

h. *Display failure*—Finally, if the display seems to be correct, except that a digit or missing segment does not light up, then the display device itself is the likely faulty component. In this case there is also the possibility that output display drivers of integrated circuit U_6 are not functioning, making U_6 the faulty device.

5. *Complete Manufacturing and Quality Review:* When the preliminary design and the draft test and fault isolation procedures are complete, they should be reviewed with the appropriate manufacturing and quality personnel. This is done to make sure that the test fixture and procedures meet the manufacturing process plan in place and that the goals of the testing are accomplished as completely and efficiently as possible.

6. *Complete Final Design and Documentation:* Any general changes as well as those that result from the review with manufacturing and quality should be incorporated into the final design. The final design should be documented,

just like any other product, so that it can be reproduced or modified in the future without difficulty. All test and calibration procedures should be released for review and the test fixture should be assembled and tested.

Step Five: Verify the Results

The test fixture and procedures will all be tested together after they are completed. The real verification of both will begin with the pilot manufacturing run of the new product and will continue throughout the initial months of production. Verifying the test fixture and procedures will involve answers to the following questions:

1. What percent of failures at the final product test step result from failures that could have been detected earlier?

2. How do the test times compare with the allotted manufacturing test time goals for the product?

In other words, we are measuring the performance of the test fixtures and procedures toward achieving the goal of complete test and calibration of the digital thermometer in an efficient manner.

Step Six: Develop a Set of Conclusions

As the results are reviewed, there are obvious ways that both the fixture and procedures can be improved. All of these should be noted, and those that are practically achievable should be implemented before the product is released. The next example shows the development of accelerated life tests for the digital thermometer.

Example 12–3, Accelerated Life Testing of a Digital Temperature Control

In this example an accelerated life test for the digital thermometer (with control outputs included) is developed. The goal of the test is to simulate two years of average operation in a life test that takes three months. The digital thermometer with a control has an analog setpoint potentiometer and an electromechanical relay output for turning a heat source on or off. After a review of the digital thermometer design, the components that are susceptible to fatigue and their degree of susceptibility can be classified as follows:

1. *Electromechanical relay*—This device is energized and de-energized every time the heat source is turned on or off. It is the device most susceptible to fatigue and eventual failure because of its inherent electromechanical nature. This device experiences most of its wear on the contacts, where their repeated switching removes the contact material over time. The electromechanical relay is also prone to the mechanical failure of the mechanism or the electrical failure of the coil.

2. *Analog set-point potentiometer*—This device will be subject to the next level of wear and fatigue. This will result from the continuous repositioning of the potentiometer, usually within a small segment of the potentiometer.

The amount of wear depends on how often the setting is readjusted, so a worst case value must be determined.

3. *All other electronic components*—All other passive or active electronic components.

A good accelerated life test will expose each of these three categories of components to tests that simulate the accelerated fatigue of that category of component. To do this, the expected maximum number of wear operations that the component would experience in 24 hours of application must be determined. For example, we estimate that the electromechanical relay is switched a maximum of 40 times in 24 hours. To simulate 24 hours of wear on the electromechanical relay, we have to switch the relay on and off 40 times with a load current equal to the maximum value in the specifications. The daily estimate for the three categories is as follows:

1. *Electromechanical relay*—Energized and de-energized 40 times daily.

2. *Analog set-point potentiometer*—The temperature setting is changed 10 times daily.

3. *All other electronic components*—Most of these will not endure cycling wear but experience wear simply when powered. It is estimated that the digital thermometer and controller are powered 24 hours daily every day in normal operation.

In order to achieve the goal of simulating two years of operation, 24 hours of wear must occur in three hours of the life test. Extending this line of thought, every day of the life test simulates eight days of actual operation of the digital thermometer and controller. Accordingly, the daily estimates for each component category must be experienced in three hours of the life test; the electromechanical relay is switched on and off 40 times, and the analog set point is repositioned 10 times. The only way to simulate accelerated wear on the basic electronic components is to operate them at a higher temperature, such as 50°C.

The accelerated life test for the digital thermometer and controller involves its operation at a temperature of 50°C, while a test fixture simulates temperature changes that switch the electromechanical relay on and off 40 times in three hours. The contacts of the relay carry the full load of their rating. A motor is connected to the analog set point, which is repositioned ten times every three hours. The controller should be examined for accuracy every week. At the end of the three months, the controller should be completely torn down, inspected, and evaluated. If the unit is still functional in terms of its ability to meet all design specifications, then the testing should continue for another three months.

The product documentation is complete for the digital thermometer. The field tests, pilot manufacturing run, and the design review are complete as well. Figure 12–7 shows the results of the design review.

The product brochure for the complete digital indicator and controller version of the digital thermometer is shown in Figures 12–8 and 12–9. It includes a picture of the product, a product description, and a brief set of specifications.

1.0 Results of Product Assurance Testing

The Product Assurance tests were successfully completed on the Digital Thermometer using Design Specifications # 12233221 Rev B. The Digital Thermometer meets the specifications in all cases. See QA Product Assurance Test Report #12233221 for details of the testing and to review the test data.

2.0 Field Test Results

The Digital Thermometer was installed in four field test sites that represent the variety of applications expected. Each Digital Thermometer operated accurately and installed without any problems. There were numerous suggestions from the different sites for changes to the operator manual, calibration procedure, shipping carton and a suggestion to change the artwork labeling on the front panel. All of these changes have been implemented except the artwork change. The artwork change has been approved but it will not take effect until the current supply of artworks is used up. It was agreed that the significance of this change is such that it can be implemented on a routine basis. See Service Engineering Field-Test Report #12233221 for details of the field-test and to review the test data.

3.0 Results of Manufacturing Pilot Run

A total of ten Digital Thermometers were manufactured as part of the Pilot Run for this project. All of the final units were tested and verified per the specified test procedures. A large number of documentation errors were found and corrected. The final test fixture and test procedures were changed significantly as a result of feedback from the Pilot Run. See Product Engineering Pilot Run Report for detailed information on the result of the Pilot Run.

4.0 Reliability Test Results

One Digital Thermometer was placed on an accelerated life test. The unit has survived the specified time period (150% of the warranty period) and the test continues. Testing will continue until there is failure. The test unit will continue to be shut down and inspected on a monthly basis.

5.0 Product Literature Status

All Product Literature is complete and ready for release.

6.0 Product Documentation Status

The documentation is complete and ready for final signoff. All the changes that were approved after the field test and the manufacturing pilot run have been implemented. After release all changes will be implemented within the ECN system.

7.0 Test Fixture/Procedure Status

All Test Fixtures and Procedures are ready for release.

8.0 Financial Review

The estimates for manufacturing cost are below the manufacturing goal for the base Digital Thermometer unit. There is some concern about the cost for product options compared to the price of the options. It was agreed that only the base Digital Thermometer unit will be released at this time until the cost of the options can be determined more accurately.

The Sales department has reviewed the project and feels confident with the current sales forecast. See the Digital Thermometer financial review for all of the financial details.

9.0 Final Summary

Based upon the successful completion of all tests and financial criteria, the base version of the Digital Thermometer is ready for formal release. The project team will remain in effect for the next six months to resolve problems and monitor the performance of the project.

▲ **FIGURE 12–7**
Digital thermometer design review

INCONTROL

the Industrial Control *SPECIAL*-ists

introducing...

the IC100 Digital Controller

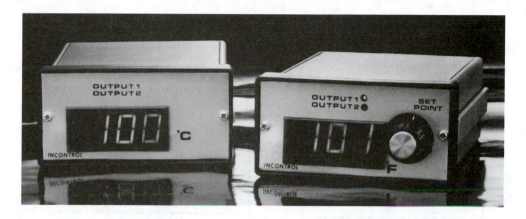

The IC100 is a temperature control. It represents the application of the latest in electronic packaging; surface mount technology. Its uniqueness is realized by its compact size and competitive price.

The standard IC100 comes as an indicator or a control, accepts thermocouple or RTD inputs and provides up to two outputs. But this is just the start. At INCONTROL, our goal is to supply OEM customers with a cost-effective, state-of-the-art controlling device, tailored to meet their requirements.

We are developing a unique line of standard products as well as working on custom OEM products and would like to review any control problem. We can provide 1/8 DIN, 1/4 DIN or custom-sized enclosures and can utilize analog, digital or microprocessor electronics as required by the application.

If the standard IC100 fits your application, why not try out an evaluation unit. If not, let us review your application to determine if modifications, or a clean sheet of paper is in order.

We encourage you to consider *INCONTROL* as an alternative to high-priced engineering charges and standard products that don't work exactly the way you would like. Let us put together a *SPECIAL* for you.

▲ **FIGURE 12–8**
Digital indicator and controller product brochure

IC100 SPECIFICATIONS

INPUT: TYPE J THERMOCOUPLE
OR 100 OHM RTD.

ACCURACY: .5% OF RANGE + OR -1 LSD.

INPUT TEMPERATURE RANGE:
MAXIMUM -200 TO 999 F OR C.

AMBIENT TEMPERATURE RANGE:
0 TO 55 DEGREES C.

POWER INPUT: 115 VAC + OR -10%
OR 230 VAC + OR -10%

DISPLAY: .56" LED DISPLAY OF
TEMPERATURE OR DISPLAY OF
SET POINT WHEN "PUSH-TO-SET"
DIAL IS DEPRESSED.

CONTROL OUTPUTS: TWO ELECTRO-
MECHANICAL RELAYS. CONTACTS
RATED AT 5 AMPS 230 VAC.

CONTROL FUNCTION: OUTPUT 1 CAN BE
ON-OFF OR TIME PROPORTIONING
WITH MANUAL RESET FUNCTION.
OUTPUT 2 CAN BE ON-OFF.

DIMENSIONS

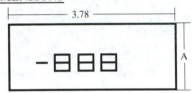

FOR 1/8 DIN DIMENSION A=1.89"
FOR 1/4 DIN DIMENSION A=3.78"

PANEL CUTOUTS: 1/4 DIN-3.62" H BY 3.62" W
1/8 DIN-1.77" H BY 3.62" W

IC100 ORDER MATRIX

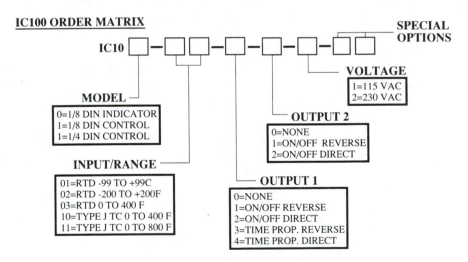

INCONTROL, INC.

1212 E. DOMINICK ST.   **TELEPHONE: 315-336-5812**
ROME, N.Y. 13440   **FAX: 315-337-9176**

▲ **FIGURE 12–9**
Digital indicator and controller product brochure

▶ References

Sommerville, I. 1992. *Software Engineering*. 5th ed. Reading, MA: Addison-Wesley.

▶ Exercises

12–1   What is the primary difference between the product assurance testing described in this chapter versus the prototype testing discussed in Chapter 11?

12–2   What is the purpose of product assurance testing, and why is it usually performed separately from engineering performance testing?

12–3   What is the difference between prelminary and released software?

12–4   What does the system release number indicate on most software version numbers? What does the version number indicate?

12–5   What are the two basic types of manufacturing processes to be selected from when determining the manufacturing process for a new product?

12–6   Define what is meant by the term *batch processing*, and discuss its advantages and disadvantages.

12–7   Define what is meant by the term *one-piece flow processing*, and discuss its advantages and disadvantages.

12–8   What is meant by the term *palletized board assembly*? What are the advantages of using the palletized board concept?

12–9   What method is used to connect test fixtures to circuit points located directly on the printed circuit board? What are the different methods of applying force to make the connections?

12–10   Define what is meant by an *infantile failure*.

12–11   What is the process called *burn-in*, and what is its purpose?

12–12   What parameter is used to describe the reliability of a design?

12–13   What are the two ways of projecting the reliability of a design?

12–14   List the primary goals of the manufacturing pilot run.

12–15   What are the primary goals of completing the field tests?

12–16   List all of the types of documents that make up the manufacturing documents for a product.

12–17   What are all of the ramifications of releasing a product or project design?

13 Step Six: Conclusion (Design Improvements and Project Performance Monitoring)

▶ Introduction

At this point in a project, we are looking back and saying: "How did we do, how well have we met the project goals?"[*] It is important to take note of what went right, what went wrong, and what can be learned from it. This is the "Conclusion" step, the last of the Six Steps. This stage of the project results in design improvements for the design that can be implemented now or later. It is also important to take the time to complete this step for the sake of future projects. There is much that is learned while developing any project. The magnitude of the learning is not apparent until time is taken to reflect on it. This last step of developing conclusions is the implementation of the continuous improvement concept.

In this chapter we will discuss the process of design improvements and project performance monitoring as it should be done in a professional project environment. This chapter includes the following specific topics:

- ▶ Performance goals
- ▶ Schedule performance

---

[*]In the senior project class that I teach every year, the students present their senior project to the class on the last class day of the semester. As with most students, they start to think about the requirements for the final report and presentation about two weeks before they are due. When they ask about this, I always refer them back to the course syllabus where all of the detail they require is available—just as it was on the first day of class. When grading the final report and their presentations, I measure how well the student reviewed and delivered the requirements stated in the syllabus. This is one of the most important objectives of the course, because it determines how well the students defined and resolved the problem posed to them at the beginning of the semester. In their presentations, each student is asked to discuss the problems encountered while completing their projects as well as their solutions to those problems. Each student encounters different problems, and this sharing of problems and solutions is another important part of the learning experience in the course.—J. S.

- ▶ Cost goals
- ▶ Quality performance
- ▶ Sales goals
- ▶ Overall economic performance

As you review this chapter, you may wonder about the importance of cost, quality, and sales goals and their importance to the electronics professional. Their relevance comes from the fact that the project's performance in each of these areas determines the success level of the project. The cost, quality, and sales performance depend on how well the design meets the goals defined in the original specifications. A good design can make all the difference in the world.

13–1 ▶ Performance Goals

The performance goals for the project represent all of the operational and environmental requirements of the specifications. We discussed these in detail in Chapter 6. At this point this question should be posed: "How much have the performance requirements changed over the course of the project?" The answer is a measure of how reasonable and accurate the initial specifications were and how closely they fit all of the other requirements for the project (i.e., manufacturing cost, size, complexity, and the like, as shown in Figure 13–1).

The project manager should request that each functional design area (electronic, mechanical, and so on) complete a technical *design performance review* of the project. This review includes a listing of the obstacles encountered in meeting the performance specifications and how they were overcome. The design

▶ **FIGURE 13–1**
Design performance goal summary

1. Quantity of specification changes

2. Primary design obstacles

3. Performance accuracy

4. Speed performance

5. Power efficiency

6. Cost effectiveness

7. Ease of use

performance review report should also evaluate the design generally in areas such as accuracy, speed, power efficiency, cost effectiveness, and ease of use. A guideline for the design performance report follows.

Specification Changes

When the specifications are changed during a project, it is usually because they were made more stringent than necessary. This is often done with products when attempting to gain an advantage over competition in the marketplace. Before accepting overly aggressive specifications, the project team should determine if there are real benefits to the improved specifications and if the higher performance level can be achieved while meeting the other project objectives. The report should indicate each specification change made, the reason for the change, and the particular problem that made the original specification unachievable. In addition, any ideas or developments that would serve to make the specifications more readily achievable should be noted.

Primary Design Obstacles

Every challenging design has obstacles that make the design difficult to achieve. It is important to note them and how they were overcome. In reflecting back on a project, it is also beneficial to reevaluate the decision process for selecting a solution to a design obstacle and determine with perfect hindsight if the correct decision was made. This section of the report should list all of these obstacles with a review of the decision process in selecting their design solution as well as a current evaluation of those decisions.

Accuracy

Accuracy is simply a measure of how well the design accomplishes a task that it is required to do. In the case of the digital thermometer project, its accuracy was determined by how correctly it displayed the actual probe temperature. The accuracy of an electronic watch is determined by how well it keeps time after it is calibrated. Most designs have requirements in their specifications that are a measure of accuracy. After the project is complete, it is important to review how well the accuracy requirements were accomplished in the design and how the design could be modified to improve accuracy.

Speed

The speed of a design is determined by the time it takes to complete a particular operation. In a computer, the speed at which instructions are performed is determined by the clock speed. In the digital thermometer project, speed is determined by how quickly the circuit can sample the analog temperature input. More speed is usually desirable—and usually comes at a cost. That cost can come in many forms: less accuracy and higher component costs are typical examples. If speed is an important aspect of a design, then the relative speed performance of the design project should be measured. How easily were the specifications met, and what future changes could yield further speed improvements?

Power Efficiency

In today's world, power efficiency should be an important factor for any design that uses electrical power. Power efficiency has always been important for battery-operated components. With the realization that there is a limit to the number of resources for the generation of power, power efficiency is becoming increasingly important. Power efficiency, like speed, usually comes at a cost. The best electronic example of this is TTL integrated circuits versus CMOS integrated circuits. TTL integrated circuits offer very high switching speeds, but they use a lot of power when compared to CMOS components, which are very power efficient yet slower. In this case the cost of power efficiency is a slower speed. The power efficiency of the design should be evaluated, and methods of improving the efficiency should be reviewed and noted.

Cost Effectiveness

Cost effectiveness is a measure of how well the design performs the required tasks when compared to the cost of its manufacture. The best way to make this comparison is against another design such as a competitive product design. A design that can perform as well as a competitor's, but costs less, has a distinct advantage. In the marketplace, a lower cost can support a lower price, which can translate into higher sales and profits. After the project has been completed, a measure of the cost effectiveness of the design should be determined. In addition, a list of the top ten component and assembly operation costs for the design should be listed as well as the top five opportunities for reducing cost without relaxing the specifications.

Ease of Use

"User friendly," a term coined in the 1980s, has been used—and overused—so much that people became tired of hearing it. The concept remains, though, and has increased in significance as a competitive advantage for all products. The primary reason for this is the increased use of microprocessors in products and the resulting increase in the number of features and options included within them. These variations in operation and features require the user to program or set up the product before use. The flashing clock readout on a VCR has become a common occurrence and is a good example of the problem, as few people can remember how to set a VCR clock correctly after a power outage.

Ease of use should be a goal for all products. Ease of use should include the sale, installation, setup, programming, and use of the product. The best way to evaluate the ease of use of a product is to compare it to another competing design. Learn to take every opportunity to compare the ease of use of products and make it a practice to evaluate the ease of use of a product whenever buying one. The following three areas are a measure of the ease of use of a product:

1. *Requirement for an operator's manual*—Can the majority of the most common operations be completed without the use of an operator's manual?

2. *Quantity and clarity of the required steps*—How many steps are required to complete a particular operation, and how clearly are these steps presented?

3. *Amount of information to be remembered*—This means the number of items about the operation that must be committed to memory in order to operate the product.

To summarize, a product that is easy to use is one that can be operated almost intuitively without an operator's manual. The product, if possible, should prompt you through the process. The steps required to set up and use the product should be minimal and clearly presented. Finally, the amount of information about the product required to perform common operations should be minimal and always noted on the product somehow.

13–2 ▶ Schedule Performance

When a project is not completed on time, major problems can develop. Sometimes if a schedule is not met, the entire project may be a failure. In other cases the expected benefit of the project may be put off indefinitely. This is the case when a product is scheduled for introduction with much fanfare at an annual trade show. When the product is not completed on time, the ability to effectively introduce the product is lost for one year. In the one-year period following the show, competition can introduce a new product and the market situation can change drastically. Schedule performance is the most obvious and easily measured performance area of a project. Its measurement is determined very simply by the actual completion date compared with the scheduled date. However, the analysis of schedule performance should entail much more than that. As important as schedules are, meeting them is one area of a project's performance most often not achieved. The primary reasons for failure to meet schedules usually stem from one of the following:

1. The requirements of the project were not clearly defined in the specifications.

2. The project was highly innovative and involved many new concepts. Consequently, task times were difficult to estimate accurately.

3. Poor schedule development and project management
   Many times a schedule is completed by first defining the desired end point instead of going through the process of determining a realistic completion date.

4. Lack of teamwork

It is important to look back at the project and determine why the schedule was not met. It is also important to analyze those rare occasions when a project schedule is exceeded. In reality, a successful schedule is one that is met almost exactly.

A schedule that is completed well ahead of schedule is one that usually was developed with too much slack time in it. This can cause problems when it interacts with other project schedules or introduction dates.

To properly evaluate schedule performance, the subproject area in which the schedule slippage occurred should be noted and analyzed. Which particular tasks within this subproject caused the project slippage? Was the slippage due to a poorly defined task, improper scheduling, or some other reason? What was the impact of the project being completed later than planned? These questions should be answered and analyzed for each area of project slippage or excess. Finally, conclusions and recommendations for use on the next project should be included in project performance report.

13–3 ▶ Cost Goals

Initially cost goals are more difficult to measure than any other goals. This is because the cost goals were developed with a specific volume level for the product. Right after a product is released, the cost goals are seldom achieved. This is because the manufacturing department is still learning how to produce the product efficiently. In addition, sales levels have not yet achieved the volume where the component costs are reduced enough to support meeting the cost goals. The initial manufacturing costs decline by as much as 25% to 35% as manufacturing learns to produce the product efficiently and as sales volume increase to the forecast levels.

The best way to measure the performance to cost goals is to take the actual costs that are incurred and apply estimates for the reduction of labor costs due to improved efficiency and lower component costs at the forecast sales levels. These estimates should be compared to the cost goals for the project. If the cost goals are not being met, it is important to determine why. Are the higher costs due to higher labor costs? To higher component costs? Once it has been determined that the manufacturing costs are not measuring up to the goals, some action must be taken: reduce costs, increase prices, accept lower profits, or discontinue the product. Sometimes, if costs are way out of line, costs are reduced, prices are increased, and lower profits are accepted rather than discontinue the newly completed project. Once the reason for the cost overage has been determined, the cause of the incorrect estimate should be determined and noted in the report so that this error is not repeated on the next project.

It is sad to see a great product that cannot meet the established cost goals even when it performs and sells well, when customers like it. The response of senior management when the sales and profit numbers are reviewed in typical monthly review meetings is usually: "Sales are nice, but where are the profits?" A product can meet every other objective and still be considered unsuccessful if unable to provide a profit or other financial benefit. After all, as discussed earlier, financial benefits are why the project was selected and initiated. There is always an economic advantage that is the basis for taking on a project. If this economic advantage is not fulfilled, the project is considered a failure regardless of how wonderful it is.

13–4 ▶ Quality Performance

In Chapter 12 we discussed how the quality department oversees the product assurance testing in an effort to provide an independent verification that the specifications have been satisfied. The development of reliability projections, in the form of mean time between failure, was also discussed. Now that the product has been released, the quality level of the product should be monitored during manufacture and after shipment to customers.

Quality Performance in Manufacturing

This is a measure of the quality of the manufacturing process and how well procedures are being followed. To accomplish this, data is taken at the various test points in the manufacturing process. When a product fails a test, it is noted in addition to the reason for the failure. The reasons for failure, presented in the order of the frequency of their occurrence, may be some of the following:

1. *Components improperly assembled in the wrong location*—This occurs with components that are the same size, such as resistors or integrated circuits with the same physical package connections. This is probably the most common error made on circuit boards that are manually assembled. It also occurs on autoinserted components when the incorrect component is placed in a feeder tray of the autoinsertion machine.

2. *Poor quality connections due to poor solder joints and improperly installed connectors*—This fault occurs in both manual and automated soldering operations. With manual soldering, the reason is simple fatigue or forgetting— or hurrying—to complete a solder joint. Automated solder failures are usually due to lack of control of the process, such as incorrect solder temperature, failure to remove flux residue periodically, or improper wave profile. Improperly applied connectors are usually due to sloppy workmanship.

3. *Procedures not properly followed*—There are many variations of this type of failure. In general, it means that the required procedure was not completed in the correct order or was not completed correctly. Many times this is seen when a test is performed improperly, indicating a failure and when, in fact, the product is well within specifications if the tests were properly completed.

4. *Faulty component*—This is the least-occurring type of failure and signifies that either a faulty component was installed in the product or that some part of the manufacturing process caused the component to fail. If the component was faulty on receipt (identified at incoming inspection), then this fact should be monitored for the frequency of occurrence. At some point incoming inspection and testing of the component may be warranted. If a failure is being induced in a component, it is critical to determine the exact cause of this. The most common defect in manufacturing processes that induces failures in manufactured products is the improper handling of MOS-type components in an environment of static electricity.

Customer Quality Performance

There are a number of ways to measure the quality performance of a product from the customer's perspective. The first is the customer's opinion about its quality. These measurements reflect the customer's expectations and therefore are subject to their point of view. How easy was the product to install? Was the operator's manual helpful and easy to understand? One customer might think that the operator's manual was comprehensive while another might think just the opposite, that the manual did not have enough information—or the right information. These measurements are best taken with a simple questionnaire that allows the customer to comment on the general appearance, function, and ease of use of the product. The performance of the product in these areas should be determined by taking an average of the respondent's replies, developing a rating for the product, and comparing the change in rating as attempts are made to improve weak areas.

The second method for measuring the quality of a product involves its failure and the reason for its failure. From the customer's perspective, a failure is a failure. However, failures that occur sooner rather than later leave a less favorable impression. These four time lengths or levels of failure have important distinctions and are explained next. Figure 13–2 shows a summary of the four levels of failure.

Out-of-Box Failures

These are the most severe failures as they cause the greatest amount of dissatisfaction for the customer. In these cases, as the name implies, the customer receives the product and unpacks it, and the product fails to perform right out of the box. Out-of-box failures can be induced by rough handling during shipment; otherwise they indicate a complete failure of the manufacturer's quality system. If the failure was induced by rough shipment, then a determination should be made as to the correct use of the packaging and any evidence of improper handling should be reviewed.

When the out-of-box failure is not due to shipment, then the exact nature of the failure must be determined. The most prevalent cause of these types of failures is incomplete testing of the entire assembly before shipment. Once identified, every effort must be made to make changes that prevent any reoccurrence of this particular problem.

▶ **FIGURE 13–2**
Product failure categories summary

1. Out-of-box failures

2. 30-day failures

3. Within-warranty failures

4. Out-of-warranty failures

Another type of out-of-box failure may not involve a failure of the product at all. An example is when an operator's manual does not explain the operation of the product to the customer to the point where the customer believes it is not functional. In this case the product failure is caused by the operator manual.

Failures Within 30 Days of Receipt

These failures are called *infantile* or *premature failures*. The product is received by the customer, set up, and verified for use. After a period of less than 30 days, the product fails. This is a very special case, because the product did function at the customer location and now it has failed. This type of failure is usually caused by a component failure or a weak electrical connection. These failures can occur after routine use or when excessive strain is placed on the product. It is therefore important to determine the failed mechanism, be it a component or connection. The application of the product should be reviewed with the customer for any specifications that are being exceeded.

If the failure occurs under normal use of the product, it is a true infantile failure. The only method available to minimize infantile failures is with a burn-in process, as discussed in Chapter 12. Infantile failures are almost as aggravating to the customer as out-of-box failures.

Failures Within the Warranty Period

These are similar in nature to infantile failures except that the product operates for a longer period of time (greater than 30 days) but less than the specified warranty period. Of course, some percentage of the product will always fail before the warranty has expired. The actual percentage level of product failures under warranty significantly affects the product's success. These failures are usually due to a weak component, connection, or a design problem. Data indicating the root cause of the failure should be maintained and monitored.

Failures Outside the Warranty Period

These failures are the least bothersome for both the customer and the supplier. From the customer's perspective, that depends on the amount of time after warranty expiration that the failure occurs. A failure that occurs one week after a one-year warranty period is viewed much differently than a failure occurring one year after a one-year warranty expiration. Although these failures are not a financial burden to the manufacturer, they do represent a problem to the customer. Therefore, the reasons for these failures should be determined. Data regarding them should be kept and reviewed regularly while efforts are made to reduce the overall MTBF for the product.

These four categories of failures represent four different aspects of the reliability of the product and the quality process that has been put in place to ensure it.

1. Out-of-box failures are a measure of how well a supplier checks out, packages, and ships the product to the customer.

2. 30-day failures are a measure of infantile failures, which can be affected by burn-in.

3. Warranty failures indicate the level of failures within the warranty period excluding 30-day failures. These are usually the result of inferior or misapplied components or other design problems.

4. Nonwarranty failures are used in conjunction with all other failures to determine the overall MTBF for the product. These are the result of inferior or misapplied components, other design problems, or the limited life of the product.

Product reliability can be improved by a combination of design improvements, test fixture and procedure changes, product packaging, information enhancements, and modified burn-in.

Example 13–1, Determination of Infantile Failures

This example will show how to analyze a failure to determine if it should be considered an infantile failure. A stereo amplifier has an output power MOSFET transistor that drives the speaker directly. The amplifier has a maximum drive capability of 8 Ω. After about 20 days of use, one channel of the amplifier fails completely and no sound can be heard from that channel. Inspection and replacement indicates that the power MOSFET no longer functions and its replacement successfully repairs the amplifier. The essential question is whether the customer used the amplifier under the specified conditions, that is, with a load of not less than 8 Ω. In this case the design of the stereo amplifier utilizes a temperature detector to sense the power MOSFET temperature and will shut it down if a predetermined temperature level is detected. The predetermined temperature level equates to slightly exceeding the maximum load. After replacing the power MOSFET, the temperature detector and temperature shut-off circuit are tested to verify their function to prove that their malfunction was not the cause of the MOSFET failure. After verifying this, the failure can be classified as an infantile failure of the MOSFET.

13–5 ▶ Sales Goals

With every design project that results in a product for sale, there is the all-important question: How well is the product selling? If all of the information put into the specifications reflected the market situation accurately, and if the product truly does meet the specifications, then the product should sell as well as expected. The sales performance depends on how well the sales team executes the sales plan and on the factory support of sales. Executing the sales plan involves introducing and training the sales force on the new product, sales promotion (advertising, trade shows, sales contests, and the like), and introductory sales calls. As part of the project team and a technical professional, you may find yourself involved in the sales training program. The factory should support the sales effort with short delivery lead times and timely application and technical support for salespeople and customers alike. As a technical expert on the product, you may also find yourself being involved with applications and technical support.

Early in the project, the sales department is asked to submit a forecast of sales for the new product, contingent on completing the project on schedule and meeting all the required specifications. These forecasts are constantly reviewed and modified as the project progresses. When the product is finally released, the forecast is a firm commitment from the sales department that this sales forecast will be met. On a monthly basis as other project performance measurements are being reviewed, the sales levels are reviewed and compared to the forecast. Upper management is always interested in these numbers and will usually request explanations for any underachievements. Accordingly, experienced salespeople submit sales forecasts that can be achieved reasonably.

As mentioned before, if the product meets the specifications and they are based on a sound market strategy, sales will usually exceed expectations. It is the magic that results from a good problem definition, a solid plan, and strong execution. Sometimes, however, unforeseen events can affect sales of new products. An example is when a competitor introduces a new product that changes the market significantly. Another case is when the economy declines just as a new product is released.

13–6 ▶ Overall Economic Performance

The overall economic performance of the product is dependent on how well the product performs in each of the performance areas already discussed. If the product was completed on time, meets all aspects of the specifications including manufacturing costs, exudes quality to the customer, and meets the projected sales levels, the overall economic performance is almost guaranteed. If one or more of these performance areas are not achieved, the entire project's success comes into question.

The economic performance is measured as profit. There are many profit measurements the finance department or group uses. Operating profit, net profit, and earnings before taxes (EBT) are some of the more common measurements. Each one of these differs in the amount of actual company costs that are applied to the product. The accounting department generally classifies the expenditures of the company as manufacturing costs, capital, or overhead. Manufacturing costs include only the component and labor costs to manufacture a product. Capital expenditures are long-term investments, a piece of equipment, a building, or improvements to a building. The only balance sheet cost a company incurs when money is spent on capital is the amount of depreciation that the particular item experiences in a given year. This is because when the capital item is purchased, it initially has a net value equal to its cost, so the balance sheet will show a capital expense, while the asset balance sheet records an asset value increase. Overhead expenses are essentially all expenses not directly related to the manufacture of the product. It includes items such as sales commissions, advertising, and administrative costs. There are many different ways that accountants allocate these costs in order to calculate the profitability of a product line. Some companies take the total overhead number and determine the amount of overhead expenses to apply to a product line by simply determining the percent of sales of the product line. The percent of sales number is multiplied times the total overhead costs to estimate overhead expense for the product line. Other companies make a significant effort to itemize the actual utilization of overhead by the various product lines, thereby giving a much more accurate picture of the profitability of the product line.

The profit review is the last part of the monthly review meeting and is usually completed by the finance department. Whichever profit measurement has been chosen as the primary measurement (operating profit, net profit and so on), it should be determined and reviewed. Any deviations plus or minus should be investigated to determine the root cause. A failure to meet profitability projections most often comes from not meeting either sales or cost goals.

▶ Summary

At the conclusion of each of the other Six Steps, a list of deliverables has been presented. For the sixth and final step, the deliverables are as follows:

1. A successful project

2. A list of ways to improve the project design in the future

3. A lot of knowledge

There are two aspects to project performance monitoring that have been discussed in this chapter. The first involves issues such as schedule performance, the design, and its cost effectiveness. These issues are reviewed once in order to make conclusions about the project, so that anything learned during the project can be utilized in the future. The second involves issues that will continue to be reported monthly for the life of the product: sales, quality, costs, and profits. Figure 13–3 summarizes all of the areas of project performance.

About six months after the product is released, the final project review meeting is scheduled at which a review of all the one-time performance areas is completed for use as future reference of the project. These include:

Performance goals:
1. The number and reason for specification changes
2. Major design obstacles and how they were overcome
3. A design evaluation that rates the accuracy, speed, power efficiency, cost effectiveness, and ease of use of the design

Schedule performance: An evaluation of the project's schedule performance, including major schedule oversights and how they were resolved, along with future recommendations.

The second category of issues will continue to be reviewed on a monthly basis as a normal part of the company's operations review should be reviewed by the project group one final time. These include:

1. Project performance:
   Number of specification changes
   Number of obstacles overcome
   Design evaluation (accuracy, speed, power
      efficiency, cost effectiveness, ease of use)

2. Schedule performance: overall evaluation

3. Quality:
   Product lead times and on-time delivery
   Manufacturing quality level
   Customer quality level
   Overall mean time between failure

4. Cost: costs compared to goals

5. Sales: sales compared to goals

6. Profit: profits compared to goals

▶ **FIGURE 13–3**
Summary of project performance areas

Quality level:
1. Product lead times and meeting customer promise dates
2. Manufacturing quality level
3. Customer quality level—out-of-box, 30-day, warranty, and nonwarranty failures, and customer questionnaires
4. Overall MTBF

Cost levels: Component and labor costs compared to goals

Sales levels: Sales compared to projections and success and failure with key accounts

Profit levels: Profit levels should be reviewed as a percentage and as a total compared to the projections for the product.

The project is now considered complete. The project team is disbanded, and there are no project meetings to attend. Projects are like many other important happenings in your life. They affect you and change you, as indeed they should. Working on a project is always a learning experience. Approach them with the mindset to do your best in meeting their objectives while learning as much as possible. In time you will look back on your career and your life in general as just a long series of projects, each one providing a different kind of knowledge, experience, and enrichment. Projects are like people—there are never two completely alike.

Digital Thermometer Example Project

The digital thermometer project is complete. If it was an actual product, it would have been released for sale, and the project team would be reviewing its performance in the areas of sales, profits, quality, reliability, and customer satisfaction. Each one of these areas are reviewed on a monthly level and compared with forecasts or goals for the product. Remember the Step One proposal, where a sales forecast was developed for use product specifications. These are the sales forecasts that are used initially to measure the product's performance. Figure 13–4 shows how actual sales and profits would be compared against forecast numbers. Figure 13–5 shows the actual quality indicators as they compare to quality goals for the product.

▶ Exercises

13–1   Why is it important to review the number of specification changes that occurred on a project?

13–2   As part of Step Six, why is it important to reflect on design decisions that are made on a project?

13–3   List the five different types of performance parameters that were discussed in this chapter that relate to measuring the performance of a design and define each.

13–4   Discuss the methods for determining the ease-of-use level for a product.

13–5   List as many negative factors as possible that can result from not meeting a product development schedule.

13–6   What are the primary reasons for not meeting a project schedule that were discussed in this chapter?

13–7   Why is it difficult to measure and compare manufacturing costs to goals on the initial production runs?

13–8   Discuss why achieving sales goals are an important part of meeting the manufacturing cost goals.

13–9   When the manufacturing cost goals are not being met, what are the two primary actions that can be taken to rectify the situation?

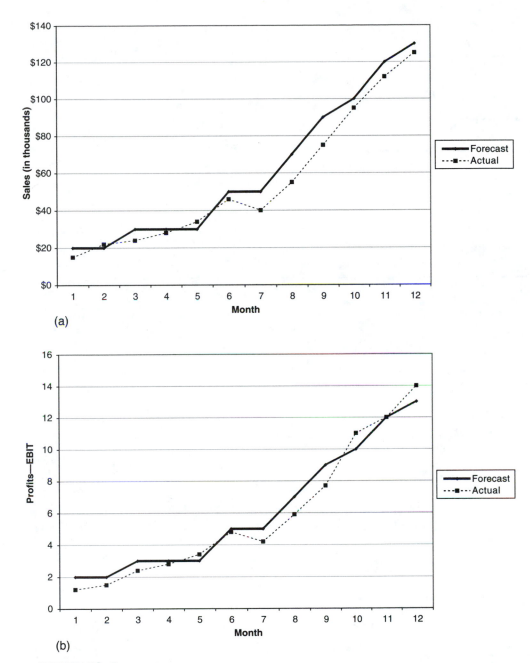

(a)

(b)

▲ **FIGURE 13–4**

Digital thermometer sales and profit reports

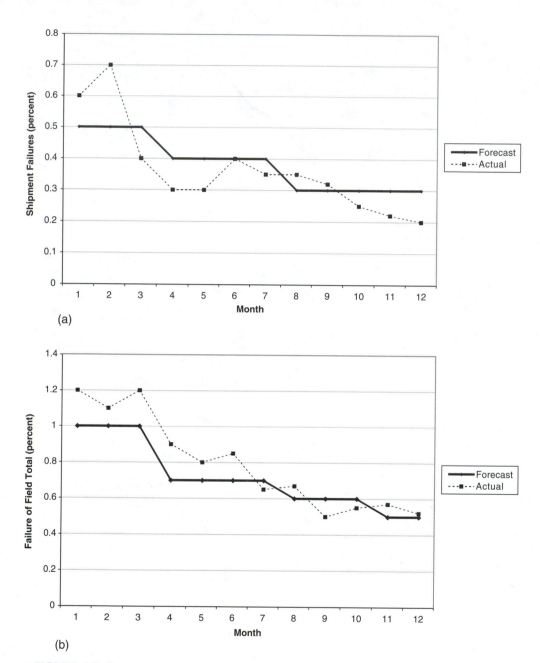

▲ **FIGURE 13-5**
Digital thermometer quality reports

13–10 List all of the ways that the customer quality level of a product can be measured that were discussed in this chapter.

13–11 Explain the difference between 30-day failures and out-of-box failures? Which type is more critical and why?

13–12 How can infantile failure levels be reduced?

13–13 How can out-of-box failure levels be reduced?

13–14 What does MTBF mean and how is it calculated?

13–15 What is the most likely reason that a newly assembled printed circuit board is not functional?

13–16 What is the least likely reason for a newly assembled printed circuit board to be nonfunctional?

13–17 Define in your own words what is meant by a successful project.

Appendix A ▶ Component Reference Information

A–1 ▶ Resistors

Resistors limit the flow of current and divide voltage in electrical circuits. If the current through a resistor is known, the voltage drop across it can be calculated using Ohm's Law ($V = I \times R$). Resistors are not frequency dependent, so they have the same resistance value for AC and DC voltage.

Resistor color codes are shown in Figure A–1. Included in Figure A–1 are the four-band code used for 5%, 10%, and 20% tolerance resistors and the five-band color code used for 1% tolerance resistors. Figure A–2 shows the standard value resistors that are available at different tolerance values.

A–2 ▶ Capacitors

Capacitors store an electrical charge on plates separated by a dielectric material. Capacitors oppose any changes in the voltage across them by giving up previously stored charge if the voltage decreases—or storing more charge if the voltage across them increases. Current never flows through a capacitor (except a small leakage current), but with AC voltages current seems to flow through because the current is constantly changing direction. In AC circuits, the capacitor is continually charging and discharging as the current flows in one direction and then reverses. Capacitive reactance, X_C, is the impedance that the capacitor offers resisting the flow of AC current and is equal to

$$X_C = 1/(2 \pi f C)$$

where f = frequency
C = capacitance value

As frequency increases, the capacitive reactance decreases. For DC circuits then, $f = 0$, so the capacitive reactance is theoretically infinite. In DC circuits,

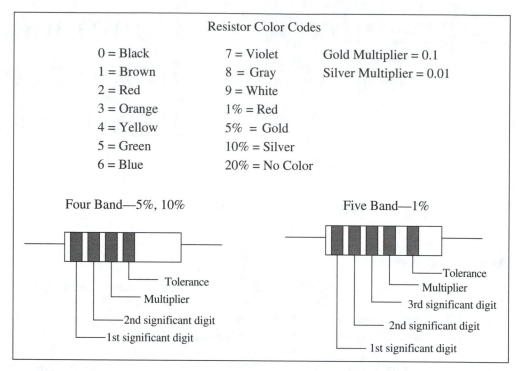

▲ **FIGURE A–1**
Resistor color codes

capacitors completely block DC current flow (except for leakage) and charge up to the DC voltage. The capacitance value determines its ability to store a charge. The larger the value, the more space it has to store charge. It makes sense that capacitance is added when you connect two capacitors in parallel. Conversely, it follows that two capacitors in series will react as two resistors in parallel would and can be calculated by the formula

$$C_t = C_1 \times C_2/(C_1 + C_2)$$

This is because the effective plate area is reduced and the distance between the plates is increased.

Capacitors are rated by their value, dielectric type, and the voltage they can withstand. These values are either stamped on the capacitor, or color codes are used. The standard capacitor values available use the same multipliers as for 10% tolerance resistors. Figure A–3 shows the standard capacitor values. It is important to note that not all of the values are available for a particular dielectric type or a specific manufacturer. For example, aluminum electrolytic capacitors are usually available only in multiples of 10, 22, 33, 47, and 68. Figure A–3 also shows the voltage ratings that are usually available for capacitors.

Capacitor color code locations are dependent on the type of dielectric used. The location and meaning of the color codes are shown in Figure A–4.

Standard Resistance Values
(the following values are available in powers of ten)

0.1% 0.25% 0.5%	1%	2% 5%	10%
10.0	10.0	10	10
10.1			
10.2	10.2		
10.4			
10.5	10.5		
10.6			
10.7	10.7		
10.9			
11.0	11.0	11	
11.1			
11.3	11.3		
11.4			
11.5	11.5		
11.7			
11.8	11.8		
12.0		12	12
12.1	12.1		
12.3			
12.4	12.4		
12.6			
12.7	12.7		
12.9			
13.0	13.0	13	
13.2			
13.3	13.3		
13.5			
13.7	13.7		
13.8			
14.0	14.0		
14.2			
14.3	14.3		
145			
14.7	14.7		
14.9			
15.0	15.0	15	15
15.2			
15.4	15.4		
15.6			
15.8	15.8		
16.0		16	
16.2	16.2		

0.1% 0.25% 0.5%	1%	2% 5%	10%
16.4			
16.5	16.5		
16.7			
16.9	16.9		
17.2			
17.4	17.4		
17.6			
17.8	17.8		
18.0		18	18
18.2	18.2		
18.4			
18.7	18.7		
18.9			
19.1	19.1		
19.3			
19.6	19.6		
19.8			
20.0	20.0	20	
20.3			
20.5	20.5		
20.8			
21.0	21.0		
21.3			
21.5	21.5		
21.8			
22.1	22.1		
22.3			
22.6	22.6		
22.9			
23.2	23.2		
23.4			
23.7	23.7		
24.0		24	
24.3	24.3		
24.6			
24.9	24.9		
25.2			
25.5	25.5		
25.8			
26.1	26.1		
26.4			

0.1% 0.25% 0.5%	1%	2% 5%	10%
26.7	26.7	27	27
27.1			
27.4	27.4		
27.7			
28.0	28.0		
28.4			
28.7	28.7		
29.1			
29.4	29.4		
29.8			
30.1	30.1	30	
30.5			
30.9	30.9		
31.2			
31.6	31.6		
32.0			
32.4	32.4		
32.8			
33.2	33.2	33	33
33.6			
34.0	34.0		
34.4			
34.8	34.8		
35.2			
35.7	35.7		
36.1			
36.5	36.5	36	
37.0			
37.4	37.4		
37.9			
38.3	38.3		
38.8			
39.2	39.2	39	39
39.7			
40.2	40.2		
40.7			
41.2	41.2		
41.7			
42.2	42.2		
42.7			
43.2	43.2		

0.1% 0.25% 0.5%	1%	2% 5%	10%
43.7		43	
44.2	44.2		
44.8			
45.3	45.3		
45.9			
46.4	46.4		
47.0		47	47
47.5	47.5		
48.1			
48.7	48.7		
49.3			
49.9	49.9		
50.5			
51.1	51.1	51	
51.7			
52.3	52.3		
53.0			
53.6	53.6		
54.2			
54.9	54.9		
55.6			
56.2	56.2	56	56
56.9			
57.6	57.6		
58.3			
59.0	59.0		
59.7			
60.4	60.4		
61.2			
61.9	61.9	62	
62.6			
63.4	63.4		
64.2			
64.9	64.9		
65.7			
66.5	66.5		
67.3			
68.1	68.1	68	68
69.0			
69.8	69.8		
70.6			

0.1% 0.25% 0.5%	1%	2% 5%	10%
71.5	71.5		
72.3			
73.2	73.2		
74.1			
75.0	75.0	75	
75.9			
76.8	76.8		
77.7			
78.7	78.7		
79.6			
80.6	80.6		
81.6			
82.5	82.5	82	82
83.5			
84.5	84.5		
85.6			
86.6	86.6		
87.6			
88.7	88.7		
89.8			
90.9	90.9	91	
92.0			
93.1	93.1		
94.2			
95.3	95.3		
96.5			
97.6	97.6		
98.8			

▲ FIGURE A–2
Standard resistor values

Standard Capacitance Values

Picofarads	Picofarads	Microfarads	Picofarads	Microfarads	Microfarads	Microfarads	Microfarads
1.8	120		12,000	0.012	1.0	120	1800
2.2	180		15,000	0.015	1.2	180	2200
2.7	220		18,000	0.018	1.8	220	2700
3.3	270		22,000	0.022	2.2	270	3300
3.9	330		27,000	0.027	2.7	330	3900
4.7	390		33,000	0.033	3.3	390	4700
5.0	470		39,000	0.039	3.9	470	5600
5.6	560		47,000	0.047	4.7	560	6800
6.8	680		56,000	0.056	5.6	680	8200
8.2	820		68,000	0.068	6.8	820	
10	1000	.001	82,000	0.082	8.2	100	
12	1200	.0012	100,000	0.1	10	120	
15	1500	.0015		0.12	12	150	
18	1800	.0018		0.15	15	180	
22	2200	.0022		0.18	18	220	
27	2700	.0027		0.22	22	270	
33	3300	.0033		0.27	27	330	
39	3900	.0039		0.33	33	390	
47	4700	.0047		0.39	39	470	
56	5600	.0056		0.47	47	560	
68	6800	.0068		0.56	56	680	
82	8200	.0082		0.68	68	820	
100	10,000	.01		0.82	82	1000	
					100	1200	

Standard Voltages

6.3	250
10	400
16	450
35	500
50	1000
63	2000
100	3000

▲ FIGURE A–3
Standard capacitor values

A–3 ▶ Inductors

An inductor is a coil of conductive material that stores electrical energy in the magnetic field created by the changing current passing through the inductor. Inductors oppose any change in the current passing through them. They accomplish this by giving up previously stored energy if the current is decreasing—or storing more energy in the magnetic field if the current is increasing. In AC circuits the inductors continually store and release energy as the current increases, reaches a maximum, and then declines and eventually flows in the other direction. Inductive reactance, X_L, is the impedance that the inductor offers to resist the flow of current and is equal to

$$X_L = 2\pi f L$$

where f = frequency
L = inductance

As frequency increases, the inductive reactance, X_L, increases. This is exactly opposite of the way capacitors function (X_C decreases as frequency increases). For

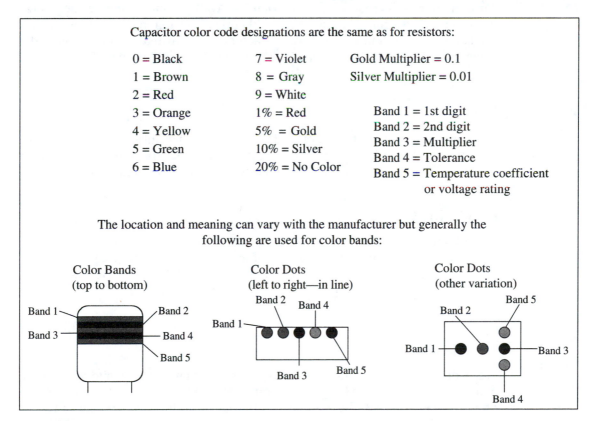

Capacitor color code designations are the same as for resistors:

0 = Black	7 = Violet	Gold Multiplier = 0.1
1 = Brown	8 = Gray	Silver Multiplier = 0.01
2 = Red	9 = White	
3 = Orange	1% = Red	Band 1 = 1st digit
4 = Yellow	5% = Gold	Band 2 = 2nd digit
5 = Green	10% = Silver	Band 3 = Multiplier
6 = Blue	20% = No Color	Band 4 = Tolerance
		Band 5 = Temperature coefficient or voltage rating

The location and meaning can vary with the manufacturer but generally the following are used for color bands:

Color Bands (top to bottom)
Band 1
Band 2
Band 3
Band 4
Band 5

Color Dots (left to right—in line)
Band 2
Band 4
Band 1
Band 3
Band 5

Color Dots (other variation)
Band 5
Band 2
Band 1
Band 3
Band 4

▲ **FIGURE A–4**
Capacitor color codes

DC circuits, $f = 0$, so inductive reactance is 0. In DC circuits inductors are a steady state short circuit. The inductance value, L, is given in units called *henries* and determines the inductor's ability to store energy. Practical ranges of inductance are millihenries (mH) or microhenries (μH). The larger the inductance value there is, the greater the capacity there is to store energy. Inductors are similar to resistors in that the equivalent of two inductors in series is simply the total of the two inductances. The equivalent of two inductors in parallel can be calculated by

$$L_T = L_1 \times L_2/(L_1 + L_2)$$

It is impossible to have a pure inductive component because of the resistance of the conductor used to make up the coil. The quality factor, Q, is a measure of this fact and is a ratio of the inductive reactance over the DC resistance:

$$Q = X_L/R_{DC}$$

Inductor values are given in henries. Inductors are rated by their value, tolerance, and maximum current. Inductor values are determined from the same standard number set as resistors and capacitors. Available inductance values range from nanohenries, microhenries, and millihenries. The nominal inductance values available are shown in Figure A–5. The actual values available depend upon the magnetic material used, the type of package, and the inductor manufacture.

A–4 ▶ Diodes

There are many different types of diodes designed for use in specific applications. Each type of diode will be discussed, along with the key parameters involved in its selection. The available package designations are shown in Figure A–6.

Rectifier Diodes

The primary purpose of rectifier diodes is simply the conduction of current in one direction, most often for the purpose of converting AC voltage to DC voltage. Generally, the key parameters for rectifier diodes are the forward biased current and the *peak inverse voltage*. The switching speed and reverse bias leakage are also important.

Zener Diodes

Zener diodes are commonly used to regulate or clamp voltages and as such are connected reverse biased. The key parameters are the zener voltage, zener current, and power rating. The tolerance and variation of the zener voltage and zener current with temperature are also important considerations. The most significant parameter for their selection is the Zener voltage.

Standard Inductance Values					
Nanohenries	Nanohenries	Microhenries	Nanohenries	Microhenries	Microhenries
1.0	120	0.12	12,000	12	1000
1.2	180	0.18	15,000	15	1200
1.8	220	0.22	18,000	18	1800
2.2	270	0.27	22,000	22	2200
2.7	330	0.33	27,000	27	2700
3.3	390	0.39	33,000	33	3300
3.9	470	0.47	39,000	39	3900
4.7	560	0.56	47,000	47	4700
5.6	680	0.68	56,000	56	5600
6.8	820	0.82	68,000	68	6800
8.2	1000	1	82,000	82	8200
10	1200	1.2	100,000	100	
12	1500	1.5		120	
15	1800	1.8		150	
18	2200	2.2		180	
22	2700	2.7		220	
27	3300	3.3		270	
33	3900	3.9		330	
39	4700	4.7		390	
47	5600	5.6		470	
56	6800	6.8		560	
68	8200	8.2		680	
82	10,000	10		820	
100					

▲ **FIGURE A–5**
Standard inductance values

Shottky Diodes

Shottky diodes feature a very high switching speed and are consequently used in high-frequency and fast-switching applications. The key parameter in their selection is the speed at which the diode goes from the forward to reverse biased operation.

Light-Emitting Diodes

Light-emitting diodes (LEDs) are used as indicators or displays or to transmit information optically. As indicators, the LEDs are usually used in individual packages. As displays, seven-segment or five-by-seven dot matrix LEDs are common. LEDs are used to transmit optical information down glass and plastic fiber in fiber optics applications. The key selection parameters are:

1. Wavelength or color of the output light

2. Intensity of the output light

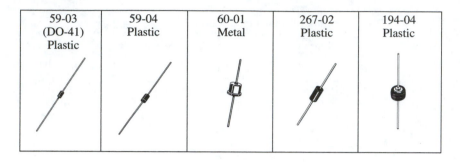

3. Switching speed

4. Efficiency (the amount of output light intensity per the input current)

Photodiodes

The photodiode is operated reverse biased. When light is applied to the reverse-biased photodiode, the leakage current increases. The key operational parameters for selecting a photodiode are:

1. Wavelength of maximum sensitivity

2. Sensitivity (the amount of reverse current per the input optical) power

A–5 ▶ Transistors

Transistor selection involves first the choosing of either bipolar junction transistors (BJTs) or unipolar field effect transistors (FET) devices. Bipolar junction transistors are characterized by faster switching speeds and poorer power efficiency. Field effect transistors are generally slower, but they draw very little current and are therefore very efficient. The available transistor packages are shown in Figure A–7.

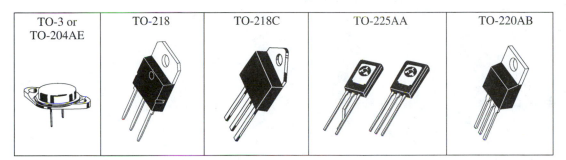

▲ **FIGURE A–7**
Transistor packages *(Copyright of Semiconductor Component Industries, LLC. Used by permission)*

Bipolar Junction Transistor

Bipolar junction transistors are generally used in small- to medium-power applications. The key parameters for selection are the maximum collector-to-emitter voltage, the maximum collector current, and the DC current gain, β.

Field Effect Transistor

The field effect transistor is also used in small- to medium-power applications where power efficiency is more important than speed. High power field effect transistors are called *MOSFETs* and are discussed next. The key field effect transistor parameters are the drain-to-source and the gate-to-source voltages, the drain-to-source current, and the forward transconductance. The forward transconductance is the change in drain-to-source current per the change in gate to source voltage. This is similar to the current gain, β, in BJTs.

Metal Oxide Semiconductor FET (MOSFET)

The primary application for MOSFETs are as power transistors called *power MOSFETs*. The power MOSFET is capable of high currents because of the very small voltage drop across the drain-to-source voltage. The key MOSFET parameters are the drain-to-source and the gate-to-source voltages, the drain-to-source current, and the forward transconductance.

A–6 ▶ Integrated Circuits

There are a wide variety of analog, digital, and hybrid (analog/digital) integrated circuits available. Their selection will involve all of the usual parameters, input and output voltage and current, power supply voltage, temperature coefficients, and frequency and speed capabilities. The different integrated circuit packages available are shown in Figures A–8 and A–9.

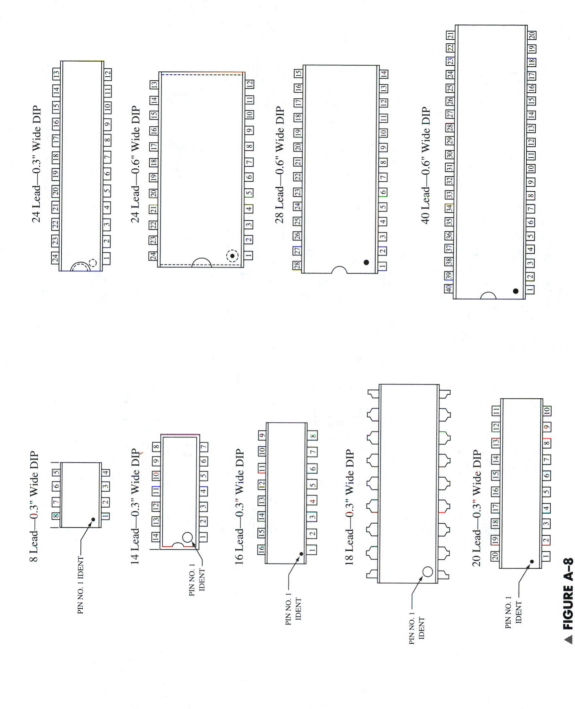

▲ **FIGURE A–8**
Integrated circuit packages

20-lead PLCC Package

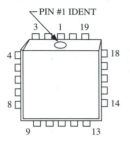

28-lead PLCC Package

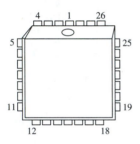

44-lead PLCC Package

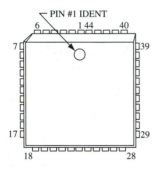

44-lead Quad Flat Package

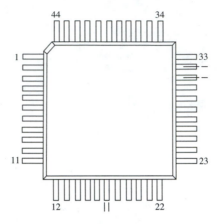

44-lead Cerquad Package

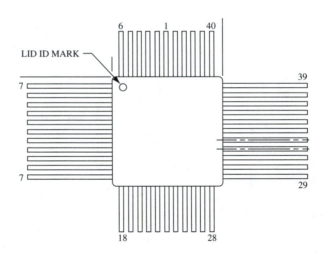

▲ **FIGURE A–9**
Integrated circuit packages

Appendix B ▶ Test Equipment

▶ Introduction

This appendix is a reference for what you have already learned and provides additional practical information on the general use of test equipment. The test equipment topics covered in this appendix are as follows:

- ▶ Multimeters
- ▶ Power supplies
- ▶ Function generators
- ▶ Oscilloscope
- ▶ Logic analyzers
- ▶ Spectrum analyzers
- ▶ Noise immunity testing
- ▶ Noise radiation testing

B–1 ▶ Multimeters

A multimeter is probably the oldest piece of existing electronic test equipment. It has also experienced the most dramatic changes. The analog dial volt-, ohm-, and milliammeter, better known as the VOM, was the mainstay of electrical and electronic test equipment for many decades. The VOM was plagued with a low-input impedance and the inaccuracy of reading the analog dial. The ideal voltmeter will have an infinite input impedance, thereby eliminating completely any loading effects. The first design improvement made to the VOM was to increase the input impedance by using electronic devices. The first of these was the vacuum tube voltmeter, better known as a VTVM. The VTVM maintained the analog dial meter, but it featured a much higher input impedance than the VOM. Next, the vacuum tubes were replaced with semiconductors and then integrated circuits in quick succession, and digital displays replaced the analog dial. The result was called a

373

digital voltmeter, or a DVM. When the ohmmeter and milliammeter functions were added to the DVM, it became a digital multimeter, or DMM for short.

Today the full-featured DMM is the most commonly used piece of electronic test equipment. The DMM is the result of many design improvements that overcome the weaknesses of the original VOM that it replaced. DMMs feature very high input impedance, a digital display, autoranging, and many other state-of-the-art features that will be discussed in this section. Figure B–1 shows a typical digital multimeter.

The operation of all of these devices, VOMs through DMMs, is very much the same except for the more advanced features included in DMMs that were not available on the VOM. These meters have two test leads: a red positive lead and a black negative lead. They all can read volts, ohms, and milliamperes. There are usually three connections for the two test leads: one ground connection that is always used, one lead for measuring current, and one for measuring volts or ohms. The current test lead is usually fused, and, in most college laboratories, this particular fuse experiences a very short life. Sometimes a fourth fused test lead is provided if the meter has an additional ampere range for measuring current (usually one ampere).

General Use

When using any DMM or equivalent for taking voltage, current, or resistance readings, it is important to use the proper terminals for the measurement being taken. The current measurement test lead is separate from the volt and ohm test lead so that the person choosing to take a current measurement must consciously think about the measurement being taken and physically change the connection on the meter. The newest DMMs sound an audible alarm whenever the test leads are connected to the current test jacks and a noncurrent function is selected on the meter.

Other than selecting the desired function of the meter, AC or DC, volts, ohms or milliamps, the range should be selected that will offer the best resolution of the measurement being taken. Meter resolution is the smallest incremental value into which the meter can break the measured signal. To achieve the best resolution available on any meter, it is desirable to spread out the reading as much as possible across one of the ranges on the meter. A simple practice that accomplishes this is to use the smallest range possible without going over range. In other words, decrease the range setting on the meter until an over range indication is given and then increase the range setting to the one just above where the over range indication was given. This range will offer the best resolution possible for the particular meter and measurement being taken.

Voltage Readings

Voltage readings are taken by connecting the proper test leads to the points in the circuit to be measured and then selecting the function and range of the meter. When selecting this function, be sure to select AC or DC volts as appropriate. One thousand volts is usually the maximum voltage reading that can be taken with this type of meter. The input impedance for voltage readings is on the order of 10 MΩ with a

▶ **FIGURE B-1**
Digital multimeter or DMM
*(Courtesy of Fluke Corporation.
Reproduced with permission)*

capacitance of less than 100 pF. For very high impedance circuits, the input impedance of the meter will determine the loading error that the meter will induce.

The voltage ratings for the meter will indicate the noise rejection levels in decibels (dB) for what is called *normal mode* and *common mode* noise. This type of specification is common and can be applied to many types of equipment. For this reason it is important to understand exactly what noise reduction levels mean. The dB measurement is a relative indication of what will happen to a signal that passes through a circuit. A signal can be amplified (positive dB) or attenuated (negative dB). In this particular case, the circuit is a noise filter. The formula for dB is:

$$dB = 20 \, \text{Log}_{10} \, (\text{output/input}) \tag{B-1}$$

The input signal in this case is the noise signal present on one or both test leads. The output is the amount of noise that remains after passing through the noise filter. The amount of noise remaining after the filter becomes part of the total meter reading. Most DMM specifications will state the Noise Rejection Specifications as follows:

DC volt normal mode rejection: > 60 dB at 50 Hz or 60 Hz
DC volt common mode rejections: > 120 dB at DC, 50 Hz or 60 Hz

Normal mode means that the noise is present on one test lead only. *Common mode* means that the same noise level is present on both test leads. Common mode noise is easier to reduce because a differential amplifier will amplify only the difference between two signals, eliminating almost completely any signal common to both test leads. To interpret the example meter specifications just presented, if 10 V of noise with a frequency between 50 Hz and 60 Hz was present on one test lead only, the remaining noise signal would be attenuated by at least 60 dB. Attenuation implies a negative dB value. Plugging into the formula shown in equation B–1:

$$-60 \text{ dB} = 20 \text{ Log}_{10} \text{ (output/10 V)}$$

$$-60 \text{ dB/20} = -3 \text{ dB} = \text{Log}_{10} \text{ (output/10 V)}$$

$$e^{-3} = \text{Output/10 V}$$

$$0.497 \text{ V} = \text{Output}$$

The calculation reveals that a 10 V input noise signal will be reduced to 0.497 V, a reduction of about 95%. If the same 10 V input noise signal was present on both leads, meaning that it is viewed as common mode noise, the noise reduction would be greater than 120 dB, which calculates out to a reduction of 99.7%. When taking DC voltage measurements, select DC volts on the meter's function selector and select the range with the best resolution, as previously discussed. For AC measurements, remember that the meter is indicating an RMS value of what it assumes to be an AC sinusoidal waveform. The RMS value for a waveform is related to the peak value of the waveform as shown below:

$$V_{RMS} = 0.707 \times V_{peak} \tag{B–2}$$

The AC voltage readings will be most accurate when the frequency of the AC waveform is between 50 Hz and 60 Hz. Later technology meters can measure AC waveforms in the range of 45 Hz to 20 kHz, although the accuracy decreases as the frequency increases above 60 Hz.

Ohm Readings

To measure resistance, select the ohm function on the meter function selector and the range that offers the best resolution. The most important thing to remember when taking ohm readings is that power should not be applied to the circuit being measured. Not only will the measurements be inaccurate, but there is the possibility of damaging the meter. Because of this fact, most ohmmeter functions have overload protection. Also, be sure to consider that when you measure the resistance between two points in a circuit, you are actually measuring all the parallel resistive paths between the two points. If other components are in parallel with the two points where a resistance measurement is to be taken, one lead of the appropriate component should be lifted to break the parallel circuit path.

Milliamp Readings

When taking milliamp measurements in a circuit, great care should be taken in selecting the measurement points. Remember that the ideal ammeter has zero resistance and the actual resistance of most DMMs is less than 1 Ω. When a meter is set up to function as an ammeter, a very small resistance (essentially a dead short) exists between the two circuit points where the ammeter test lead connections are made. In most cases the ammeter must be inserted into a circuit by breaking the series conductor path and inserting the ammeter in the conductor branch where the current is to be measured. In order to measure current, the red test lead must be placed in the appropriate range test jack (milliamps or amps) and the range offering the best resolution should be selected.

The most current DMMs offer many outstanding features and functions that will be discussed in the following subsections.

Diode Test

This is a special feature added to improve the diode testing capabilities of most multimeters. Previously, on most VOMs and DVMs, diode tests were performed with the ohmmeter function by measuring the very high reverse impedance of the diode compared to the very small forward impedance. Most of these meters do not have a high enough internal battery voltage to sufficiently forward bias some diodes. In these cases the test results were misleading.

In order to test a diode with a DMM that possesses the diode test function, look for the small diode symbol on the DMM function selector. Connect the red lead to the diode's anode and the black lead to the cathode. The DMM will indicate the forward voltage drop across the diode, which should be in the range from 0.5 to 0.9 V, depending on the type of diode. When the leads are reversed, creating a reversed biased diode, the voltage reading should equal the internal circuit test voltage of the particular meter, indicating that diode has a very high impedance. If either—or both—of the readings are incorrect, the diode will not function properly. When a diode fails, it can fail completely open, shorted, or exhibit a small resistance in both directions.

Maximum and Minimum Values

The current technology DMMs not only take measurements and indicate them. They can also be set up to store the maximum and minimum values for anything that can be measured on the meter. The DMM acts like a peak signal data recorder when used in this way. With this function the technician can walk away from the test site and leave the DMM in place to monitor and store the maximum and minimum values. These can be retrieved later by the technician. The true average of all of the values sampled is calculated and stored as well.

Frequency Counter

An extended feature of many modern DMMs is frequency counting. To determine the frequency of a signal, connect the test leads as you would to take a voltage measurement and select the hertz or frequency counter function on the DMM function selector. The DMM will indicate the frequency in Hz over a range of about 0.5 Hz to 200 kHz. This is an exceptional feature because this reading once required the use of another meter. Also within the frequency counter function is the ability to determine the duty cycle of the waveform. *Duty cycle* is the amount of time the signal is in the high state divided by the period of the waveform, and it is important for communication system testing when a receiver operates over a limited range of duty cycle. The DMM can be connected to the signal while the duty cycle function is selected with the maximum and minimum option also selected. The DMM will store the maximum and minimum duty cycle measured.

Capacitance Readings

With the increased use of SMT components and chip capacitors, the ability to measure capacitance for the purpose of determining a capacitor's value is increasingly important. The newer DMMs offer this capability over a range of 10 pF to 5 µF usually as an extension of the ohmmeter function. The operation is identical to the ohmmeter function.

B–2 ▶ DC Power Supplies

DC power supplies are used to energize circuits by supplying the required voltage sources. In some cases, off-the-shelf power supplies are the permanent DC power source for a circuit. Most often a custom-designed power supply is the permanent DC power source for a circuit while a bench DC power supply is used for the initial testing only. In this section power supplies are discussed in general so that all of the specifications and requirements are clearly understood.

There are two basic types of power supplies: linear and switching power supplies. Linear power supplies (see Figure B–2) utilize a variable series resistance in some form, usually a semiconductor, to regulate the output voltage to the proper value. The variable resistance is located between the unregulated input voltage and the output voltage. A control circuit senses the output voltage and develops a signal that adjusts the collector emitter resistance of the semiconductor to provide just the right output voltage.

Switching power supplies (see Figure B–3) also use a semiconductor that is placed in between the unregulated input voltage and the output voltage. Switching regulators switch the semiconductor completely on or off to regulate the output voltage. The duty cycle (on time versus off time) of the on/off switching is what determines the ultimate output voltage. The switched voltage signal must be flattened out with a filter circuit as shown in Figure B–3. The output voltage is measured and fed back to a control circuit that adjusts the duty cycle accordingly to provide the proper output voltage.

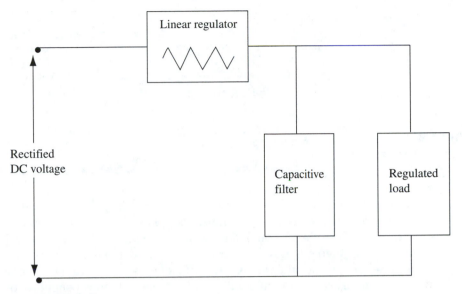

▲ **FIGURE B-2**
Linear power supply block diagram

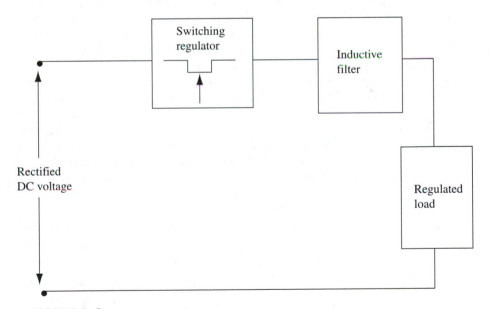

▲ **FIGURE B-3**
Switching power supply block diagram

The basic difference between the design philosophy of the linear and switching power supplies results in many differences in their performance. Efficiency, regulation, radiated noise, complexity, size, and cost are the important performance issues. As can be seen in Figure B–4, the switching power supply is generally much

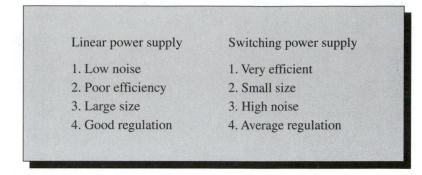

Linear power supply	Switching power supply
1. Low noise	1. Very efficient
2. Poor efficiency	2. Small size
3. Large size	3. High noise
4. Good regulation	4. Average regulation

▶ **FIGURE B–4**
Linear and switching power supply comparison

more efficient, complex yet smaller, and lower in cost with good regulation and higher levels of radiated noise. Linear supplies offer the best regulation, very little noise, and less circuit complexity while being physically larger with poor efficiency and are higher in cost. The primary issues that usually determine the selection of one type or the other are usually efficiency, size, and radiated noise.

There are many configurations of off-the-shelf power supplies, but this discussion focuses on bench-test power supplies. These are usually the linear-type power supply because of the low noise requirements and size, which is not an important factor. Bench-top power supplies are available with one, two, three, or more separate supplies. Some of the voltages are fixed, while others are adjustable. Often, when testing linear op amp circuits, plus and minus voltages that track together are needed for optimum performance. Many power supplies have the option of connecting two power supplies to a common ground, one with the negative side grounded, while the positive side is grounded on the other supply. The two supplies can be selected to track together, which means that they will regulate to the same absolute value voltage.

Usually individual power supplies are completely isolated. This means that they can be connected together with either the positive or negative sides grounded as desired. It is important to make sure that the power supplies are isolated and not connected to earth ground at any point before making ground connections. Many supplies have a grounding strap that allows connection of the isolated power supply to AC ground. This connection must be removed if the supply is to have a ground other than AC ground.

Many newer power supplies have adjustable current limit protection circuits. When the adjusted current limit is exceeded, the supply is shut down while a fault indication is lit up on the power supply front panel. An example of a power supply is shown in Figure B–5.

B–3 ▶ Function Generators

Function generators are used to simulate a variety of waveforms to a circuit. There are two basic levels of these devices: a basic unit that generates sine, square, and triangle waveforms or a more sophisticated unit that can create very specific pulses. The latter are sometimes called *pulse generators*. The basic type of function generator is shown in Figure B–6. There is usually a frequency range selector, frequency

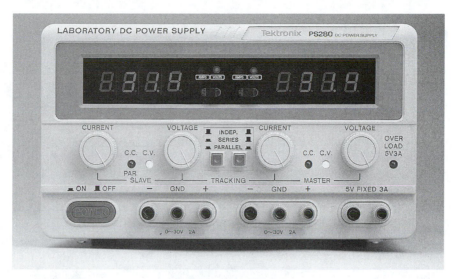

▲ **FIGURE B–5**
Power supply *(Reprinted with permission. Copyright Tektronix, Inc. All rights reserved.)*

▲ **FIGURE B–6**
Basic function generator *(Courtesy of Fluke Corporation. Reproduced with permission)*

adjustment, waveform type selector, duty cycle, and amplitude adjustment. All of the waveforms develop a positive and negative cycle. In other words the waveform increases until a positive peak is reached, and then the waveform begins decreasing and approaching the negative peak value. In order to have just a positive or negative signal, most function generators have the capability of adding a DC offset voltage that shifts the selected waveform either positively or negatively. In this case the waveform rides on top of a DC offset voltage.

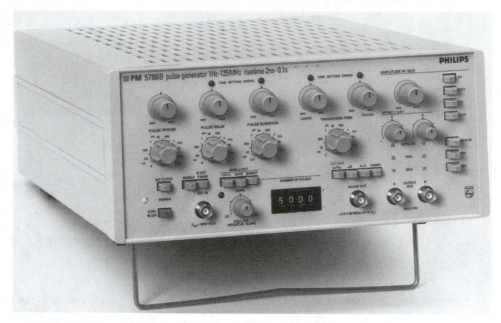

▲ **FIGURE B–7**
Pulse generator *(Courtesy of Fluke Corporation. Reproduced with permission)*

The type of waveforms available on function generators include the sine, square, triangle, and sawtooth waveforms. An additional feature available on many of the basic function generators is the ability to modulate a waveform with a separate sine wave frequency. This function is used to simulate an amplitude modulated signal.

Pulse generators (see Figure B–7) are used mostly in digital circuits, where a very specific pulse must be applied to a circuit. The degree requirements for the pulse may involve specifying the rise-and-fall time along with the pulse duration and overall frequency. Pulse generators have a high degree of accuracy and consequently are much more expensive than the basic function generator.

B–4 ▶ Oscilloscopes

The oscilloscope is an electronic test instrument, the general use of which is second only to the digital multimeter. Oscilloscopes duplicate the waveform connected to their input terminals and present it on a screen with a time base much slower than it actually is so that our eyes can see it. The oscilloscope is indispensable when taking measurements and testing analog and digital circuits, because it displays the actual waveform in a circuit. The oscilloscope can be viewed as a voltmeter that graphs voltage measurements taken in quick succession. The instantaneous values of voltage are plotted in the vertical direction with time denoting the horizontal position. The os-

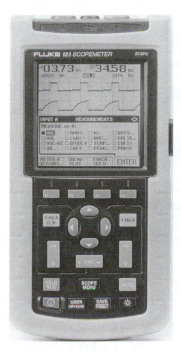 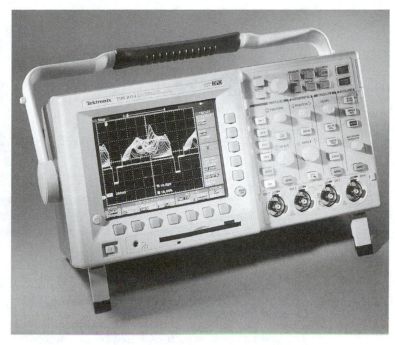

▲ FIGURE B–8

Examples of recent oscilloscope technology *(Left: Courtesy Fluke Corporation. Reproduced with permission. Right: Reprinted with permission. Copyright Tektronix, Inc. All rights reserved.)*

cilloscope measures voltage, but the voltage can represent many other parameters, such as resistance, current, or other transducer outputs. The number of channels included on an oscilloscope determines the number of voltages that can be processed and displayed simultaneously. The basic oscilloscope has undergone much change as advances in technology have been applied to the basic requirements. The latest oscilloscopes feature a variety of very fast digital sampling, hand-held LCD screens and circuit board oscilloscope modules that plug into a personal computers. Figure B–8 shows examples of the latest in oscilloscope technology.

The complete operation of an oscilloscope is somewhat complicated. In industry and college laboratories, you see technicians and students alike wildly turning the scope selection knobs attempting to reproduce a waveform picture on an oscilloscope screen. This exemplifies a lack of understanding of proper oscilloscope operation. There are many different shapes and sizes of oscilloscopes, but they all operate much the same. The analog and digital oscilloscopes are the two primarily different types available and the following discussion applies to the operation of both. (Digital oscilloscope capabilities are discussed specifically at the end of this section.) A general layout for an oscilloscope front panel is shown in Figure B–9.

The general operation of an oscilloscope involves the selection and adjustment of three categories of control settings: overall display screen controls, vertical and voltage amplifier adjustments for each channel, and horizontal and time base selections.

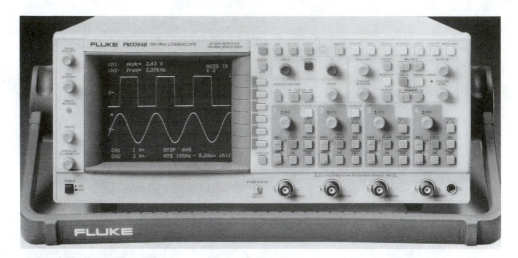

▲ **FIGURE B–9**
General oscilloscope *(Courtesy of Fluke Corporation. Reproduced with permission)*

Overall Display Screen Controls

The overall controls included on most all oscilloscopes are usually grouped together and include the display intensity and display focus controls. Display intensity affects the brightness of the displayed waveform for all waveforms being displayed while the display focus adjusts the clarity or fineness of the beam producing the waveform.

Vertical and Voltage Amplifier Controls

The vertical and voltage amplifiers adjust the level of amplification applied to the input signal as it is applied to the vertical scale on the display. Each channel has selectors for setting the volts and major division (*major divisions* are grid lines that extend across the display screen), vertical position adjustment, and a signal conditioning selector for AC, DC, or ground. These are the key adjustments specific to each channel.

Vertical Position Control

This control positions the zero reference point for each vertical amplifier. The normal zero position is in the middle of the screen, but it can be adjusted anywhere. In order to set the vertical zero position, select the ground selection on the signal conditioner and adjust the zero trace line to the desired spot on the screen. When displaying more than one channel on the screen, it is often desirable to have different zero references for each channel so that each waveform can be viewed separately and with respect to each other (see Figure B–10).

Signal Conditioner Selector

The signal conditioner selector is usually a three-way selector switch that determines whether the vertical and voltage amplifier is connected either to ground, or a signal conditioner called *DC* that passes direct currents and alternating currents, or a signal conditioner called *AC* that passes only alternating currents.

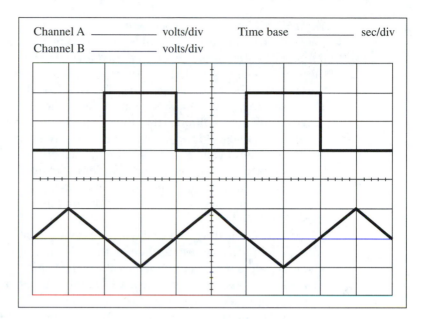

Channel A ————— volts/div   Time base ————— sec/div
Channel B ————— volts/div

▶ **FIGURE B–10**
Dual trace scope display with different zero reference

The signal conditioner ground selection disconnects the positive test lead from the test circuit and connects the positive input of the vertical and voltage amplifier to ground. This is to allow adjustment of the zero reference for that channel. This selection does not ground the positive test lead. When the signal conditioner selector is in the DC position, the vertical and voltage amplifier is connected directly to the positive test lead and allows all voltages, DC or AC, to pass into the amplifier for measurement and display. The AC signal conditioner position places a capacitor in between the vertical and voltage amplifier and the positive test lead. This allows only AC signals to pass into the amplifier for measurement and display as the capacitor charges up to and blocks any constant DC level. Figure B–11 shows the actual circuit connections that exist in the signal conditioner selector.

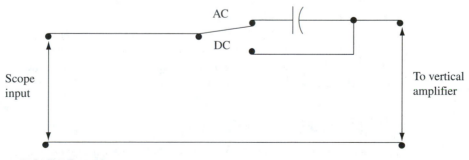

▲ **FIGURE B–11**
Signal conditioner circuit

Vertical and Voltage Amplifier Gain

This selector adjusts the gain of the vertical and voltage amplifiers that change the number of volts per division of the waveform displayed on the screen. A division in this case is one major division. All major divisions extend across the entire face of the screen while minor divisions are shown as small hash marks (see Figure B–10). The wider the range of adjustment of the vertical amplifier gain, the broader the range of voltage amplitudes that the oscilloscope can display. A typical range for a general type of oscilloscope is from 1 mV per division to 5 V per division. The smaller the number of volts per division on the selector, the larger the gain setting is on the vertical and voltage amplifier. In order to achieve the best resolution on the displayed waveform, the gain should be set as high as possible while still being able to view the entire waveform. Usually in the middle of the vertical gain selector is a variable adjustment with a position labeled "CAL." When the CAL adjustment is in the CAL position, the calibrated volts-per-division values shown on the vertical and voltage gain selector are in effect. Otherwise the calibrated values do not apply. The CAL adjustment is provided to allow the user to achieve the best view of the waveform as possible, when the actual amplitude accuracy is not important. When making amplitude measurements with an oscilloscope, it is important to verify that all the CAL adjustments are in the CAL position. Many scopes have a notch in the adjustment mechanism so that you can mechanically feel the CAL position when it is in place, while others have an LED that indicates that CAL is in effect.

The final controls for the vertical and voltage channels are selectors that determine which channels are displayed on the screen and the form in which they are to be displayed. There is usually one selector that will select either channel 1, channel 2, or both channels for display on the screen. Another selector determines whether channel 2 is shown in its normal form or inverted. Finally a third selector determines whether the two waveforms are added together or shown in the "alternate" or "chop" modes. The "invert channel 2" and "add channels 1 and 2 together" selections together are commonly used to subtract two voltage measurements, as described later in this section.

In the alternate mode, channel 1 is selected for display for one sweep of the screen, and then channel 2 is selected on the succeeding sweep, while the previous sweep display of channel 1 is maintained. In the chop mode, the selector switches back and forth from channel 1 to channel 2 during each sweep at a rate that is asynchronous with the input waveforms. The result is short gaps in both the waveform displays that result from the momentary switching to the other channel during the sweep. If the chop rate (the rate at which the channels are alternately selected during the sweep) is sufficiently fast when compared to the input signals, the gaps are not visible. The alternate mode works best at fast sweep and signal repetition rates, while the chop mode performs better at slower sweep and signal repetition rates.

Horizontal and Time Base Controls

The horizontal and time base controls determine the time scale presented on the horizontal axis and the horizontal zero position of the waveform on the screen. These are much like the selectors for the vertical and voltage gain amplifiers, except that there is only one horizontal time base available for the entire instrument. The adjustment of the horizontal and time base selects the seconds per division represented in the displayed waveform. Again, to achieve the best resolution of one waveform, the time base should be adjusted to spread one cycle of the waveform out as much as possible across the display screen. The horizontal and time base control also has a separate CAL adjustment that functions exactly as the CAL described for the vertical and voltage amplifier. In the time base area, there is a horizontal position adjustment that can move all waveforms displayed back and forth in order to locate a key point in the waveform exactly at a major time division. The time base setting is used to determine the waveform rise and fall times, frequency, period, pulse width, and duty cycle.

Trigger Controls

The trigger controls are usually separate from the horizontal and time base controls, but they are strongly tied to them. For this reason the trigger controls are usually adjacent to the horizontal and time base controls. The trigger determines the point in the waveform that the oscilloscope begins its sweep across the display. In other words, the trigger determines the point of the waveform that will be shown at the farthest left side of the display screen. The effect and adjustment of the trigger controls are the least understood of all the oscilloscope controls.

Trigger Slope and Level Selection

The trigger point is determined by the edge of the waveform: the point at which it begins in either a positive or negative direction. The positive slope is said to be a *positive-going transition*, while the negative slope is a *negative-going transition*. The oscilloscope's trigger controls allow the selection of either. In addition, the trigger level can be adjusted that determines the voltage threshold—or point—on the waveform where the trigger point will be established. In Figure B–12 the trigger point is shown for two different trigger levels with the positive slope trigger selected.

▶ **FIGURE B–12**
Trigger level adjustment

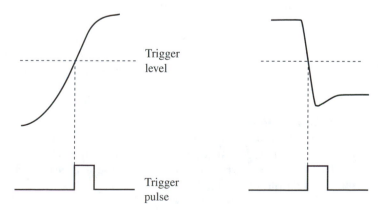

Trigger Source

With the selection of the desired trigger slope and level, a selection must be made of the signal source that will drive the trigger. The trigger source is the signal that will be monitored to determine the trigger point selected by the trigger slope and level adjustments. There are many potential sources for the trigger source, either of the waveforms being measured, the 60 Hz power AC waveform, or some other external source. One of these must be selected as a trigger source. The 60 Hz AC power trigger source is useful for signals that are synchronous with the AC power source being measured. More typically, one of the input channel waveforms is used as the trigger source. External clock sources are used when an ideal trigger source exists that is not being measured, such as the clock signal in a computer circuit. In this case the clock signal is connected to the external clock input connection located in the trigger control area and the external trigger source is selected.

Sweep Control

Once the first horizontal sweep has been triggered for a waveform, there are typically three different ways that the succeeding sweeps will occur. These are selectable with the sweep control selections: single, normal, or automatic. In the single sweep mode, the occurrence of the selected trigger will initiate one sweep of the screen, after which signal acquisition ceases. Depressing the reset button allows one more single sweep of the display to occur. The normal setting initiates a new sweep every time the trigger source retriggers. In the normal mode, if the trigger fails to occur, the sweep does not start and the display vanishes. The automatic sweep control mode provides for an automatic trigger to occur after a certain time, that is, if the trigger source signal selected does not happen as it should. When the trigger source is intermittent, the resulting display may show waveforms that are not synchronous, but a display is always present. For this reason the automatic sweep control mode is favored for first-time measurements.

Trigger Source Signal Conditioner

The trigger source also has different signal conditioners that are available similar to the vertical and voltage amplifier. In this situation the intent is to reject signals that might make the trigger source intermittent. As before, either DC or AC signal conditioners can be selected. The DC selection includes both direct current and alternating current components of the trigger waveform, while AC will include only the alternating current component. Other selections are for high-frequency pass, low-frequency pass, and band pass signal conditioners that again are selected to reject spurious signals that may cause the trigger source to be unstable.

Hold-Off Adjustment

Many oscilloscopes have a variable hold-off adjustment that prevents a trigger from reoccurring for an adjustable period of time. The purpose of this is to impede spurious triggering from occurring as well. Adjusting the hold-off value to zero eliminates hold off altogether.

Delayed Sweep

Delayed sweep is a feature that is included on some oscilloscope models. This feature is necessary when a time-magnified view of a portion of the measured waveform is needed. Delayed sweep works by the generation of a second sweep generator and trigger. The second sweep generator is initiated by the initial trigger after a delay that is adjusted by the operator. When the delayed sweep is selected, the time base of the displayed waveform should be read from the delayed sweep time base setting.

Negative Test Lead—Ground Connection

One important fact about an oscilloscope that makes it different from a voltmeter is that the oscilloscope negative test lead is actually connected to AC ground. Subsequently, the standard oscilloscope is said to be a single-ended instrument: It measures voltages with respect to AC ground only. Therefore, when using an oscilloscope, the negative test lead can only be connected to AC ground. Connection to any other point in a circuit will force that point to be AC ground, which usually disrupts the circuit operation. In the worst case, a dead short is created across a power supply, which sometimes causes it to smoke profusely. The connection between the negative scope test lead and AC ground (see Figure B–13) can be defeated by isolating the scope from AC ground, with an isolation transformer, or by disconnecting the ground terminal from the AC plug. This practice is not considered safe, because the scope chassis is not connected to ground, causing a hazard if the scope chassis were somehow connected to a live voltage.

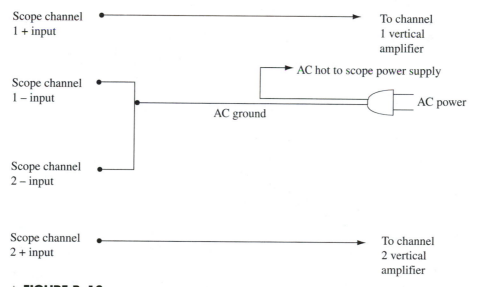

▲ **FIGURE B–13**
AC ground connections for an oscilloscope

The single-ended nature of the standard oscilloscope poses a problem to measuring waveforms between two points when neither are AC ground. There are two solutions: the subtraction method or use of a differential mode oscilloscope.

Subtraction Method

The subtraction method uses both channels of the standard oscilloscope to measure one waveform between two points where neither are AC ground. In the circuit shown in Figure B–14, channel 1 of the oscilloscope is connected from the top side of R_1 to AC ground. Channel 2 measures the waveform across resistor R_2 to ground. In order to show the waveform of the voltage across R_1, the channel 2 voltage must be inverted and then added to channel 1. The resulting waveform is the desired measurement across R_1. Most oscilloscopes have either an "invert channel 2 and add to channel 1" or simply a "subtract channel 2 from channel 1" mode selection provided specifically for this purpose.

Differential Oscilloscope

The differential oscilloscope (see Figure B–15) is specifically designed to make measurements in the differential mode: measuring and displaying the voltage waveform between two points that are not necessarily connected to ground. The differential oscilloscope is considered to be a more precise instrument for use in determining analog-oriented measurements, where the accuracy of the waveform amplitude is more critical. The differential oscilloscope has two scope probe connections that represent the two points to be measured between. The two positive scope probe connections are connected to the points to be measured. The negative scope probe clips are not used.

Bandwidth

The bandwidth specification of an oscilloscope is most important. Bandwidth defines the highest frequency waveform that the oscilloscope can measure and display. A 100-MHz oscilloscope can display waveforms over the range DC to 100 MHz. More importantly, the bandwidth also determines the fastest rise time on a pulse waveform that can be shown on the oscilloscope display. A 100-MHz

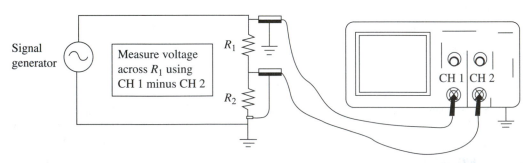

▲ **FIGURE B–14**
Oscilloscope subtraction method

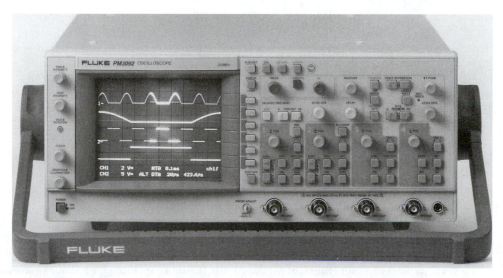

▲ **FIGURE B–15**
Differential oscilloscope *(Courtesy of Fluke Corporation. Reproduced with permission)*

oscilloscope, when compared to a 200-MHz scope, cannot display shorter-duration transient waveforms. The fastest rise time that is displayed on the oscilloscope can be approximated by the following relationship:

$$\tau_r = 0.35/f_{BW}$$

where τ_r = rise time
f_{BW} = scope bandwidth

   The price paid for an oscilloscope is determined largely by the bandwidth capabilities.

Test Probe

A test probe is used to make the connection between the oscilloscope and the circuit test points to be displayed. A variety of 1:1 and attenuating 10:1, both passive and active, test probes are available.

1:1 Passive Probe

This probe is simply a coaxial cable connected to the oscilloscope vertical and voltage channel input. With the 1:1 passive probe in place, the input impedance connected to the test circuit is the oscilloscope input impedance (approximately 1 MΩ and 25 pF) plus the additional capacitance of the coaxial cable (approximately 50 pF). Waveforms with fast rise times will be severely affected by capacitance values that could total 75 pF. For these reasons the 10:1 test probe that is discussed next is preferred.

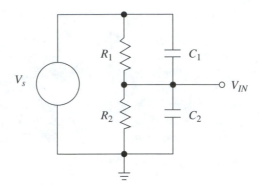

▶ **FIGURE B–16**
10:1 probe equivalent circuit

10:1 Passive Probe

This probe is the most popular test probe used with oscilloscopes because it minimizes the capacitive loading on the test circuit and, therefore, allows the display and measurement of signals with faster rise times. A parallel resistor–capacitor circuit is included within the probe assembly that forms a circuit with the oscilloscope input impedance, as shown in Figure B–16.

In order for the DC attenuation to be a factor of ten, the relationship between the probe resistance R_p and the scope input resistance R_{IN} is $R_p = 9 \times R_{IN}$. At high frequencies, to maintain the 10:1 attentuation, then C_{IN} must $= 9 \times C_p$. The net input impedance of the oscilloscope, combined with the 10:1 probe assembly, is a resistance that is 10 times the input resistance of the scope, plus a capacitance that is one-tenth the cable capacitance plus the input capacitance of the oscilloscope. The capacitor, located in the 10:1 probe assembly, is usually adjustable to ensure that $R_p\,C_p = R_{IN}\,C_{IN}$. Most oscilloscopes provide a square wave that can be used to calibrate the 10:1 passive probe capacitance. Figure B–17 shows a representation of the calibrator waveform when the probe capacitance is adjusted improperly.

10:1 Active Probe

These probes further minimize the input capacitance of the oscilloscope and probe assembly by locating a small amplifier very close to the probe tip. These probes are necessary when measuring very high-frequency waveforms. Active probes are available in a range of attenuation from 1:1 to 10:1.

▶ **FIGURE B–17**
10:1 probe capacitance
adjustment

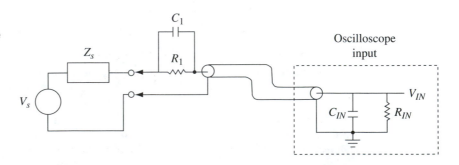

Digital Oscilloscopes

The digital oscilloscope differs from its analog counterpart by the fact that the measured waveform is actually sampled and converted to a digital number. The series of digital numbers that represent the waveform can then be stored in memory, sent to a computer or printer, or saved on a disc. The quality of the waveform sampled depends on the resolution of the analog-to-digital converter sampling the waveform and how often a sample is taken. The digital oscilloscope offers many advantages over an analog scope, but it comes at a much higher cost. Its primary benefits are as follows:

1. Multiple channels

2. Ability to play back a waveform

3. Ease of transient signal capture

4. Ease of saving test data

The high cost and complexity of the digital oscilloscope favors the use of the analog oscilloscope for most basic test applications. Figure B–18 shows an example of a digital oscilloscope.

B–5 ▶ Logic Analyzers

The development and growth of digital electronics has brought about a need for a different type of instrument to display pulse waveforms on a screen. Digital electronics required the display of many pulse waveforms, switched between two logic levels, simultaneously. The resulting instrument was called a *logic analyzer*, which extended the channel capacity and speed of the oscilloscope while

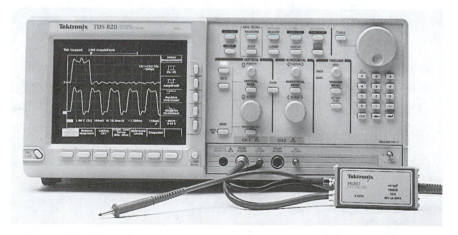

▲ **FIGURE B–18**
Digital scope *(Reprinted with permission. Copyright Tektronix, Inc. All rights reserved.)*

limiting the amplitude of the waveform to various standard logic levels. Because the logic analyzer is only required to determine whether the waveform is at either logic level, the number of samples that must be taken to reconstruct the waveform is minimal, and the sample speed can be very fast. Logic analyzers are used to debug both hardware and software, and data can be stored and played back as desired.

The logic analyzer begins sampling when an event, defined as the "trigger event," begins the waveform sampling process. Once triggered, the logic analyzer, like the oscilloscope, generates a sweep signal that can be initiated by an internal clock or an external clock supplied by the circuit under test. The logic analyzer is said to be in the asynchronous mode with the internal clock or the synchronous mode when an external clock is applied. The logic analyzer displays either a high- or low-logic level for a waveform with respect to whichever clock is used. Waveform variations above or below either logic level are ignored. The samples are stored in a high-speed memory that can be looked at as a continuous loop of the waveform data. Figure B–19 shows an example of the logic analyzer display of a typical time-varying logic circuit waveform. The basic logic analyzer is much like an oscilloscope that can display many waveform channels, 8 to more than 100, but at just two specified logic levels. There are many different types and manufacturers of logic analyzers, so the following discussion focuses on the primary parameters that must be specified for their use and their most common applications.

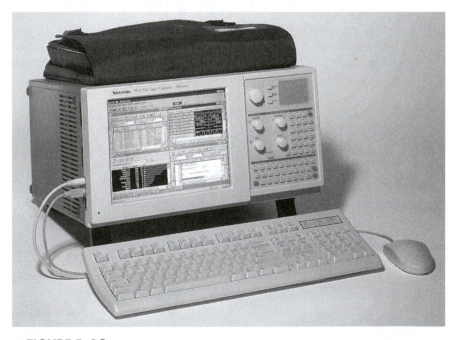

▲ **FIGURE B–19**
Logic analyzer *(Reprinted with permission. Copyright Tektronix, Inc. All rights reserved.)*

Basic Operation

The setup and use of most logic analyzers involves the use of a portion of the display screen to select various operating modes and parameters. Push buttons are provided for cursor movement and mode selection. Many analyzers provide pull-down menus similar to computer programs. The primary concerns are to select the logic levels and the clock-group channels under one label or name and to define the display format.

Probes

The input connections to the logic analyzer are usually separated into groups of 8 or 16 that are called *pods* (see Figure B–20). Each pod is connected to the instrument with a flat-cable connector. The pod end, where the measurement connections are made, incorporates separate channel clip-on connections.

The pod number and the connection (bit) number are then used to identify the signal that is being measured. The first input channel would be pod 1, bit 0, then pod 1, bit 1, and so on.

▶ **FIGURE B–20**

Pod clip-on connections
(Reprinted with permission. Copyright Tektronix, Inc. All rights reserved.)

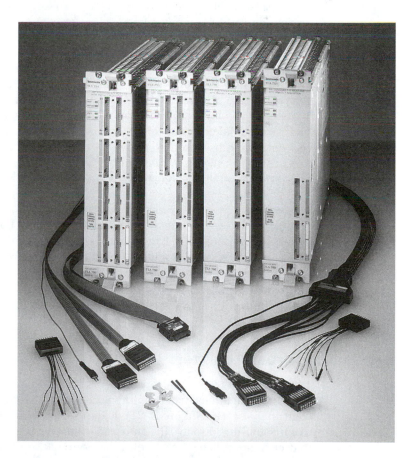

Logic Level Specification

The type of logic being measured should be specified to the analyzer, that is, TTL or CMOS.

Clock Selection

For asynchronous operation the internal clock should be selected. The rate of the internal clock is usually shown in the form of its period instead of its frequency. The fastest sample rate should be selected, unless the amount of data that must be stored per sample exceeds the available memory in the instrument. In the synchronous mode, one or more input sources can be identified as the clock source. A clock can be generated on just one input source, or when two or more input source conditions are combined. This is usually accomplished in the menu setup to configure the clock.

Assigning Channel Labels

The use of a logic analyzer can be made much simpler with the use of channel labels. The large number of input channels provided make it difficult to remember which one represents which waveform. The assignment of channel labels overcomes this problem by allowing the user to key in names that represent the particular channels. The channel names are presented in the display format along with the waveform in question.

Display Format

The primary display format decision is to determine if the display screen shows the waveform data in a graphical or text format (ones and zeros). When the text format is selected, the logic analyzer can convert the ones and zeros to a number system such as decimal or hexadecimal. It is also possible to have the logic analyzer convert data streams into the equivalent microprocessor coded *mnemonic instructions*. This is called a *disassembler function*, because it reverses the process of converting assembling code into the machine code on which the computer operates.

Triggering

The trigger event is the event that begins sampling the data that is loaded into the high-speed memory. The trigger can be defined as an event that encompasses many input values logically. A trigger could be defined for the specific situation: When the microprocessor clock goes high, the address-latch enable goes low, and address location 1000 Hex is on the address bus. Once the trigger event has occurred, the defined clock signal strobes the data into the high-speed memory.

B–6 ▶ Spectrum Analyzers

The increased use of microprocessors in equipment, combined with advances in telecommunications, has promoted the continued growth of applications for spectrum analyzers (see Figure B–21). The spectrum analyzer measures the selected electronic signal and displays the amplitude of the signal versus the frequency component that corresponds to that amplitude. Fourier theory states that all signals can be

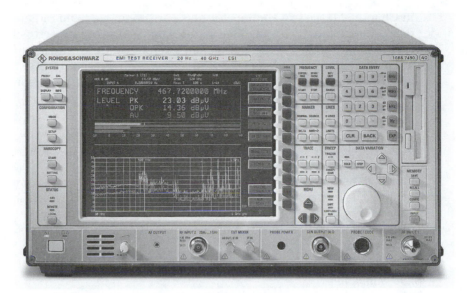

▲ FIGURE B–21
Spectrum analyzer *(Reprinted with permission. Copyright Tektronix, Inc. All rights reserved.)*

broken down into a summation of sine and cosine functions at various frequencies. The spectrum analyzer breaks a signal down into these various frequencies and shows their amplitudes. Spectrum analyzers look similar to an oscilloscope, but that is as far as the similarity goes. An oscilloscope shows the actual signal waveform, amplitude versus time, while the spectrum analyzer shows the amplitude versus frequency of the components that make up the signal. Spectrum analyzers can measure very small signals over a wide range in frequency and, by doing so, can indicate signal amplitudes, distortion levels, stability, modulation types, and so on. The latest spectrum analyzers can analyze signals that range in frequency from a few hertz to 100 GHz.

The basic use of a signal analyzer involves connecting the signal and selecting the range in frequency over which the signal is to be analyzed. The display screen indicates the amplitude of all of the component signals plotted against their corresponding frequencies. A typical use of a spectrum analyzer is performing testing, similar to that described in Chapter 11, to determine the radiated EMI levels to determine if equipment meets current FCC levels. In this case a spectrum analyzer is connected to an antenna, which is located at the prescribed distances from the unit under test, to determine the strength and the frequency of the radiated emissions. There are many other more specific uses in the telecommunications area.

B–7 ▶ Noise Immunity Testing Equipment

In order to complete most noise immunity testing, a controlled source of noise must be generated. Once generated, the impact of the application of the noise source is monitored in a manner appropriate for the unit being tested. As discussed in Chapter 11, the two sources for noise are conducted and radiated noise. Conducted noise can enter equipment through the AC line or an electrostatic discharge (ESD) to the surface of the equipment. Radiated noise is simply radio waves that are picked up by the equipment under test. In order to create these noise sources, an AC line spike generator, an ESD simulator, and an RFI generator is needed.

AC Line Spike Generator

This piece of test equipment can generate large voltage spikes, of an amplitude and duration that is specified by the user, and apply them to the AC power line that is, in turn, connected to the unit under test. In addition, the exact phase angle in which the spike should be injected can be specified by the user as well. Transient analysis yields the fact that each circuit has a phase angle at which an applied transient will result in the largest impact on a circuit. The AC line spike generator can be used to find the phase angle of maximum effect and continually test the circuit for immunity at that phase angle.

ESD Simulator

This device can generate thousands of volts to simulate an ESD pulse. The ESD simulator device is usually packaged in the form of a gun where the probe tip at the end of the gun is the point of contact to the unit under test. The use of an ESD simulator involves selecting the ESD voltage level and the duration of the spike. The technician then fires away over the surface of the unit under test as required in the test specifications while monitoring the results on the operation of the unit under test. Test specifications usually call for the application of an ESD spike for every two square inches of surface area of the unit being tested, while special attention is paid to any keypad devices or switches.

RFI Generator

An RFI generator simply generates and radiates EMI signals at a selectable level and frequency. The resulting signal is radiated and adjusted to develop the proper EMI field conditions in the area of the unit under test. This test setup was discussed in Chapter 7.

B–8 ▷ Impedance Measurement

Years ago, the impedance bridge was one of the first test instruments introduced to the electronics student. Today, digital technology has replaced the impedance bridge with the digital impedance meter. The digital impedance meter resembles a digital ohmmeter, except that it measures and calculates the total impedance of an unpowered circuit. It is a simple instrument to use, similar to the operation of a DVM. One important difference is the potential for error when taking precise measurements. Digital impedance meters offer a variety of methods of connection to the test points. The simplest connection is the two-terminal connection. The excitation current supplied by the impedance meter travels from one terminal to the test circuit and then back to the other terminal. The impedance meter takes its measurement at the

▲ **FIGURE B–22**
Impedance meter *(Courtesy of Fluke Corporation. Reproduced with permission)*

two terminals. This method is adequate for many situations but inherently introduces two sources for error, the contact resistance of the contact points and stray impedance in parallel with the test points. Some digital impedance meters utilize a third terminal, a guard terminal that eliminates the effect of the stray impedance. In order to eliminate the contact resistance error, two more terminals can be added to separate the source current and the measurement circuits. Digital impedance meters incorporate various combinations of the terminal arrangements just discussed. Figure B–22 shows an example of an impedance meter.

B–9 ▶ Simulators, Emulators, and Microtesters

Microprocessors and microcomputers have initiated many new requirements for test equipment. One of the most significant is a category of instruments used primarily for debugging software but which also have many applications debugging and troubleshooting hardware problems. A testing problem arises because of the inherent nature of microprocessors: they execute program instructions very quickly in a sequential fashion. The only way to determine if the device under test is operating correctly is to slow down its operation so that it can be monitored and determined if the instructions are being executed properly.

Simulators
Simulators are software programs written to simulate a specific microprocessor or microcomputer. When the computer executes the simulator program, it acts like the processor in question and executes the program code by changing the data in memory locations and registers just as the actual processor would. The program operation can be checked by examining the contents of memory locations and registers after program code execution. A simulator is useful in debugging software only.

Emulators
An emulator operates in a similar fashion as the simulator, except that it utilizes the actual circuit hardware, verifying both hardware and software operation. An emulator allows the unit under test to execute instructions step by step, while the

results of the operation are verified by the hardware. The program can be run in single-step mode or operated in real time from point A to point B and then halted, while the hardware is examined to verify proper operation.

Microtesters

These are a family of devices that are primarily intended to troubleshoot microprocessor boards in a manufacturing test environment. Faulty microprocessor-based circuit boards experience the same troubleshooting problems as the initial hardware and software designs: there must be some means of taking control of the microprocessor and have it execute instructions while the results are monitored. A microtester device provides this and a number of built-in tests. Some examples are built-in bus tests, memory tests, and ROM sum tests and the ability to write and read I/O ports.

▶ Summary

In this appendix you have reviewed the operation and capabilities of many of the available electronic instruments used in designing and testing electronic projects. This discussion is by no means complete, but it does include the most commonly used instruments. It is important to develop the knowledge and skills necessary to utilize electronic test instrumentation and to keep up with new developments in technology as they are applied to the instruments. Electronic test instruments are simply additional tools that assist the electronics professional in testing and verifying electronic designs in a complete and efficient manner.

▶ References

Coombs, C., Jr. 1992. *Basic Electronic Instrument Handbook.* New York: McGraw-Hill.

Appendix C ▶ Selected Contact Information

C–1 ▶ Periodicals

The following lists are the better-known periodicals that the electronics profession will be interested in receiving. Many are available for free if the subscriber meets certain qualifications. These qualifications are fulfilled when employed by a company that purchases some of the items advertised in the magazine. A free subscription can be obtained by filling out a free subscription questionnaire.

Communications News
Profiles the people, products, and technologies for today's
  communications solutions.
Published monthly

Nelson Publishing
2500 Tamiami Trail North
Nokomis, FL 34275-3482
Web address: www.comnews.com

Electronic Design
Reviews technology, applications, products, and solutions.
Published biweekly

Penton Media, Inc.
1100 Superior Ave.
Cleveland, OH 44114-2543
Tel: (201) 393-6060
Web address: www.elecdesign.com

Electronic Design News
Published biweekly

Cahners Business Information
8773 South Ridgeline Blvd.
Highlands Ranch, CO 80126-2329
Web address: www.ednmag.com

Electronic Systems — Technology and Design
Published monthly

Embedded Systems Programming
600 Harrison St.
San Francisco, CA 94107
Web address: www.embedded.com

Electronic Engineering Times
An industry newspaper for engineers and technical management.
Published weekly

600 Community Drive
Manhasset, NY 11030
Tel: (516) 562-5000
Web address: www.eet.com

Integrated System Design
Reviews tools and technologies for electronic designers.
Published monthly

411 Borel Ave.
Suite 100
San Mateo, CA 94402
Tel: (650) 513-4300
Web address: www.isdmag.com

Printed Circuit Design
A professional publication for engineers and designers of printed circuits
and related technologies.
Published monthly

P.O. Box 1277
Skokie, IL 60076-8277
Tel: (847) 676-9745
Web address: www.pcdmag.com

Software Development
Published monthly

600 Harrison St.
San Francisco, CA 94107
Tel: (415) 905-2200
Web address: www.sdmagazine.com

Test and Measurement World
Published monthly

8773 S. Ridgeline Blvd.
Highlands Ranch, CO 80126
Tel: (617) 558-4671
Web address: www.tmworld.com

C–2 ▶ Approval and Standards Agencies

These are Web sites and, in some instances, mailing addresses for the approval
and standards agencies discussed in Chapter 3.

ANSI—The American National Standards Institute
Develops nationally recognized technical standards.

1819 L Street, N.W.
Washington, DC 20036
Tel: (202) 293-8027
Web address: www.ansi.org

ASTM—American Society for Testing and Materials
100 Bar Harbor Drive
West Conshohocken, PA 19428-2959
Tel: (610) 832-9585
Web address: www.astm.org

CE—Standards for the European Union
To determine and obtain the appropriate standards for testing equipment to comply with European standards and allow application of the CE mark, consult with
one or more of the following agencies:

European Committee for Standardization (CEN)
Works with the CENELEC to voluntary harmonized standards for the Euro-
pean Union. Located in Brussels, Belgium.
36 rue de Stassart
B–1050 Brussels
Tel: +32 2 519 6871
Web Address: www.cenorm.be

European Committee for Electro-technical Standardization (CENELEC)
Provides a contact page for answering any questions. Located in Brussels,
Belgium.
36 rue de Stassart
B–1050 Brussels
Tel: +32 2 519 6871
Web Address: www.cenelec.be

European Telecommunications Standards Institute (ETSI)
The central secretariat is located in Sophia Antipolis, France.

06921 Sophia Antipolis
Cedex, France
Tel: +33 (0)4 92 944 222
Web address: www.etsi.fr

CSA—Canadian Standards Association International
178 Rexdale Blvd.
Etobicoke, ON M9W1R3
Canada
Tel: (416) 747-4058
Web address: www.csa-international.org

FCC—Federal Communications Commission
Federal agency.

445 12th St., S.W.
Washington, DC 20554
Tel: (202) 418-0190
Web address: www.fcc.gov

FDA—Food and Drug Administration
Federal agency.

UFI40
Rockville, MD 20857
Tel: (888) INFO-FDA
Web address: www.fda.gov

IEC—The International Electo-technical Commission
Develops international technical standards. Located in Geneva, Switzerland.

3 rue de Varembé
P.O. Box 131
CH-1211 Geneva 20
Switzerland
Tel: +41 22 919 0211
Web address: www.iec.ch

ISO—International Standards Organization
Develops general industrial international standards. Located in Geneva,
  Switzerland.

1 rue de Varembé
Case postale 56

CH-1211 Geneva 20
Switzerland
Tel: +41 22 74901
Web Address: www.iso.ch

ITS — Intertek Testing Services (formerly ETL)
A nationally recognized testing laboratory.

70 Codman Hill Road
Boxborough, MA 01719
Tel: (800) 967-5352
Web address: www.etlsemko.com

NEC — The National Electrical Code
Published by the National Fire Protection Association.

National Fire Protection Association
One Battery March
Quincy, MA 02269
Tel: (617) 770-3000
Web address: www.nfpa.org

UL — Underwriters Laboratory
333 Pfingsten Road
Northbrook, IL 60062
Tel: (847) 272-8800
Web address: www.ul.com

VDE — Verband der Elektronotechnik
Stresemannallee 15
60596 Frankfurt, Germany
Tel: ++ 4969/6308-0
Web address: www.vde.de

C–3 ▶ Professional and Manufacturers Organizations

IEEE — Institute for Electrical and Electronic Engineers
Professional association.

3 Park Avenue
17th Floor
New York, NY 10016-5997
Tel: (212) 419-7900
Web address: www.ieee.org

IPC—The Institute for Interconnecting and Packaging Electronic Circuits
Manufacturer's association.

2215 Sanders Road
Northbrook, IL 60062-6135
Tel: (847) 509-9700
Web address: www.ipc.org

ISA—The Instrument Society of America
A professional organization for instrumentation professionals.

P.O. Box 3561
Durham, NC 27702
Tel: (919) 549-8411
Web address: www.isa.org

NEMA—National Electrical Manufacturers Association
Manufacturer's association that develops standards for electrical equipment.

11300 North 17th Street
Suite 1847
Rosslyn, VA 22209
Tel: (703) 841-3200
Web address: www.nema.org

Index